AF804066

Advances in Experimental Medicine and Biology

Volume 1365

Series Editors

Wim E. Crusio, Institut de Neurosciences Cognitives et Intégratives d'Aquitaine, CNRS and University of Bordeaux, Pessac Cedex, France

Haidong Dong, Departments of Urology and Immunology, Mayo Clinic, Rochester, MN, USA

Heinfried H. Radeke, Institute of Pharmacology & Toxicology, Clinic of the Goethe University Frankfurt Main, Frankfurt am Main, Hessen, Germany

Nima Rezaei, Research Center for Immunodeficiencies, Children's Medical Center, Tehran University of Medical Sciences, Tehran, Iran

Ortrud Steinlein, Institute of Human Genetics, LMU University Hospital, Munich, Germany

Junjie Xiao, Cardiac Regeneration and Ageing Lab, Institute of Cardiovascular Science, School of Life Science, Shanghai University, Shanghai, China

Advances in Experimental Medicine and Biology provides a platform for scientific contributions in the main disciplines of the biomedicine and the life sciences. This series publishes thematic volumes on contemporary research in the areas of microbiology, immunology, neurosciences, biochemistry, biomedical engineering, genetics, physiology, and cancer research. Covering emerging topics and techniques in basic and clinical science, it brings together clinicians and researchers from various fields.

Advances in Experimental Medicine and Biology has been publishing exceptional works in the field for over 40 years, and is indexed in SCOPUS, Medline (PubMed), Journal Citation Reports/Science Edition, Science Citation Index Expanded (SciSearch, Web of Science), EMBASE, BIOSIS, Reaxys, EMBiology, the Chemical Abstracts Service (CAS), and Pathway Studio.

2020 Impact Factor: 2.622

Xiao-Hong Sun
Editor

Innate Lymphoid Cells

Springer

Editor
Xiao-Hong Sun
Program in Arthritis and Clinical Immunology
Oklahoma Medical Research Foundation
Oklahoma City, OK, USA

ISSN 0065-2598 ISSN 2214-8019 (electronic)
Advances in Experimental Medicine and Biology
ISBN 978-981-16-8386-2 ISBN 978-981-16-8387-9 (eBook)
https://doi.org/10.1007/978-981-16-8387-9

© The Editor(s) (if applicable) and The Author(s), under exclusive license to Springer Nature Singapore Pte Ltd. 2022
This work is subject to copyright. All rights are solely and exclusively licensed by the Publisher, whether the whole or part of the material is concerned, specifically the rights of translation, reprinting, reuse of illustrations, recitation, broadcasting, reproduction on microfilms or in any other physical way, and transmission or information storage and retrieval, electronic adaptation, computer software, or by similar or dissimilar methodology now known or hereafter developed.
The use of general descriptive names, registered names, trademarks, service marks, etc. in this publication does not imply, even in the absence of a specific statement, that such names are exempt from the relevant protective laws and regulations and therefore free for general use.
The publisher, the authors and the editors are safe to assume that the advice and information in this book are believed to be true and accurate at the date of publication. Neither the publisher nor the authors or the editors give a warranty, expressed or implied, with respect to the material contained herein or for any errors or omissions that may have been made. The publisher remains neutral with regard to jurisdictional claims in published maps and institutional affiliations.

This Springer imprint is published by the registered company Springer Nature Singapore Pte Ltd.
The registered company address is: 152 Beach Road, #21-01/04 Gateway East, Singapore 189721, Singapore

Preface

This book covers the current concepts on innate lymphoid cells written by leading experts in the field. Since the recognition of the family of innate lymphoid cells over 10 years ago, a wealth of knowledge has accumulated, and new information continues to emerge. Innate lymphoid cells bridge the function of innate and adaptive immunity and shape immune responses in a variety of tissues by sensing environmental cues. The purpose of this book is to apprise immunologists and biologists in other disciplines and to alert clinicians as they treat diseases like pulmonary, gastroenterological, and neurological illnesses, as well as cancers.

Oklahoma City, OK, USA Xiao-Hong Sun

Contents

1 **Overview: Themes in Innate Lymphoid Cell Biology** 1
Marco Colonna

2 **ILC Differentiation from Progenitors in the Bone Marrow** . . 7
Arundhoti Das, Christelle Harly, Yi Ding, and Avinash Bhandoola

3 **ILC Differentiation in the Thymus** 25
Xiao-Hong Sun and Sandra Bajana

4 **NK Cells and ILC1s in Cancer Biology** 41
Anna Sliz and Wayne M. Yokoyama

5 **Finding a Niche: Tissue Immunity and Innate Lymphoid Cells** 57
Haerin Jung, Do-Hyun Kim, Yilin Wang, and Steven J. Van Dyken

6 **ILCs and Allergy** 75
Hiroki Kabata, Yasutaka Motomura, Tsuyoshi Kiniwa, Tetsuro Kobayashi, and Kazuyo Moro

7 **Innate Lymphoid Cells and Inflammatory Bowel Disease** 97
Vincent Peng, Natalia Jaeger, and Marco Colonna

8 **Coordination of Mucosal Immunity by Innate Lymphoid Cells** 113
Jordan Z. Zhou and Gregory F. Sonnenberg

9 **Interaction Between Innate Lymphoid Cells and the Nervous System** 135
Yuanyue Zhang, Rachel Grazda, and Qi Yang

10 **Development of αβ T Cells with Innate Functions** 149
José Alberola-Ila

11 **The Road from Mouse to Human ILCs: A Perspective of Understanding the Roles of ILCs in Disease** 161
Hergen Spits

Overview: Themes in Innate Lymphoid Cell Biology

1

Marco Colonna

Abstract

Since their initial identification more than 10 years ago, innate lymphoid cells (ILCs) have emerged as denizens of an immune realm parallel to that of T cells. Here I highlight basic similarities shared by all and underscore features related to development, tissue residency and regulation that distinguish ILCs from T cells. I further discuss the potential of ILCs as promising targets for therapeutic intervention in human diseases.

Keywords

Innate lymphoid cells · Cytokines · Regulation · Human · Therapy

1.1 ILCs and T Cells: Identification of Two Parallel Universes

Until 2009, immunologists knew two types of innate lymphoid cells (ILCs): natural killer (NK) cells and lymphoid tissue inducer (LTi) cells. Extensive characterization of NK cells demarcated their innate capacity to kill tumor cells and virally infected cells without previous exposure [1–4]. Recruitment of LTi cells to embryonal anlagen was found to be essential for the development of lymph nodes and Peyer's patches [5]. The limited diversity of ILCs was in stark contrast with that of T cells, which encompassed cytotoxic CD8 T cells and a variety of helper CD4 T cells, each with a distinct functional polarization, including T_H1, T_H2, T_H17, T_{FH}, and Treg. After 2009, a number of groups began to identify novel subsets of ILCs, each also defined by a discrete functional polarization, that are now called ILC1, ILC2, and ILC3. Remarkably, this diversity mirrors that of T_H1, T_H2, and T_H17, respectively [6]. Assuming that cytotoxic NK cells are the innate counterparts of CD8 T cells, it became evident that T cells and ILCs represented two parallel universes of cell types with the same functional modules: cytotoxic (perforin and granzymes), type 1 (IFNγ), type 2 (IL-4, IL-5, IL-13), and type 3 (IL-17, IL-22). T cells differ from ILCs mainly in their capacity to recognize specific antigens through the T cell receptor. In retrospect this makes perfect sense: ILCs serve as tissue "first responders" by sensing the release of soluble inflammatory mediators during infections and tissue damage and rapidly communicating danger via cytokine secretion; in T cells these host defense modules are further equipped with the T cell receptor, which provides specificity and memory for a targeted pathogen.

M. Colonna (✉)
Washington University School of Medicine, St Louis, MO, USA
e-mail: mcolonna@wustl.edu

© The Author(s), under exclusive license to Springer Nature Singapore Pte Ltd. 2022

X.-H. Sun (ed.), *Innate Lymphoid Cells*, Advances in Experimental Medicine and Biology 1365,
https://doi.org/10.1007/978-981-16-8387-9_1

1.2 The Importance of Tissue Residency

Why did immunologists miss ILCs for such a long time? At least two reasons contribute to this: first, ILCs comprise very small cell populations that were difficult to detect with the available tools years ago; second and perhaps more importantly, most of these cells reside in peripheral tissues rather than in blood and lymphoid organs. Despite encompassing crucial sites for immune responses, peripheral tissues fell below the radar of immunologists before 2009, especially in humans. Thus, another important outcome of the discovery of ILCs has been appreciating the relevance and diversity of immune cells in peripheral tissues.

ILC diversity in tissues extends far beyond the strict definitions of NK cells, ILC1, ILC2, and ILC3. Single-cell RNA sequencing of ILCs in different tissues has unveiled substantial transcriptional heterogeneity, as each major ILC group presents distinct tissue-specific features that reflect the influence of the tissue microenvironment on phenotype and function [7–12]. Many tissue factors imprint ILCs: cytokines produced in steady state, such as TGFβ, or during tissue damage, particularly pro-inflammatory cytokines [13–15]; microbiota, bacterial products, and nutrients to which the tissue is exposed, such as short-chain fatty acids, ligands of the aryl hydrocarbon receptor, and vitamin A [16–20]; oxygen tension [21]; growth factors, like IGF-1 [22]; lipid mediators, such as prostaglandins [23, 24]; and neurotransmitters released by nerve fibers [25–27]. ILCs also express quorum-sensing molecules and chemokine receptors that further regulate their function based on cell density and spatial localization [28]. Overall, these observations emphasize the impact of tissue microenvironment on controlling ILC diversity and, conversely, the ability of ILCs to functionally adapt to local stimuli and tailor their responses to the tissue niche.

One of the latest developments in this regard is the recent observation that ILCs, particularly ILC3 in the gastrointestinal tract, adapt not only to their location, but also to circadian fluctuations of the tissue [29, 30]. The gastrointestinal tract, as many organs, is attuned to a circadian rhythm that primarily ensues in response to light and dark. ILC3 in the gastrointestinal tract express clock genes that undergo circadian oscillations and control the expression of genes encoding for cytokines. Moreover, ILC3 circadian oscillations are coordinated with the central clock in the suprachiasmatic nucleus of the central nervous system. As ILC3 expresses receptors for peptide neurotransmitters, the suprachiasmatic nucleus can connect long distance with ILC3 through enteric neurons secreting these peptides. Moreover, the daily cycle of nutrition can further contribute to entrain the ILC3 clock by food-induced circadian stimulation of enteric neurons [31]. It is likely that circadian rhythms control ILC functions in many other tissues to ensure timely coordination of ILC functions with those of the tissue.

1.3 ILC Versus T Cell Commitment: Shared and Unique Pathways

The parallel between ILC and T cell functional models extends to their development. Similar to T cells, all ILC subsets and NK cells originate from common lymphoid progenitors (CLPs). The commitment of CLPs towards ILCs rather than T cells occurs early on at the level of the common innate lymphoid progenitor CILP, the foremost progenitor with restricted potential to generate ILCs and NK cells [32, 33]. CILPs then differentiate into progenitors with more restricted potential, which together give rise to NK cells and all ILCs: the NK progenitor that gives rise to NK cells, the common helper lymphoid progenitor (CHILP), and the innate lymphoid cell progenitor (iLCP). The transcriptional repressor Id2 is a primary switch that propels the differentiation of CLPs towards NK cells and ILCs by blocking E2 family transcription factors that prompt T cell development [34–36]. However, the final delineation of NK cells, ILC1, ILC2, and ILC3 lineages from innate precursors largely depends on the same transcription factors that mediate polarization of T cells, i.e., Eomes, Tbet, GATA3, and Rorγt.

Epigenetic studies have confirmed the lineage and functional kinship of ILCs with their T cell counterparts. Analyses of gene regulatory circuitries of ILC-T pairs have revealed that each ILC type shares a circuitry devoted to lineage commitment and functional polarization with its Th counterpart [37]. However, epigenetic differences do exist between ILCs and their T cell counterparts, which are mainly related to their activating signals. T cell activation requires signals from TCR and co-stimulatory receptors, which drive expansion, differentiation, and cytokine expression. In contrast, ILC effector responses chiefly depend on tissue stimuli, endowing them with more rapid response profiles. Thus, ILCs and T cells employ both shared and divergent enhancers to express genes dependent on activating signals. The fundamental nature of these epigenetic differences in ILCs and T cells is still not well understood. Transcriptomic and chromatin accessibility studies on ILC subsets have been instrumental in comprehending the complexity of transcriptional modules in these cells. To build upon these initial findings, we should begin interrogating the epigenetic mechanisms that establish these modules, such as histone modifications, DNA methylation, and 3D chromatin conformation, as well as identifying ILC-specific regulatory elements.

Finally, it should be noted that ILCs have been shown to be rather plastic and can toggle their functional polarization in order to adapt their responses to disparate tissues and diverse pathogenic stimuli [13, 38]. Thus, it is possible that ILC lineage commitment is somewhat changeable and that gene regulatory circuitries may be flexible or reversible in certain contexts. It will be important to determine whether lineage commitment and plasticity are governed by distinct epigenetic modifications.

1.4 Where Do ILCs Develop?

Recent studies on macrophage development have identified two types of progenitors [39, 40]. Tissue-resident macrophages derive mainly from embryonal progenitors, which colonize developing tissues and persist throughout life by self-renewing. Monocyte-derived macrophages are generated from bone marrow progenitors during definitive hematopoiesis and populate tissues during inflammation and remodeling. Although ILCs have been extensively shown to develop from bone marrow progenitors, some ILC subsets are predominantly present in fetal tissues and tend to decline with age. Thus, it is possible that ILCs, like macrophages, develop in part from embryonal or fetal progenitors and populate peripheral tissues, generating a subset of cells capable of self-maintenance [41]. ILC development has also been observed in the thymus when T cell progenitors deviate from their developmental trajectory and become ILCs rather than mature T cells [42]. Human studies have also suggested the possibility that ILCs and NK cells may develop in part from fetal liver and thymus. Thus, it is likely that ILC diversity depends not only on tissue localization, but also on developmental origin. Future studies should address the life span of ILCs originating from disparate origins in both steady state and disease. Moreover, it is important to see whether human ILCs generated in vitro from various hematopoietic and/or lymphopoietic sources can be eventually exploited in cell-based therapies.

1.5 Tregs and T_{FH}: Why in T Cells Only?

It is of note that the similarity between ILCs and T cells seems limited to effector modules. In contrast to T cells, ILCs have not developed a separate lineage, such as Foxp3+ Tregs, dedicated to limiting the effector subsets. However, some ILCs can behave as regulatory cells. One subset of ILC2 can produce IL-10, acting as a regulatory cell in contexts in which ILC2 is exposed to IL-2 and IL-10 [43, 44]. ILC3 can induce tolerogenic T cells [45]. Through the expression of MHC class II, ILC3 can present antigens to T cells in a modality that induces tolerogenic rather than activated T cells, as ILC3 lacks costimulatory molecules. Additionally, ILC3 produces IL-2, which sustains Tregs [46]. These observations open interesting questions: Which mechanisms mediate antigen endocytosis and processing in ILC3? How does ILC3 antigen presentation differ from that of dendritic cells? What is the relative contribution of each to T cell responses? Do ILC3 and DC directly cooperate in T cell activation?

No direct counterpart of T_{FH} has been identified in ILCs. However, ILCs can stimulate B cell responses through multiple mechanisms. Embryonal LTi cells promote the ontogenesis of secondary lymphoid organs, which are essential for B cell responses [5]. Similarly, LTi-like ILC3 promotes the postnatal generation of intestinal cryptopatches, the antecedents of B cell-rich lymphoid follicles that produce IgA [47] and enhance IgA production in lamina propria and Peyer's patches by interacting with DCs [48, 49]. LTi cells and ILC3 also express lymphotoxins that promote B cell expansion and differentiation [50]. Thus, although ILCs have not developed a unique counterpart of T_{FH} cells, they do provide help to B cell responses.

1.6 ILCs: A Target for Disease Treatment?

Since their discovery, ILCs have been implicated in the defense against intracellular bacteria, extracellular bacteria, fungi, parasites, and viruses [6]. On the other hand, their uncontrolled activation has been implicated in various autoimmune diseases and allergies [51, 52]. More recent studies have shown that ILCs control lipid absorption and metabolism in the gut and adipose tissue [31, 53–55] and can participate in immune responses to cancer, with either pro-tumorigenic or anti-tumorigenic effects [56–58]. Beyond mouse models, ILCs have also been implicated in human diseases, with beneficial effects in infections and detrimental effects in autoimmune and allergic diseases [59]. Given the broad impact of ILCs in diseases, can we consider ILCs as suitable targets for immunotherapy? What tools are available for modulating ILCs? There is a major barrier that prevents a satisfying answer to these questions. ILCs share many programs and molecules with T cells. Because of this original sin, it is difficult to dissect the role of ILCs from that of T cells as well as to target ILCs independently of T cells. There are only few animal models in which ILCs can be selectively investigated. Indeed, to date, the most frequently used mouse model is the *Rag* knockout mouse in which adaptive responses are missing. Better models should be developed to analyze the impact of ILCs in the context of intact adaptive responses. No antibodies have been developed that can selectively deplete ILCs without impacting T cells in mice or humans. Available drugs (antibodies and small molecules) targeting ILC cell surface molecules [60], cytokines (IFNγ, IL-5, IL-13, IL-17, IL-22), signaling mediators (such as JAKs), and transcription factors (such as Rorγt) equally target T cells. One essential direction for future studies is the development of more sophisticated approaches to specifically modify ILCs in the context of mouse models and human diseases. Additionally, given the development of NK cell adoptive therapies for treating cancer [61] it is important to determine whether ILCs can also be effectively generated in vitro and used for adoptive transfer therapies to enhance innate immune responses [62] or, conversely, to modulate adaptive responses, depending on the context. Nonetheless, ILCs do have the potential to become the focus of a new generation of immunotherapies.

Acknowledgements I would like to thank Susan Gilfillan, Vincent Peng, and Marina Cella for helpful discussions.

References

1. Morvan MG, Lanier LL. NK cells and cancer: you can teach innate cells new tricks. Nat Rev Cancer. 2016;16(1):7–19.
2. Geiger TL, Sun JC. Development and maturation of natural killer cells. Curr Opin Immunol. 2016;39:82–9.
3. Lam VC, Lanier LL. NK cells in host responses to viral infections. Curr Opin Immunol. 2017;44:43–51.

4. Hammer Q, Ruckert T, Romagnani C. Natural killer cell specificity for viral infections. Nat Immunol. 2018;19(8):800–8.
5. Bar-Ephraim YE, Mebius RE. Innate lymphoid cells in secondary lymphoid organs. Immunol Rev. 2016;271 (1):185–99.
6. Vivier E, Artis D, Colonna M, Diefenbach A, Di Santo JP, Eberl G, et al. Innate lymphoid cells: 10 years on. Cell. 2018;174(5):1054–66.
7. Gury-Ben Ari M, Thaiss CA, Serafini N, Winter DR, Giladi A, Lara-Astiaso D, et al. The spectrum and regulatory landscape of intestinal innate lymphoid cells are shaped by the microbiome. Cell. 2016;166 (5):1231–46 e13.
8. Yu Y, Tsang JC, Wang C, Clare S, Wang J, Chen X, et al. Single-cell RNA-seq identifies a PD-1(hi) ILC progenitor and defines its development pathway. Nature. 2016;539(7627):102–6.
9. Xu H, Ding J, Porter CBM, Wallrapp A, Tabaka M, Ma S, et al. Transcriptional atlas of intestinal immune cells reveals that neuropeptide alpha-CGRP modulates group 2 innate lymphoid cell responses. Immunity. 2019;51(4):696–708 e9.
10. Bielecki P, Riesenfeld SJ, Hutter JC, Torlai Triglia E, Kowalczyk MS, Ricardo-Gonzalez RR, et al. Skin-resident innate lymphoid cells converge on a pathogenic effector state. Nature. 2021;592(7852):128–32.
11. McFarland AP, Yalin A, Wang SY, Cortez VS, Landsberger T, Sudan R, et al. Multi-tissue single-cell analysis deconstructs the complex programs of mouse natural killer and type 1 innate lymphoid cells in tissues and circulation. Immunity. 2021;54(6): 1320–37 e4.
12. Hernandez DC, Juelke K, Muller NC, Durek P, Ugursu B, Mashreghi MF, et al. An in vitro platform supports generation of human innate lymphoid cells from CD34(+) hematopoietic progenitors that recapitulate ex vivo identity. Immunity. 2021;54(10): 2417–32.
13. Colonna M. Innate lymphoid cells: diversity, plasticity, and unique functions in immunity. Immunity. 2018;48(6):1104–17.
14. Domingues RG, Hepworth MR. Immunoregulatory sensory circuits in group 3 innate lymphoid cell (ILC3) function and tissue homeostasis. Front Immunol. 2020;11:116.
15. Meininger I, Carrasco A, Rao A, Soini T, Kokkinou E, Mjosberg J. Tissue-specific features of innate lymphoid cells. Trends Immunol. 2020;41(10):902–17.
16. Eberl G. Development and evolution of RORgammat+ cells in a microbe's world. Immunol Rev. 2012;245(1): 177–88.
17. Belkaid Y, Harrison OJ. Homeostatic immunity and the microbiota. Immunity. 2017;46(4):562–76.
18. Britanova L, Diefenbach A. Interplay of innate lymphoid cells and the microbiota. Immunol Rev. 2017;279(1):36–51.
19. Chun E, Lavoie S, Fonseca-Pereira D, Bae S, Michaud M, Hoveyda HR, et al. Metabolite-sensing receptor Ffar 2 regulates colonic group 3 innate lymphoid cells and gut immunity. Immunity. 2019;51(5): 871–84 e6.
20. Fachi JL, Secca C, Rodrigues PB, Mato FCP, Di Luccia B, Felipe JS, et al. Acetate coordinates neutrophil and ILC3 responses against *C. difficile* through FFAR2. J Exp Med. 2020;217(3):20190489.
21. Pral LP, Fachi JL, Correa RO, Colonna M, Vinolo MAR. Hypoxia and HIF-1 as key regulators of gut microbiota and host interactions. Trends Immunol. 2021;42(7):604–21.
22. Zhou L, Sonnenberg GF. In situ support of ILC precursors. Immunity. 2020;52(2):207–9.
23. Kabata H, Moro K, Koyasu S. The group 2 innate lymphoid cell (ILC2) regulatory network and its underlying mechanisms. Immunol Rev. 2018;286(1): 37–52.
24. Konya V, Mjosberg J. Lipid mediators as regulators of human ILC2 function in allergic diseases. Immunol Lett. 2016;179:36–42.
25. Veiga-Fernandes H, Mucida D. Neuro-immune interactions at barrier surfaces. Cell. 2016;165(4): 801–11.
26. Klose CS, Artis D. Neuronal regulation of innate lymphoid cells. Curr Opin Immunol. 2019;56:94–9.
27. Godinho-Silva C, Cardoso F, Veiga-Fernandes H. Neuro-immune cell units: a new paradigm in physiology. Annu Rev Immunol. 2019;37:19–46.
28. Secca C, Bando JK, Fachi JL, Gilfillan S, Peng V, Di Luccia B, et al. Spatial distribution of LTi-like cells in intestinal mucosa regulates type 3 innate immunity. Proc Natl Acad Sci. 2021;118:23.
29. Burrows K, Mortha A. Going green with solar-powered ILC3 homeostasis. Sci Immunol. 2019;4 (40):0433.
30. Jacquelot N, Belz GT, Seillet C. Neuroimmune interactions and rhythmic regulation of innate lymphoid cells. Front Neurosci. 2021;15:657081.
31. Talbot J, Hahn P, Kroehling L, Nguyen H, Li D, Littman DR. Feeding-dependent VIP neuron-ILC3 circuit regulates the intestinal barrier. Nature. 2020;579 (7800):575–80.
32. Yu X, Wang Y, Deng M, Li Y, Ruhn KA, Zhang CC, et al. The basic leucine zipper transcription factor NFIL3 directs the development of a common innate lymphoid cell precursor. Elife. 2014;3:e04406.
33. Ishizuka IE, Constantinides MG, Gudjonson H, Bendelac A. The innate lymphoid cell precursor. Annu Rev Immunol. 2016;34:299–316.
34. Bedoui S, Gebhardt T, Gasteiger G, Kastenmuller W. Parallels and differences between innate and adaptive lymphocytes. Nat Immunol. 2016;17(5):490–4.
35. Qian L, Bajana S, Georgescu C, Peng V, Wang HC, Adrianto I, et al. Suppression of ILC2 differentiation from committed T cell precursors by E protein transcription factors. J Exp Med. 2019;216(4):884–99.
36. Spinner CA, Lazarevic V. Transcriptional regulation of adaptive and innate lymphoid lineage specification. Immunol Rev. 2021;300(1):65–81.

37. Fernando N, Sciume G, O'Shea JJ, Shih HY. Multidimensional gene regulation in innate and adaptive lymphocytes: a view from regulomes. Front Immunol. 2021;12:655590.
38. Bal SM, Golebski K, Spits H. Plasticity of innate lymphoid cell subsets. Nat Rev Immunol. 2020;20(9):552–65.
39. Lavin Y, Mortha A, Rahman A, Merad M. Regulation of macrophage development and function in peripheral tissues. Nat Rev Immunol. 2015;15(12):731–44.
40. Bleriot C, Chakarov S, Ginhoux F. Determinants of resident tissue macrophage identity and function. Immunity. 2020;52(6):957–70.
41. Turner JE, Gasteiger G. Innate lymphoid cells: key players in tissue-specific immunity. Semin Immunopathol. 2018;40(4):315–7.
42. Wang HC, Qian L, Zhao Y, Mengarelli J, Adrianto I, Montgomery CG, et al. Downregulation of E protein activity augments an ILC2 differentiation program in the thymus. J Immunol. 2017;198(8):3149–56.
43. Seehus CR, Kadavallore A, Torre B, Yeckes AR, Wang Y, Tang J, et al. Alternative activation generates IL-10 producing type 2 innate lymphoid cells. Nat Commun. 2017;8(1):1900.
44. Bando JK, Gilfillan S, Di Luccia B, Fachi JL, Secca C, Cella M, et al. ILC2s are the predominant source of intestinal ILC-derived IL-10. J Exp Med. 2020;217(2):e20191520.
45. Sonnenberg GF, Hepworth MR. Functional interactions between innate lymphoid cells and adaptive immunity. Nat Rev Immunol. 2019;19(10):599–613.
46. Zhou L, Chu C, Teng F, Bessman NJ, Goc J, Santosa EK, et al. Innate lymphoid cells support regulatory T cells in the intestine through interleukin-2. Nature. 2019;568(7752):405–9.
47. Tsuji M, Suzuki K, Kitamura H, Maruya M, Kinoshita K, Ivanov II, et al. Requirement for lymphoid tissue-inducer cells in isolated follicle formation and T cell-independent immunoglobulin A generation in the gut. Immunity. 2008;29(2):261–71.
48. Kruglov AA, Grivennikov SI, Kuprash DV, Winsauer C, Prepens S, Seleznik GM, et al. Nonredundant function of soluble LTalpha3 produced by innate lymphoid cells in intestinal homeostasis. Science. 2013;342(6163):1243–6.
49. Reboldi A, Arnon TI, Rodda LB, Atakilit A, Sheppard D, Cyster JG. IgA production requires B cell interaction with subepithelial dendritic cells in Peyer's patches. Science. 2016;352(6287):aaf 4822.
50. Magri G, Miyajima M, Bascones S, Mortha A, Puga I, Cassis L, et al. Innate lymphoid cells integrate stromal and immunological signals to enhance antibody production by splenic marginal zone B cells. Nat Immunol. 2014;15(4):354–64.
51. Ebbo M, Crinier A, Vely F, Vivier E. Innate lymphoid cells: major players in inflammatory diseases. Nat Rev Immunol. 2017;17(11):665–78.
52. Friedrich M, Pohin M, Powrie F. Cytokine networks in the pathophysiology of inflammatory bowel disease. Immunity. 2019;50(4):992–1006.
53. O'Sullivan TE, Rapp M, Fan X, Weizman OE, Bhardwaj P, Adams NM, et al. Adipose-resident group 1 innate lymphoid cells promote obesity-associated insulin resistance. Immunity. 2016;45(2):428–41.
54. Guendel F, Kofoed-Branzk M, Gronke K, Tizian C, Witkowski M, Cheng HW, et al. Group 3 innate lymphoid cells program a distinct subset of IL-22BP-producing dendritic cells demarcating solitary intestinal lymphoid tissues. Immunity. 2020;53(5):1015–32.
55. Cardoso F, Klein Wolterink RGJ, Godinho-Silva C, Domingues RG, Ribeiro H, da Silva JA, et al. Neuro-mesenchymal units control ILC2 and obesity via a brain–adipose circuit. Nature. 2021;597(7876):410–4.
56. Panda SK, Colonna M. Innate lymphoid cells: a potential link between microbiota and immune responses against cancer. Semin Immunol. 2019;41:101271.
57. Bald T, Wagner M, Gao Y, Koyasu S, Smyth MJ. Hide and seek: plasticity of innate lymphoid cells in cancer. Semin Immunol. 2019;41:101273.
58. Stamatiades EG, Li MO. Tissue-resident cytotoxic innate lymphoid cells in tumor immunosurveillance. Semin Immunol. 2019;41:101269.
59. Klose CS, Artis D. Innate lymphoid cells as regulators of immunity, inflammation and tissue homeostasis. Nat Immunol. 2016;17(7):765–74.
60. Guia S, Fenis A, Vivier E, Narni-Mancinelli E. Activating and inhibitory receptors expressed on innate lymphoid cells. Semin Immunopathol. 2018;40(4):331–41.
61. Myers JA, Miller JS. Exploring the NK cell platform for cancer immunotherapy. Nat Rev Clin Oncol. 2021;18(2):85–100.
62. Cobb LM, Verneris MR. Therapeutic manipulation of innate lymphoid cells. JCI Insight. 2021;6(6):e146006.

2 ILC Differentiation from Progenitors in the Bone Marrow

Arundhoti Das, Christelle Harly, Yi Ding, and Avinash Bhandoola

Abstract

Innate lymphoid cells (ILCs) are a family of immune cells that possess similar functions as T cells. We review steps of central ILC development in the bone marrow of adult mice and discuss recent evidence for peripheral ILC development suggesting extramedullary sites of ILC development. We also assess the contribution of development during different phases of life towards shaping the composition of the adult ILC pool. Finally, we briefly review the local cues that lead to heterogeneity of ILCs between tissues. We propose that tissue-resident ILC progenitors may economically allow tissues to elicit rapid expansion of specific ILC types that are needed based on the nature of antigenic assaults in tissues.

Keywords

Innate lymphoid cells · Central ILC progenitors · Local precursors

A. Das · Y. Ding · A. Bhandoola (✉)
T-Cell Biology and Development Unit, Laboratory of Genome Integrity, Center for Cancer Research, National Cancer Institute, National Institute of Health, Bethesda, MD, USA
e-mail: avinash.bhandoola@nih.gov

C. Harly
Université de Nantes, CNRS, INSERM, CRCINA, Nantes, France

LabEx IGO "Immunotherapy, Graft, Oncology", Nantes, France

2.1 Introduction

Innate lymphoid cells (ILCs) are immune cells that lack specific antigen receptors but possess similar effector functions as T cells. Concordantly, ILCs express many transcription factors known to be important for T cell effector functions. For these reasons, ILCs are considered innate counterparts of effector T cells.

Based on their functions and their expression of key transcription factors, ILCs are broadly classified into group 1 ILCs comprising NK cells and ILC1s, group 2 ILCs (ILC2s), and group 3 ILCs that comprise ILC3s and lymphoid tissue inducer (LTi) cells [1–3]. Another ILC subset analogous to T regulatory cells, regulatory ILC (ILC_{reg}), was identified in the mouse and human intestine based on its ability to secrete IL-10 and suppress cytokine production by ILC1s and ILC3s in intestinal inflammation [4]; however, this subset is not observed in all strains of laboratory mice [5]. Recently, ILC2s were identified as a major source of IL-10 in the intestine [5]. Additionally, recent evidence indicates the existence of functionally intermediate states between the main ILC groups [6–8].

ILCs display a wide array of receptors on their cell surface that enable responses to local cues and facilitate interaction with adaptive immune cells [9–11]. These receptors are distinct from the clonally expressed TCR on T cells; however, they trigger ILC immune effector functions that are similar to T cell effector functions

© The Author(s), under exclusive license to Springer Nature Singapore Pte Ltd. 2022

X.-H. Sun (ed.), *Innate Lymphoid Cells*, Advances in Experimental Medicine and Biology 1365,
https://doi.org/10.1007/978-981-16-8387-9_2

[12, 13]. Like $CD8^+$ T cells, NK cells are cytotoxic to tumor cells and virus-infected cells. The canonical cytokines produced by Th1, Th2, and Th17 cells are secreted by ILC1s, ILC2s, and ILC3s, respectively [14]. ILCs and T cells share transcriptional networks that are likely to be responsible for the similarities of their effector functions [15]. ILCs are located in various lymphoid and nonlymphoid tissues [16]. Despite many similarities between T cell and ILC, ILCs possess some key features that distinguish ILCs from T cells. They do not express antigen-specific receptors on the cell surface. As they are part of the innate immune system, ILCs were thought to lack memory. However, recent studies indicate that ILC2s as well as NK cells do possess memory features [17, 18].

Initially, studies on ILC development were largely focused at primary hematopoietic sites, which are the liver in fetal mice or the bone marrow in adult mice [19, 20]. However, many recent reports identified ILC progenitors in peripheral tissues in mice and humans, suggesting the presence of peripheral ILC development [21–24]. However, the extent to which ILC development occurs centrally versus at peripheral sites is still unclear.

In this chapter, we present an overview of the steps of central ILC development in the bone marrow of adult mice, and we also present recent evidence for peripheral ILC development.

2.2 ILC Subsets and Their Sites of Abundance

ILC subsets are distributed widely throughout the body [25]. ILCs are found in primary lymphoid organs such as bone marrow and thymus, secondary lymphoid organs, as well as nonlymphoid tissues such as skin, liver, lung, small intestine, colon, uterus, salivary gland, and fat [26–28].

2.2.1 Group 1 ILC

Group 1 ILCs include both ILC1 and conventional NK (cNK) cells. They are found in liver, spleen, gut, skin, uterus, and salivary glands [29, 30]. cNK cells recirculate between blood and tissues, and are cytotoxic effectors that can kill virus-infected cells [31, 32], whereas ILC1s are mainly tissue resident and are weakly cytotoxic [33]. ILC1s function as a first line of defense against infections with viruses and certain bacteria such as *T. gondii* by secreting IFN-γ at the local sites of infections [34]. Both ILC1s and cNK cells produce IFN-γ and TNF-α, and both express the transcription factor T-bet. T-bet is indispensable for ILC1 development and important for terminal maturation of NK cells [35]. In addition to T-bet, NK cells also require transcription factor Eomesodermin (Eomes) for development, and Eomes is required for the expression of perforin and granzymes that promote cytotoxic functions of NK cells [36]. Thymic NK cells express IL-7Rα and need GATA-3 for development unlike cNK cells, and are less cytotoxic [37–39]. Tissue-resident NK cells in liver display distinct markers and transcription factors as compared to thymic and conventional splenic NK cells [40]. Recently, a comprehensive single-cell analysis identified the transcriptional programs of ILC1s and NK cells from multiple tissues, including blood, spleen, liver, salivary glands, uterus, small intestines, and adipose tissues. $NK1.1^+$ $NKp46^+$ cells are classified into ILC1s with $EOMES^-$ $TCF\text{-}1^{lo}$ $HOBIT^+$ phenotype and NK cells with $EOMES^+$ $TCF\text{-}1^{lo/hi}$ $HOBIT^-$ phenotype [41, 42]. The study allowed better discrimination of tissue-specific ILC1s from tissue-specific and circulating NK cells. However, there is at present no cell surface marker that reliably differentiates NK cells from ILC1s in all tissues.

2.2.2 Group 2 ILC

Group 2 ILCs are dominant in lung, skin, white adipose tissues, and small intestinal lamina propria. ILC2s express transcription factor GATA-3 and retinoic acid receptor (RAR)-related orphan receptor (RORα) that are indispensable for ILC2 development. They do not highly express transcription factors Eomes,

T-bet, and RORγt that are characteristic of other ILC subsets [43]. ILC2s are stimulated by the cytokines IL-33, TSLP, and IL-25 [44]. They produce IL-5 and IL-13 which induce eosinophilic lung inflammation [45], as well as amphiregulin [46], IL-2, IL-9, and IL-10 [5, 47, 48]. ILC2s can express KLRG1, GITR/ligand, and ICOS/ligand that enable ILC2s to interact with a variety of immune cell types and participate in a broad range of immune responses [49]. ILC2s can also produce granulocyte-macrophage-colony-stimulating factor (GM-CSF) that facilitates eosinophil recruitment and enhances antitumor immunity [50]. ILC2s co-localize with cholinergic neurons that express the neuropeptide neuromedin U (NMU). ILC2s express NMU receptor 1 (NMUR1) and respond to NMU both in vitro and in vivo. NMU-NMUR1 signaling triggers type 2 cytokine responses characterized by ILC2 activation, proliferation, and secretion of type 2 cytokines including IL-5, IL-9, and IL-13 [51–54].

There are two distinct subsets of ILC2s. One subset is inflammatory ILC2s that are IL-25 responsive, express high levels of KLRG1, and produce large amounts of IL-5 and IL-13. The second subset is termed natural ILC2s that express ST2 (IL-33 receptor) and are IL-33 responsive. Natural ILC2s produce amphiregulin in addition to IL-5 and IL-13, and are involved in tissue repair [22, 55–58].

2.2.3 Group 3 ILC

Group 3 ILCs are abundant in intestinal crypts, lamina propria, Peyer's patches (PPs), and lymph nodes [59, 60]. They provide immunity against extracellular bacteria and maintain the integrity of the intestinal barrier [61]. ILC3s are also present within marginal zone areas of spleen where they mediate antibody production through secreting B cell helper factors APRIL and BAFF and by signaling through CD40L [62]. ILC3s are heterogenous based on the expression of surface markers (NKp46, CCR6, CD4), and on their appearance during ontogeny, but all ILC3s express the transcription factor RORγt (RAR-related orphan receptor gamma) [61]. ILC3s are further subdivided into subsets based on the cell surface expression of NKp46 and CCR6. The subset with NKp46 expression is termed natural cytotoxicity receptor-positive (NCR^+) ILC3s, the subset expressing CCR6 on its surface is termed $CCR6^+$ ILC3s, and the subset lacking NKp46 and CCR6 expression on the cell surface is called NCR^- ILC3s [63, 64]. Further, $CCR6^+$ ILC3s that are CD4 expressing are called LTi in fetal life and LTi-like ILC3s in adult life [65]. $CCR6^+$ $ROR\gamma t^+$ ILC3s arise earlier during ontogeny. Postnatally emerging $CCR6^-$ $ROR\gamma t^+$ ILCs upregulate the expression of T-bet in response to cues from microbiota and IL-23 [66, 67]. T-bet expression in $CCR6^-$ $ROR\gamma t^+$ ILC3s increases from the NCR^- to NCR^+ ILC3 subset [64].

The predominant cytokine produced by ILC3s is IL-22 which induces epithelial cells to produce antimicrobial peptides that kill bacteria [68, 69], promotes the proliferation of intestinal stem cells [68–72], and maintains intestinal homeostasis by inducing T cell tolerance to bacterial antigens through MHC class II presentation [9, 73]. In addition to IL-22, NCR^+ ILC3s produce IFN-γ, GM-CSF, and TNF [74–77], whereas $CCR6^+$ ILC3s express IL-17A and IL-17F [64, 78].

LTi cells are considered group 3 ILCs because they express transcription factor RORγt, a lineage defining transcription factor for group 3 ILCs [79]. LTi cells are found in spleen, blood, and lymph node (LN) analgen by E12.5 and in Peyer's patch (PP) analgen by E16. They continue to be present in very low numbers after birth in cryptopatches of small intestine, and colon and cortex of LN and PPs [80]. LTi cells are required for fetal organogenesis of LN and PP through the action of lymphotoxin [81]. LTi-like cells in adults play a role in the restoration of secondary lymphoid organ architecture including spleen after acute phase of LCMV infection [82].

2.3 ILC Progenitors and Steps of ILC Differentiation at Central Sites

ILCs differentiate from hematopoietic stem cells (HSCs) via lymphoid progenitors, to yield diverse ILC progenitors in fetal liver and adult bone marrow. Several transcription factors and signaling pathways regulate this differentiation process in mice and humans; however, human ILC development is less well characterized.

ILCs develop from lymphoid progenitors that reside in fetal liver and adult bone marrow (BM) [83–88]. ILCs, like B cells and T cells, arise from all lymphoid progenitors (ALP) that contain IL-7Rα-expressing lymphoid-primed multipotent progenitors ($IL7R\alpha^+$ LMPP) and $Ly6D^-$ common lymphoid progenitors (CLP) [85, 89–95]. The differentiation of ILCs from ALPs occurs via intermediate stages initially identified by cell surface expression of proteins, and subsequently by staged expression of transcription factors that drive ILC development (Table 2.1).

Pioneering studies found that the progenitor potential for ILCs resides downstream of all lymphoid progenitors (ALP) [91, 93, 97]. The earliest progenitors have been further characterized by the expression of transcription factors TCF-1, TOX, NFIL3, and GATA-3, and by the expression of cell surface integrin $\alpha_4\beta_7$ and Flt3 [83, 96]. One intermediate called common helper-like innate lymphoid progenitor (CHILP) defined as $lineage^-$ $Id2^{high}$ $IL\text{-}7R\alpha^+$ $Flt3^-$ $\alpha_4\beta_7{}^{high}$ $PLZF^{+/-}$ was identified that generated ILC subsets in vitro and in vivo [85, 86]. A more committed ILC precursor (ILCP) expressing PLZF (encoded by *Zbtb16*) was identified that generated all ILCs except LTi cells [84]. Gene expression analysis identified PD-1 to be specifically expressed by ILCP, thus allowing identification of BM ILCP without requiring PLZF reporter mice [102, 103]. $\alpha_4\beta_7$-expressing lymphoid progenitors (αLPs), another intermediate of ILC development, are defined as $lineage^-$ $IL7R\alpha^+$ $\alpha_4\beta_7{}^+$ kit^{lo} $Flt3^{+/-}$ and are a heterogenous population that contain CHILP and ILCP. They can generate all ILC subsets but retain some T cell potential. Among them, $CXCR6^+$ αLP cells lack T and B cell potentials but differentiate into all ILC subsets [97, 104]. However, these progenitors are quite rare in bone marrow and also this definition misses key ILC progenitors that are $IL7R\alpha^{low}$.

More recently, early innate lymphoid progenitors (EILPs) were identified as $IL7R\alpha^{neg/low}$ TCF-1-expressing cells that contained progenitor potential for all adult ILC subsets [83]. EILPs are intermediate precursors between ALP and ILCP that transiently downregulate IL7Rα [96]. Recent work identified transcriptional and functional heterogeneity within EILP in bone marrow using single-cell approaches and characterized two successive steps of development within EILP that were termed "specified EILP" (sEILP) that generated dendritic cells (DC) along with

Table 2.1 Current definitions for early ILC progenitors

Name	Definition	References
ALP	$Lineage^-$ $Ly6D^-$ kit^+ $CD122^-$ $Flt3^+$ $IL7R\alpha^+$ $\alpha_4\beta_7{}^-$	[94]
αLP	$Lineage^-$ kit^{lo} $Sca\text{-}1^{lo}$ $Flt3^-$ $IL7R\alpha^+$ $\alpha_4\beta_7{}^+$ (includes $CXCR6^+$ subset lacking T cell potential)	[95]
EILP	$Lineage^-$ kit^+ $TCF\text{-}1^+$ $Thy\text{-}1^-$ $IL7R\alpha^{lo\text{-}neg}$ $\alpha_4\beta_7{}^+$	[81]
CHILP	$Lineage^-$ $Id2^+$ $Thy\text{-}1^+$ $IL7R\alpha^+$ $\alpha_4\beta_7{}^+$ (includes both $PLZF^+$ and $PLZF^-$ progenitors)	[83]
ILCP	$Lineage^-$ $PLZF^+$ $Thy\text{-}1^+$ $IL7R\alpha^+$ $\alpha_4\beta_7{}^+$	[82]
NKP	$Lineage^-$ $CD122^+$ $NK1.1^-$ $DX5^-$ (heterogenous and refined as rNKP: $Lineage^-$ $CD27^+$ $CD244^+$ $CD122^+$ $Flt3^-$)	[96, 97]
LTiP	$Lineage^-$ $IL7R\alpha^+$ $\alpha_4\beta_7{}^{+\cdot}$ $Flt3^-$ $Ror\gamma t\text{-}EGFP^+$ $CXCR5^+$	[77, 90]
ILC1P (immature ILC1)	$Lineage^-$ $ROR\gamma t\text{-}fm^-$ $CD49a^+$ $IL7R\alpha^+$	[83, 98]
ILC2P	$Lineage^-$ $Sca1^{hi}$ $CD25^+$ $\alpha4\beta7^+$ $Flt3^-$	[83, 99]
ILC3P	$Lineage^-$ $Id2^+$ $IL7R\alpha^+$ $Bcl11b\text{-}tdTomato^+$ $GATA\text{-}3\text{-}hCD2^{lo}$ $Ror\gamma t\text{-}kat^+$	[85]

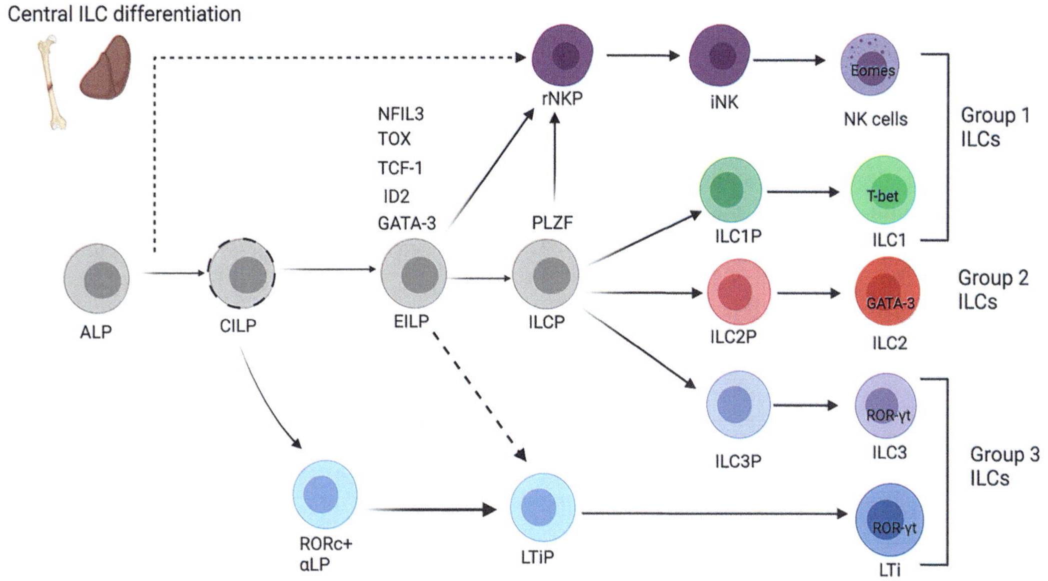

Fig. 2.1 Model for central ILC development. All lymphoid progenitors (ALP) give rise to common lymphoid innate lymphoid progenitors (CILP) that are not well identified. Bifurcation into ILC and LTi lineage is identified by the expression of *Rorc* and low *Id2* expression. EILPs, identified by *Tcf7* expression, arise downstream of CILP that can generate all ILC lineages and myeloid cells. Downstream of EILPs contain precursors with ILC and LTi lineage potential. ILCP is identified by PLZF expression and can give rise to ILC1/ILC2/ILC3 and NK cells. RORc+ αLPs are the earliest LTi progenitors that express RORγt but not TCF-1. LTiPs are identified by Rorc expression along with high expression of TCF-1 and ID2 and generate LTi cells. Lineage-biased ILC progenitors, putative ILC1P, ILC2P, and ILC3P, are identified in bone marrow that give rise to ILC1, ILC2, and ILC3, respectively. NK cells arise from different pathways as shown in the figure via NK-committed progenitors in bone marrow. The black lines show successive stages of ILC differentiation and maturation that are supported by published data. The dashed lines represent proposed steps of ILC differentiation, but these are not yet established by published data. Some key transcription factors are shown where they are expressed

ILC subsets and "committed EILP" (cEILP) that lacked DC potential and only gave rise to ILC lineages, similar to ILCP. TCF-1 was found to be dispensable for sEILP development whereas it was essential for generating cEILP and later developmental stages of ILCP [105].

Fetal LTi cells are generated from common lymphoid progenitors (CLPs) via αLP in the fetal liver [104, 106]. Fetal αLP more efficiently generated mature LTi compared to their adult counterparts in bone marrow [91, 107]. In fetal mice, the *Tcf7*-expressing EILP population was discovered to contain a population of *Rorc*-expressing LTi progenitors that was much rarer in adult mice [88]. A novel LTi-specified precursor (*Rorc*$^+$ αLP) was identified that expressed *Rorc* but not *Tcf7*, and so was distinct from EILP (Fig. 2.1). Cell fate mapping indicated that LTi cells belong to a separate lineage from ILC3s [84].

In humans, Kit$^+$ multipotent ILC progenitors were identified in peripheral blood [21, 108]. Recently, Lin$^-$ CD34$^+$ CD127$^+$ IL3RA$^+$ progenitors were identified in human fetal liver from 8–12 PCW that generated NK cells, ILC1s, and ILC2s efficiently but not ILC3s [109].

Lineage-biased precursors are also present in adult BM in mice. ILC2-restricted progenitors have been identified in BM [45, 101], and studies also indicate the existence of immature ILC1s or putative ILC1P in fetal liver and bone marrow [85, 100]. A recent study generated "5X polychromILC" transcription factor reporter mouse model that allowed examination of previously unappreciated ILC precursor heterogeneity in bone marrow and identification of rare Bcl11b$^+$

ILC3 progenitors in bone marrow. Further, single-cell analysis confirmed the divergence of ILC2P from other lineage-committed ILC precursors [87]. Another study showed that Id2$^+$ Zbtb16$^-$ Bcl11b$^-$ ILC precursor populations in BM largely harbored cells with ILC3 potential, and upregulation of Bcl11b and/or Zbtb16 was associated with loss of ILC3 potential. These data suggest that the emergence of ILC3s from Id2$^+$ ILCP may represent a branch point in ILC development that separates an ILC3 pathway from ILC1, ILC2, and/or NK cell pathways via upregulation of *Zbtb16* [86].

Immature NK cells and NK progenitors are present in BM [110]. NK cell development progresses from ALP to a heterogeneous population of NK progenitors (NKP), defined as Lin$^-$ CD122$^+$ NK1.1$^-$ DX5$^-$, to Lin$^-$ CD122$^+$ NK1.1$^+$ DX5$^-$ immature NK (iNK) cells and then to Lin$^-$ CD122$^+$ NK1.1$^+$ DX5$^+$ mature NK (mNK) cells [98, 111]. Subsequent studies identified earlier NK lineage-committed progenitors, the pre-NKP cells and refined NKP (rNKP) which gave rise to NK cells, although their ability to generate other ILC subsets was not assessed [99, 112]. CHILPs, identified using Id2-GFP reporter mice, were thought to lack NK potential [85]; however, recently described Id2-RFP reporter mice allowed identification of Id2$^+$ progenitors that gave rise to all helper ILCs and NK cells efficiently in vitro and in vivo [86]. These differences may be explained by the better discrimination of progenitors in Id2-RFP reporter mice, allowing the isolation of ILCP and NK cell progenitors that expressed Id2. Furthermore, fate mapping data using a *Zbtb16*-Cre mice showed that a minor fraction of NK cells are labelled by the expression of *Zbtb16* suggesting that the majority of NK cells may develop through another pathway which is upstream of ILCP [84, 100]. It is not fully understood how and when this split occurs and what are the underlying mechanisms that dictate the helper versus killer lineages.

2.4 Transcription Factor Requirement During Central ILC Development

BM progenitors differentiating along the ILC lineages gradually upregulate key transcription factors NFIL3 (encoded by *Nfil3*), TCF-1 (*Tcf7*), TOX (*Tox*), ID2 (*Id2*), GATA-3 (*Gata3*), and PLZF [15, 92, 96, 113, 114] (Fig. 2.1). There are many transcription factors that are important at early phases of ILC development [96, 114, 115] whereas others are important for the development of specific ILC subsets such as RORα and Bcl11b that are specifically required for ILC2 development [45, 116–118].

2.4.1 NFIL3

NFIL3 (also called E4BP4) is a basic leucine zipper transcription factor that is indispensable for conventional NK cell development [119]. *Nfil3*$^{-/-}$ mice have defects in early ILC development and lack mature ILC subsets [97, 120, 121]. *Nfil3*-null mice have greatly reduced or complete absence of PLZF$^-$ and PLZF$^+$ CHILP [122]. NFIL3 transcriptionally controls *Tox* expression [97, 123] as well as EOMES and ID2 by directly binding to the *Eomes* and *Id2* loci (see ID2 in Sect. 2.4.4) [122, 123].

2.4.2 TOX

TOX (thymocyte selection-associated HMG bOX protein) is essential for CD4$^+$ T cell development. TOX also plays an essential role in ILC development. *Tox*$^{-/-}$ mice have greatly reduced number of EILP and lack ILCP in bone marrow [96, 124]. *Tox*$^{-/-}$ mice lacked lymph nodes, and the number of Peyer's patches was significantly reduced [123].

2.4.3 TCF-1

T cell factor 1 (TCF-1), encoded by gene *Tcf7*, is a sequence-specific high-mobility-group transcription factor. TCF-1 is not expressed in ALPs, but its upregulation occurs at the EILP stage. It continues to be expressed at the later ILCP stage and is downregulated in ILC2 in bone marrow. *Tcf7*$^{-/-}$ mice lacked cEILP, ILCP, ILC2P, and NKP but ALP and sEILP were unaffected [83, 105].

2.4.4 ID2

ID2 (inhibitor of DNA binding) belongs to the family of helix-loop-helix (HLH) proteins that form heterodimers with E proteins, thereby preventing their transcriptional activities [125, 126]. Id2 plays an important role in NK cell maturation, in part via its effects on TCF-1 [127]. High levels of TCF-1 expression promoted immature NK cell expansion and inhibited terminal maturation. Id2 inhibited the expression of TCF-1 during late NK development to support the terminal differentiation of immature NK cells [128]. Id2 is required early during ILC development. *Id2*$^{-/-}$ mice showed twofold reduction in EILP, and downstream ILCP and ILC2P in bone marrow were absent. Also, *Id2*$^{-/-}$ EILPs showed increased expression of Id1 and Id3, suggesting that other Id proteins may compensate for the deficiency of Id2 during ILC development [83]. The effects of sustained E protein activity were assessed using mice expressing ET-2, which prevents ID proteins from inhibiting E protein function. In this model, ILC2 development was inhibited [129]. *Id2*$^{-/-}$ mice also lack LTi cells, and thus lack lymph nodes and Peyer's patches [130].

2.4.5 GATA-3

GATA-3 is another transcription factor required for ILC development. GATA3 is expressed by all ILC progenitors and mature ILC subsets, although at different levels. *Gata3*$^{-/-}$ mice showed reduction in EILP population whereas the ILCP subset was completely absent in adult bone marrow [96]. Recent work showed that levels of GATA-3 expression controlled the bifurcation of non-LTi lineages and LTi cells. GATA3 was dispensable for LTi cells but was essential for the generation of other ILCs. Further, a low level of GATA-3 expression mediated by a transgene was sufficient to restore LTi functions including lymph node formation, but failed to rescue the development of ILCP in bone marrow and mature ILC1s, ILC2s, and NCR$^+$ ILC3s in peripheral tissues [131].

2.4.6 PLZF

PLZF, encoded by gene *Zbtb16*, is a zinc finger protein that is highly expressed in bone marrow ILC progenitors. PLZF is important for ILC2P development but is dispensable for EILP and ILCP development in BM [84, 96]. Competitive chimeras revealed the requirement of PLZF in ILC2s' and liver ILC1s' development but not for the development of ILC3s, fetal LTi, and NK cells [84].

2.5 ILC Development at Extramedullary Sites

ILCs are relatively abundant in nonlymphoid tissues such as lung, skin, liver, adipose tissues, and gut. It is possible that extramedullary maturation may contribute for the abundance and diversity of ILCs at different anatomical locations. ILCs arise from hematopoietic progenitors in fetal liver and adult bone marrow. Fetal RORγt$^+$ progenitors seed the intestine at embryonic day E12.5–13.5, and support the development of lymphoid structures via expression of lymphotoxins [91, 92, 107]. T-bet$^+$, GATA-3$^+$, and RORγt$^+$ ILCs are present in the fetal gut and liver at E15.5 [104, 132]. BM ILCP expresses homing molecules suggesting that they may exit the bone marrow and home to different peripheral sites [96]. Furthermore, identification of ILCP in tissues of mice and humans suggests that some steps of ILC differentiation and maturation might happen locally [21–24, 108, 133, 134].

Transcription factors and cytokines that are required during early central ILC development are also required for the generation of mature peripheral ILC subsets, indicating that ILCs in tissues either originate from central ILC precursors or rely on similar transcriptional programs for development. However, there are some exceptions. One puzzle is that some transcription factors appear essential for the development of early ILC progenitors in bone marrow, but only mild defects are detected in mature ILCs at peripheral sites. For example, TCF-1 is indispensable for BM ILC precursors, and EILP, ILCP, and NK progenitors were nearly absent in *Tcf7*$^{-/-}$ mice [83, 105], but some mature NCR^{-} ILC3s, ILC1s, and NK cells were present [135]. However, competitive chimeras with WT and *Tcf7*$^{-/-}$ HSCs exhibited a near-complete lack of *Tcf7*$^{-/-}$ mature ILCs, suggesting that *Tcf7* might be required at early stages of ILC development but is dispensable in mature ILCs. Possibly, early defects in development are obscured by proliferation at later stages of ILC development.

There are other studies where proliferation does not easily explain the defects seen in early progenitors and mature ILC populations. *Nfil3*$^{-/-}$ mice lack central ILCs and NK progenitors and mature NK cells in bone marrow whereas NK cells remain unaffected in salivary glands [123, 136]. Tissue-resident NK cells are present in liver, uterus, and skin of *Nfil3*$^{-/-}$ mice [137]. Additionally, ILC1s and ILC3s are also present in the uterus of *Nfil3*$^{-/-}$ pregnant females [138, 139] and ILC1s are intact in the thymus of *Nfil3*$^{-/-}$ mice [38]. The presence of these mature cells in peripheral tissues even in competitive chimeras hints at the possibility that there may be differential requirement of transcription factors in central versus peripheral ILC development [140]. A related puzzle appears from the study of *Tox*-deficient mice where central ILC progenitors as well as many mature ILC subsets are greatly reduced or absent; however, populations of mature ILC3s (NCR^{+} ILC3s and NCR^{-} ILC3s) remain numerically intact in the gut [96, 124, 141].

Furthermore, environmental factors can shape the effector response of ILC progenitors. Lung ILCP and BM ILCP defined as *Zbtb16*-expressing precursors possess very similar transcriptomes, but they differ in the response to IGF1 growth factor signaling and chemotaxis [24]. Another study showed that ILC progenitors were 20% fate-mapped positive for *Il5* in lungs but not in bone marrow of *Il5-Cre* lineage-tracing mice, indicating that ILCP gained *Il5* expression locally and matured towards ILC2s within lung tissue [23].

These studies suggest that there is a differential requirement of some transcription factors for ILC development and maturation in central versus peripheral sites. Such tissue-resident progenitors may differentiate into mature ILC subsets locally in response to specific environmental cues to meet local demand.

2.6 ILC Progenitors in Tissues

Lineage-tracing experiments using PLZF$^{GFPcre+/-}$ mice carrying a ROSA26-floxstop-YFP fate-mapping allele showed nearly 70% labelling of lamina propria ILC2s whereas YFP labelling was significantly lower in lamina propria ILC3s, suggesting that distinct progenitors may contribute to the adult ILC pool in tissues. Furthermore, different ILC3 subsets from lamina propria were differentially fate mapped; approximately 40% of NCR^{+} ILC3s, 10% of CD4^{-}, and 1–2% of CD4^{+} LTi populations were fate mapped suggesting that different pathways in the course of ILC development may lead to the generation of different subsets of ILC3s [84]. Several reports have identified multipotent ILC progenitors in tissues in mice and humans indicating that some steps of ILC maturation may occur locally, at peripheral sites. ILC progenitors were identified in fetal mice by E13.5 in proximal gut. These were heterogenous and express varying amounts of Rorγt, T-bet, and GATA-3. However, these fetal gut precursors were largely ILC3 lineage-restricted in clonal assays [132]. Upstream hematopoietic progenitors including HSCs were present in lung and repopulated bone marrow under conditions of stem cell deficiency [142]. In the thymus, multipotent progenitors and committed T cell precursors (DN3) could differentiate into thymic ILC2s in vitro and in vivo. Furthermore, ablation of E proteins greatly enhanced the ILC fate while impairing B and T cell development [143, 144]. Future work should investigate whether ILC development normally occurs in the thymus.

IL18Rα^+ ST2$^-$ ILC progenitors were identified in neonatal lung using RORα lineage tracer mice that efficiently gave rise to multiple ILC lineages both in vitro and in vivo [145]. Recent evidence suggests that *Igfr1*-expressing ILC progenitors were present in neonatal lung. Conditional deletion of IGF1 in alveolar fibroblasts or specific deletion of IGF-1 receptor from ILC precursors in lungs significantly reduced ILC3 biogenesis and rendered newborn mice susceptible to pneumonia despite having normal ILCP in bone marrow [24]. Adult mouse liver also contained progenitors, Lin$^-$ Sca-1$^+$ Mac-1$^+$ (LSM), that were able to differentiate into multiple hematopoietic lineages and preferentially generated ILC1s rather than cNK cells in liver [146].

Several ILC progenitors are also described in humans. Multipotent Kit$^+$ ILC progenitors were detected in peripheral blood and different organs in humans suggesting that circulating Kit$^+$ ILCP is able to migrate to tissues and differentiate into mature ILC subsets [21]. This Kit$^+$ population was shown to be heterogeneous based on the surface expression of NKp46, CD56, and KLRG1. The KLRG1$^+$ subset preferentially differentiated into ILC2s but could also differentiate into ILC3s in appropriate conditions. The NKp46$^+$ CD56$^-$ subset possessed minimal ILC2 lineage potential, and instead gave rise to ILC3s and also NK cells [108]. Specified ILC3P identified as RORγt$^+$ CD34$^+$ hematopoietic progenitors were independently detected in human tonsils and intestine [133, 134].

These studies show that ILCP or other multipotent ILC progenitors seed various peripheral tissues and support ILC development locally. ILC progenitors in tissues may contribute to ILC-poiesis locally in response to infection and inflammation [147].

2.7 Tissue-Homing Features of Central ILC Progenitors

Extramedullary sites such as spleen, liver, blood, and peripheral organs harbor hematopoietic stem and progenitor cells [148–153]. The abundance of ILCs in tissues may depend on homing of ILC progenitors from central sites to tissues based on their surface expression of homing molecules. ALP and other downstream progenitors express chemokine receptors including CCR7, CCR9, and CXCR4 that can facilitate the migration of these precursors to different tissues [21, 96]. The adhesion molecule $\alpha_4\beta_7$ is highly expressed in EILP and ILCP that can help in homing to intestine and perhaps to other tissues. ILCP highly expresses many homing molecules such as CXCR5, CXCR6, CCR2, CD41, CD61, and CD226, suggesting that ILCP may possess migratory properties towards many different tissues [96, 102]. CXCR6 plays an important role in the migration of ILC progenitors as *Cxcr6*$^{-/-}$ mice showed accumulation of ILCP in bone marrow whereas mature ILC numbers were significantly reduced in the periphery [154].

Various lineage-committed ILC precursors are generated centrally and may home to specific tissues. ILC2P expresses CCR9 which supports migration to small intestine [101]. Additionally, IL-33 signaling is important for the exit of ILC2s from bone marrow. IL-33 signaling downregulates CXCR4 expression, a chemokine receptor that is essential for retaining ILC2s in the bone marrow, and thus IL-33 promotes egress of ILC2s from bone marrow to peripheral tissues [155]. RORγt$^+$ progenitors are preferentially enriched in human tonsils and intestinal lamina propria as compared to bone marrow, and are capable of differentiating into ILC3s [133, 134]. This suggests that ILC3s may develop in tissues such as tonsils and intestinal LP rather than centrally [104].

Much remains to be further investigated regarding migration of early ILC precursors. It is unknown whether all ILC progenitors are generated centrally, and then migrate from bone marrow to different organs during ILC development. Alternatively, some ILC progenitors may develop from upstream progenitors in tissues.

2.8 Multiple Developmental Waves Contributing to Adult Tissue ILC Pool

How the local ILC pool in different tissues is maintained and renewed is not fully understood. Parabiotic and shield chimera experiments suggest that mature ILCs are maintained in tissues during adulthood at steady state and there is minimal exchange once ILC populations are established [23, 156].

It is hypothesized that ILCs in mice arise during a wave of fetal liver-derived hematopoiesis from E13.5 until birth [132], in close concordance with fetal liver monocyte dissemination, and these ILCs seed peripheral tissues through unknown developmental cues [157]. A second wave of expansion extending from afterbirth through weaning is presumed to occur by regional expansion and maturation in response to cues from the niche [158]. Nevertheless, the anatomical location of ILC precursors in developing tissues, the identity of niches that support local precursors, and the knowledge of specific factors sustaining the local expansion of tissue ILCs are largely unknown.

The perinatal period is a critical window for shaping the distribution of ILCs within developing organs. Schneider et al. extensively characterized the contribution of different stages of life—fetal, postnatal, and adult—in shaping the adult tissue ILC2 pools using fate-mapping approaches and reporters for ILC2 activation. The adult ILC2 pool was mainly established within the first few weeks of life. The contribution by fetal and adult BM progenitors to the adult ILC2 pool was minimal in many tissues like lung and fat. This perinatal period was also important for ILC2 priming and acquisition of tissue-specific signature [159].

2.9 Local Niche Instructs Tissue ILC Development and ILC Subset Heterogeneity

The external factors that direct commitment to specific ILC lineages and the stromal cells that constitute the optimal microenvironment for ILC development in tissues are not fully understood. However, there are some studies that identified the role of stromal cells in instructing the ILC differentiation process. Insulin growth factor 1 (IGF1), produced by PDGFRα^+ alveolar fibroblasts, is important for the maturation and expansion of ILC precursors in the neonatal lung [24]. IGF1 also promoted the development of human NK cells [160]. Another study suggested that CD31$^-$ CD45$^-$ PDGFRα^+ gP38$^+$ mesenchymal cells in fetal and adult mesentery were essential for providing an optimal microenvironment for the terminal differentiation of ILC2s in peripheral tissues [158]. Recently, an adventitial stromal cell niche was implicated in the expansion of pulmonary ILC2s [161].

Recent evidence indicates the extensive heterogeneity based on tissue origins of ILC2s. The single-cell profiling for adult ILC2s from different tissues revealed that ILC2s isolated from different tissues like lung, fat, bone marrow, gut, and skin appeared transcriptionally distinct [162]. This transcriptional heterogeneity of tissue-resident ILC2s allowed the authors to designate tissue-specific transcriptional signature for each tissue-specific ILC2 [163]. Another analysis of human ILC2s by mass cytometry showed extensive heterogeneity among individuals and between different tissues [164]. Heterogeneity based on tissue origin was also shown in ILC3s by using bulk RNA sequencing. The bulk RNA sequencing data for ILC3s isolated from spleen and gut based on surface markers appeared distinct [165].

The expression of surface molecules and function of ILCs can be altered by different environmental cues (covered in Chap. 5). Local pools of ILCs may receive input of ILCs from distant cellular sources during inflammation and local changes in ILC subsets and abundance correlate with the kind of inflammation setting [166, 167]. Recent studies have established the phenotypic and functional heterogeneity of ILCs across different tissues in humans and mice [164, 168]. For example, splenic ILC3s showed distinct properties when compared to gut ILC3s. Splenic ILC3s inhibited tumor growth, whereas small intestinal ILC3s did not. Adoptive transfer experiments established that transferred ILCs gained the phenotypic and functional properties of the particular tissue where they settled, suggesting a crucial role of tissue microenvironment in shaping specialization of ILCs [169]. Multiple transcriptionally distinct states within known mature ILC subsets were identified using single-cell transcriptomics in gut. Under steady-state conditions, the expression profile of ILC1 and ILC2-specific genes and their chromatin landscapes showed significant alteration upon antibiotic treatment [170]. Deficiency of nutrients also greatly influences ILC composition. Deficiency of vitamin A caused significant reduction in ILC3s in small intestine and increase in ILC2s. This suggests that ILCs behave as a sensor for dietary stress at barrier surfaces [171]. Local tissue signals greatly impact the terminal differentiation of ILC2s and their T cell counterpart, Th2 cells. Indeed, both ILC2s and Th2 cells acquired similar terminal effector functions when they were exposed to exogenous cytokine signals like IL-25, IL-33, and TSLP [172].

These studies indicate that ILC subsets are more heterogenous than previously thought. The physiological relevance of ILC heterogeneity remains largely unknown. These data also suggest that tissue adaption and existence of separate source of ILCP in tissues and during different stages of life can lead to distinct transcriptional profiles of the same ILC subset at distinct sites (Fig. 2.2).

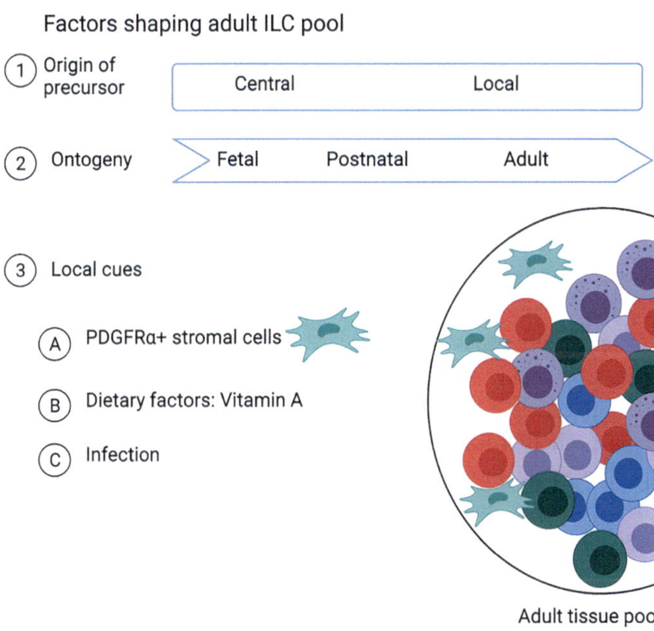

Fig. 2.2 Factors shaping the composition of adult ILC pool. The heterogeneity in ILC subsets in tissues can be due to multiple factors: (1) The developmental origin of precursors: The adult pool can be established by the expansion of BM progenitors or from local precursors. (2) Ontogeny: The adult ILC pool can be continuously replaced by different waves of ILC during life (fetal and adult life) across tissues. (3) Local cues like signals from stromal cells that may differ across the tissues. Other external factors like dietary factors and infection can also contribute to the heterogeneity across tissues

2.10 Conclusion

The origins of ILCs from lymphoid precursors in the fetal liver and adult BM are relatively well characterized. It is hypothesized that ILC precursors seed the peripheral tissues and subsequently differentiate and expand at peripheral sites as needed. Such precursors are identified at multiple sites including adult human blood, small intestine in fetal mice, and lung in postnatal mice. ILCs detected in each tissue may be generated from different cellular sources and at different times during ontogeny. The infiltrating cells can originate from bone marrow ILCP, local precursors, or mature cells exiting from other inflamed tissues. Many studies have identified local progenitors of tissue ILC, highlighting in situ differentiation as a mechanism of ILC maintenance and phenotypic diversification.

Despite extensive characterization of steps and intermediates of ILC development centrally, tools for tissue-specific deletion of ILC precursors are lacking, due to the shared surface molecules and transcription factors by the ILC precursors at different sites. Identification of organ-specific cues that impose tissue-specific signatures in ILCP at distinct sites needs further validation.

Further studies using appropriate fate-mapping tools will determine whether tissue-specific expression profiles of a particular ILC subset are acquired during fetal or postnatal development, when ILCs become established in tissues. Such studies will establish if these programs are engaged during differentiation of bone marrow precursors or instead established after migration to tissues. They will determine whether there are local precursors that shape the adult pool independent of bone marrow precursors that may meet local demand at steady state and after antigenic assault in tissues.

References

1. Klose CSN, Artis D. Innate lymphoid cells control signaling circuits to regulate tissue-specific immunity. Cell Res. 2020;30(6):475–91.
2. Vivier E, Artis D, Colonna M, Diefenbach A, Di Santo JP, Eberl G, et al. Innate lymphoid cells: 10 years on. Cell. 2018;174(5):1054–66.
3. Hazenberg MD, Spits H. Human innate lymphoid cells. Blood. 2014;124(5):700–9.
4. Wang S, Xia P, Chen Y, Qu Y, Xiong Z, Ye B, et al. Regulatory innate lymphoid cells control innate intestinal inflammation. Cell. 2017;171(1):201–16. e18
5. Bando JK, Gilfillan S, Di Luccia B, Fachi JL, Sécca C, Cella M, et al. ILC2s are the predominant source of intestinal ILC-derived IL-10. J Exp Med. 2019;217(2)
6. Huehn J, Siegmund K, Lehmann JCU, Siewert C, Haubold U, Feuerer M, et al. Developmental stage, phenotype, and migration distinguish naive- and effector/memory-like CD4+ regulatory T cells. J Exp Med. 2004;199(3):303–13.
7. Gronke K, Kofoed-Nielsen M, Diefenbach A. Innate lymphoid cells, precursors and plasticity. Immunol Lett. 2016;179:9–18.
8. Zhang K, Xu X, Pasha MA, Siebel CW, Costello A, Haczku A, et al. Cutting edge: notch signaling promotes the plasticity of Group-2 innate lymphoid cells. J Immunol. 2017;198(5):1798–803.
9. Hepworth MR, Fung TC, Masur SH, Kelsen JR, McConnell FM, Dubrot J, et al. Group 3 innate lymphoid cells mediate intestinal selection of commensal bacteria–specific CD4+ T cells. Science. 2015;348 (6238):1031–5.
10. Zhou L, Chu C, Teng F, Bessman NJ, Goc J, Santosa EK, et al. Innate lymphoid cells support regulatory T cells in the intestine through interleukin-2. Nature. 2019;568(7752):405–9.
11. Sonnenberg GF, Hepworth MR. Functional interactions between innate lymphoid cells and adaptive immunity. Nat Rev Immunol. 2019;19(10): 599–613.
12. Colonna M. Innate lymphoid cells: diversity, plasticity, and unique functions in immunity. Immunity. 2018;48(6):1104–17.
13. Klose CSN, Artis D. Innate lymphoid cells as regulators of immunity, inflammation and tissue homeostasis. Nat Immunol. 2016;17(7):765–74.
14. Spits H, Artis D, Colonna M, Diefenbach A, Di Santo JP, Eberl G, et al. Innate lymphoid cells — a proposal for uniform nomenclature. Nat Rev Immunol. 2013;13(2):145–9.
15. De Obaldia ME, Bhandoola A. Transcriptional regulation of innate and adaptive lymphocyte lineages. Annu Rev Immunol. 2015;33(1):607–42.
16. Fan X, Rudensky AY. Hallmarks of tissue-resident lymphocytes. Cell. 2016;164(6):1198–211.
17. Paust S, Gill HS, Wang B-Z, Flynn MP, Moseman EA, Senman B, et al. Critical role for the chemokine receptor CXCR6 in NK cell–mediated antigen-specific memory of haptens and viruses. Nat Immunol. 2010;11(12):1127–35.
18. Martinez-Gonzalez I, Mathä L, Steer CA, Ghaedi M, Poon GFT, Takei F. Allergen-experienced group

2 innate lymphoid cells acquire memory-like properties and enhance allergic lung inflammation. Immunity. 2016;45(1):198–208.
19. Yang Q, Bhandoola A. The development of adult innate lymphoid cells. Curr Opin Immunol. 2016;39:114–20.
20. Das A, Harly C, Yang Q, Bhandoola A. Lineage specification in innate lymphocytes. Cytokine Growth Factor Rev. 2018;42:20–6.
21. Lim AI, Li Y, Lopez-Lastra S, Stadhouders R, Paul F, Casrouge A, et al. Systemic human ILC precursors provide a substrate for tissue ILC differentiation. Cell. 2017;168(6):1086–100e10.
22. Ghaedi M, Shen ZY, Orangi M, Martinez-Gonzalez I, Wei L, Lu X, et al. Single-cell analysis of RORα tracer mouse lung reveals ILC progenitors and effector ILC2 subsets. J Exp Med. 2019;217(3)
23. Zeis P, Lian M, Fan X, Herman JS, Hernandez DC, Gentek R, et al. In situ maturation and tissue adaptation of type 2 innate lymphoid cell progenitors. Immunity. 2020;53(4):775–92. e9
24. Oherle K, Acker E, Bonfield M, Wang T, Gray J, Lang I, et al. Insulin-like growth factor 1 supports a pulmonary niche that promotes type 3 innate lymphoid cell development in newborn lungs. Immunity. 2020;52(2):275–94. e9
25. Spits H, Cupedo T. Innate lymphoid cells: emerging insights in development, lineage relationships, and function. Annu Rev Immunol. 2012;30:647–75.
26. Meininger I, Carrasco A, Rao A, Soini T, Kokkinou E, Mjösberg J. Tissue-specific features of innate lymphoid cells. Trends Immunol. 2020;41(10):902–17.
27. Robinette ML, Colonna M. Immune modules shared by innate lymphoid cells and T cells. J Allergy Clin Immunol. 2016;138(5):1243–51.
28. Lim AI, Verrier T, Vosshenrich CAJ, Di Santo JP. Developmental options and functional plasticity of innate lymphoid cells. Curr Opin Immunol. 2017;44:61–8.
29. Diefenbach A, Colonna M, Koyasu S. Development, differentiation, and diversity of innate lymphoid cells. Immunity. 2014;41(3):354–65.
30. Cortez VS, Colonna M. Diversity and function of group 1 innate lymphoid cells. Immunol Lett. 2016;179:19–24.
31. Yokoyama WM, Plougastel BFM. Immune functions encoded by the natural killer gene complex. Nat Rev Immunol. 2003;3(4):304–16.
32. Vivier E, Raulet DH, Moretta A, Caligiuri MA, Zitvogel L, Lanier LL, et al. Innate or adaptive immunity? The example of natural killer cells. Science. 2011;331(6013):44–9.
33. Jiao Y, Huntington ND, Belz GT, Seillet C. Type 1 innate lymphoid cell biology: lessons learnt from natural killer cells. Front Immunol. 2016;7:426.
34. Weizman O-E, Adams NM, Schuster IS, Krishna C, Pritykin Y, Lau C, et al. ILC1 confer early host protection at initial sites of viral infection. Cell. 2017;171(4):795–808e12.
35. Zhang J, Marotel M, Fauteux-Daniel S, Mathieu A-L, Viel S, Marçais A, et al. T-bet and EOMES govern differentiation and function of mouse and human NK cells and ILC1. Eur J Immunol. 2018;48(5):738–50.
36. Cooper MA, Colonna M, Yokoyama WM. Hidden talents of natural killers: NK cells in innate and adaptive immunity. EMBO Rep. 2009;10(10):1103–10.
37. Vosshenrich CAJ, García-Ojeda ME, Samson-Villéger SI, Pasqualetto V, Enault L, Goff OR-L, et al. A thymic pathway of mouse natural killer cell development characterized by expression of GATA-3 and CD127. Nat Immunol. 2006;7(11):1217–24.
38. Gabrielli S, Sun M, Bell A, Zook EC, de Pooter RF, Zamai L, et al. Murine thymic NK cells are distinct from ILC1s and have unique transcription factor requirements. Eur J Immunol. 2017;47(5):800–5.
39. Ribeiro VSG, Hasan M, Wilson A, Boucontet L, Pereira P, Lesjean-Pottier S, et al. Cutting edge: Thymic NK cells develop independently from T cell precursors. J Immunol. 2010;185(9):4993–7.
40. Aw Yeang HX, Piersma SJ, Lin Y, Yang L, Malkova ON, Miner C, et al. Cutting edge: human CD49e-; NK cells are tissue resident in the liver. J Immunol. 2017;198(4):1417–22.
41. McFarland AP, Yalin A, Wang S-Y, Cortez VS, Landsberger T, Sudan R, et al. Multi-tissue single-cell analysis deconstructs the complex programs of mouse natural killer and type 1 innate lymphoid cells in tissues and circulation. Immunity. 2021;54(6):1320–1337e4.
42. Mackay LK, Minnich M, Kragten NAM, Liao Y, Nota B, Seillet C, et al. Hobit and Blimp1 instruct a universal transcriptional program of tissue residency in lymphocytes. Science. 2016;352(6284):459–63.
43. Koyasu S, Moro K. Innate Th2-type immune responses and the natural helper cell, a newly identified lymphocyte population. Curr Opin Allergy Clin Immunol. 2011;11(2):109–14.
44. Van Dyken SJ, Mohapatra A, Nussbaum JC, Molofsky AB, Thornton EE, Ziegler SF, et al. Chitin activates parallel immune modules that direct distinct inflammatory responses via innate lymphoid type 2 and γδ T cells. Immunity. 2014;40(3):414–24.
45. Halim TYF, MacLaren A, Romanish MT, Gold MJ, McNagny KM, Takei F. Retinoic-acid-receptor-related orphan nuclear receptor alpha is required for natural helper cell development and allergic inflammation. Immunity. 2012;37(3):463–74.
46. Monticelli LA, Sonnenberg GF, Abt MC, Alenghat T, Ziegler CGK, Doering TA, et al. Innate lymphoid cells promote lung-tissue homeostasis after infection with influenza virus. Nat Immunol. 2011;12(11):1045–54.
47. Howard E, Lewis G, Galle-Treger L, Hurrell BP, Helou DG, Shafiei-Jahani P, et al. IL-10 production by ILC2s requires Blimp-1 and CMAF, modulates cellular metabolism, and ameliorates airway

hyperreactivity. J Allergy Clin Immunol. 2021;147 (4):1281–1295e5.
48. Neill DR, Fallon PG. Innate lymphoid cells and parasites: ancient foes with shared history. Parasite Immunol. 2018;40(2):e12513.
49. Kabata H, Moro K, Koyasu S. The group 2 innate lymphoid cell (ILC2) regulatory network and its underlying mechanisms. Immunol Rev. 2018;286 (1):37–52.
50. Jacquelot N, Seillet C, Wang M, Pizzolla A, Liao Y, Hediyeh-zadeh S, et al. Blockade of the co-inhibitory molecule PD-1 unleashes ILC2-dependent antitumor immunity in melanoma. Nat Immunol. 2021;22(7): 851–64.
51. Cardoso V, Chesné J, Ribeiro H, García-Cassani B, Carvalho T, Bouchery T, et al. Neuronal regulation of type 2 innate lymphoid cells via neuromedin U. Nature. 2017;549(7671):277–81.
52. Wallrapp A, Riesenfeld SJ, Burkett PR, Abdulnour R-EE, Nyman J, Dionne D, et al. The neuropeptide NMU amplifies ILC2-driven allergic lung inflammation. Nature. 2017;549(7672):351–6.
53. Klose CSN, Mahlakõiv T, Moeller JB, Rankin LC, Flamar A-L, Kabata H, et al. The neuropeptide neuromedin U stimulates innate lymphoid cells and type 2 inflammation. Nature. 2017;549(7671):282–6.
54. Nagashima H, Mahlakõiv T, Shih H-Y, Davis FP, Meylan F, Huang Y, et al. Neuropeptide CGRP limits group 2 innate lymphoid cell responses and constrains type 2 inflammation. Immunity. 2019;51 (4):682–695e6.
55. Huang Y, Guo L, Qiu J, Chen X, Hu-Li J, Siebenlist U, et al. IL-25-responsive, lineage-negative KLRG1(hi) cells are multipotential "inflammatory" type 2 innate lymphoid cells. Nat Immunol. 2015;16(2):161–9.
56. Price AE, Liang H-E, Sullivan BM, Reinhardt RL, Eisley CJ, Erle DJ, et al. Systemically dispersed innate IL-13–expressing cells in type 2 immunity. Proc Natl Acad Sci. 2010;107(25):11489–94.
57. Moro K, Yamada T, Tanabe M, Takeuchi T, Ikawa T, Kawamoto H, et al. Innate production of T(H) 2 cytokines by adipose tissue-associated c-kit(+) Sca-1(+) lymphoid cells. Nature. 2010;463(7280): 540–4.
58. Neill DR, Wong SH, Bellosi A, Flynn RJ, Daly M, Langford TKA, et al. Nuocytes represent a new innate effector leukocyte that mediates type-2-immunity. Nature. 2010;464(7293):1367–70.
59. Mackley EC, Houston S, Marriott CL, Halford EE, Lucas B, Cerovic V, et al. CCR7-dependent trafficking of RORγ+ ILCs creates a unique microenvironment within mucosal draining lymph nodes. Nat Commun. 2015;6(1):5862.
60. Satoh-Takayama N. Heterogeneity and diversity of group 3 innate lymphoid cells: new cells on the block. Int Immunol. 2016;28(1):29–34.
61. Melo-Gonzalez F, Hepworth MR. Functional and phenotypic heterogeneity of group 3 innate lymphoid cells. Immunology. 2017;150(3):265–75.
62. Magri G, Miyajima M, Bascones S, Mortha A, Puga I, Cassis L, et al. Innate lymphoid cells integrate stromal and immunological signals to enhance antibody production by splenic marginal zone B cells. Nat Immunol. 2014;15(4):354–64.
63. Eberl G, Colonna M, Di Santo JP, McKenzie ANJ. Innate lymphoid cells: a new paradigm in immunology. Science. 2015;348(6237):aaa6566.
64. Klose CSN, Kiss EA, Schwierzeck V, Ebert K, Hoyler T, d'Hargues Y, et al. A T-bet gradient controls the fate and function of CCR6-RORγt+ innate lymphoid cells. Nature. 2013;494(7436): 261–5.
65. van de Pavert SA. Lymphoid tissue inducer (LTi) cell ontogeny and functioning in embryo and adult. Biom J. 2021;44(2):123–32.
66. Satoh-Takayama N, Vosshenrich CAJ, Lesjean-Pottier S, Sawa S, Lochner M, Rattis F, et al. Microbial flora drives interleukin 22 production in intestinal NKp46+ cells that provide innate mucosal immune defense. Immunity. 2008;29(6):958–70.
67. Mortha A, Chudnovskiy A, Hashimoto D, Bogunovic M, Spencer SP, Belkaid Y, et al. Microbiota-dependent crosstalk between macrophages and ILC3 promotes intestinal homeostasis. Science. 2014;343(6178):1249288.
68. Zheng Y, Valdez PA, Danilenko DM, Hu Y, Sa SM, Gong Q, et al. Interleukin-22 mediates early host defense against attaching and effacing bacterial pathogens. Nat Med. 2008;14(3):282–9.
69. Guo X, Qiu J, Tu T, Yang X, Deng L, Anders RA, et al. Induction of innate lymphoid cell-derived interleukin-22 by the transcription factor STAT3 mediates protection against intestinal infection. Immunity. 2014;40(1):25–39.
70. Cella M, Fuchs A, Vermi W, Facchetti F, Otero K, Lennerz JKM, et al. A human natural killer cell subset provides an innate source of IL-22 for mucosal immunity. Nature. 2009;457(7230):722–5.
71. Lindemans CA, Calafiore M, Mertelsmann AM, O'Connor MH, Dudakov JA, Jenq RR, et al. Interleukin-22 promotes intestinal-stem-cell-mediated epithelial regeneration. Nature. 2015;528(7583):560–4.
72. Keir ME, Yi T, Lu TT, Ghilardi N. The role of IL-22 in intestinal health and disease. J Exp Med. 2020;217 (3)
73. Hepworth MR, Monticelli LA, Fung TC, Ziegler CGK, Grunberg S, Sinha R, et al. Innate lymphoid cells regulate CD4+ T-cell responses to intestinal commensal bacteria. Nature. 2013;498(7452):113–7.
74. Buonocore S, Ahern PP, Uhlig HH, Ivanov II, Littman DR, Maloy KJ, et al. Innate lymphoid cells drive interleukin-23-dependent innate intestinal pathology. Nature. 2010;464(7293):1371–5.
75. Vonarbourg C, Mortha A, Bui VL, Hernandez PP, Kiss EA, Hoyler T, et al. Regulated expression of

nuclear receptor RORγt confers distinct functional fates to NK cell receptor-expressing RORγt(+) innate lymphocytes. Immunity. 2010;33(5):736–51.
76. Pearson C, Thornton EE, McKenzie B, Schaupp A-L, Huskens N, Griseri T, et al. ILC3 GM-CSF production and mobilisation orchestrate acute intestinal inflammation. Elife. 2016:e10066.
77. Song C, Lee JS, Gilfillan S, Robinette ML, Newberry RD, Stappenbeck TS, et al. Unique and redundant functions of NKp46+ ILC3s in models of intestinal inflammation. J Exp Med. 2015;212(11):1869–82.
78. Zeng B, Shi S, Ashworth G, Dong C, Liu J, Xing F. ILC3 function as a double-edged sword in inflammatory bowel diseases. Cell Death Dis. 2019;10(4):315.
79. Eberl G, Marmon S, Sunshine M-J, Rennert PD, Choi Y, Littman DR. An essential function for the nuclear receptor RORgamma(t) in the generation of fetal lymphoid tissue inducer cells. Nat Immunol. 2004;5 (1):64–73.
80. Eberl G. Development and evolution of RORγt+ cells in a microbe's world. Immunol Rev. 2012;245 (1):177–88.
81. Mebius RE, Rennert P, Weissman IL. Developing lymph nodes collect CD4+CD3- LTbeta+ cells that can differentiate to APC, NK cells, and follicular cells but not T or B cells. Immunity. 1997;7 (4):493–504.
82. Scandella E, Bolinger B, Lattmann E, Miller S, Favre S, Littman DR, et al. Restoration of lymphoid organ integrity through the interaction of lymphoid tissue-inducer cells with stroma of the T cell zone. Nat Immunol. 2008;9(6):667–75.
83. Yang Q, Li F, Harly C, Xing S, Ye L, Xia X, et al. TCF-1 upregulation identifies early innate lymphoid progenitors in the bone marrow. Nat Immunol. 2015;16(10):1044–50.
84. Constantinides MG, McDonald BD, Verhoef PA, Bendelac A. A committed precursor to innate lymphoid cells. Nature. 2014;508(7496):397–401.
85. Klose CSN, Flach M, Möhle L, Rogell L, Hoyler T, Ebert K, et al. Differentiation of type 1 ILCs from a common progenitor to all helper-like innate lymphoid cell lineages. Cell. 2014;157(2):340–56.
86. Xu W, Cherrier DE, Chea S, Vosshenrich C, Serafini N, Petit M, et al. An Id2RFP-reporter mouse redefines innate lymphoid cell precursor potentials. Immunity. 2019;50(4):1054–1068e3.
87. Walker JA, Clark PA, Crisp A, Barlow JL, Szeto A, Ferreira ACF, et al. Polychromic reporter mice reveal unappreciated innate lymphoid cell progenitor heterogeneity and elusive ILC3 progenitors in bone marrow. Immunity. 2019;51(1):104–118e7.
88. Kasal DN, Bendelac A. Multi-transcription factor reporter mice delineate early precursors to the ILC and LTi lineages. J Exp Med. 2020;218(2)
89. Inlay MA, Bhattacharya D, Sahoo D, Serwold T, Seita J, Karsunky H, et al. Ly6d marks the earliest stage of B-cell specification and identifies the branchpoint between B-cell and T-cell development. Genes Dev. 2009;23(20):2376–81.
90. Ding L, Morrison SJ. Haematopoietic stem cells and early lymphoid progenitors occupy distinct bone marrow niches. Nature. 2013;495(7440):231–5.
91. Possot C, Schmutz S, Chea S, Boucontet L, Louise A, Cumano A, et al. Notch signaling is necessary for adult, but not fetal, development of RORγt(+) innate lymphoid cells. Nat Immunol. 2011;12(10):949–58.
92. Ishizuka IE, Chea S, Gudjonson H, Constantinides MG, Dinner AR, Bendelac A, et al. Single-cell analysis defines the divergence between the innate lymphoid cell lineage and lymphoid tissue-inducer cell lineage. Nat Immunol. 2016;17(3):269–76.
93. Cherrier M, Sawa S, Eberl G. Notch, Id2, and RORγt sequentially orchestrate the fetal development of lymphoid tissue inducer cells. J Exp Med. 2012;209 (4):729–40.
94. Ghaedi M, Steer CA, Martinez-Gonzalez I, Halim TYF, Abraham N, Takei F. Common-lymphoid-progenitor-independent pathways of innate and T lymphocyte development. Cell Rep. 2016;15(3):471–80.
95. Yang Q, Saenz SA, Zlotoff DA, Artis D, Bhandoola A. Cutting edge: natural helper cells derive from lymphoid progenitors. J Immunol. 2011;187 (11):5505–9.
96. Harly C, Cam M, Kaye J, Bhandoola A. Development and differentiation of early innate lymphoid progenitors. J Exp Med. 2018;215(1):249–62.
97. Yu X, Wang Y, Deng M, Li Y, Ruhn KA, Zhang CC, et al. The basic Leucine zipper transcription factor NFIL3 directs the development of a common innate lymphoid cell precursor. Elife. 2014;3
98. Rosmaraki EE, Douagi I, Roth C, Colucci F, Cumano A, Di Santo JP. Identification of committed NK cell progenitors in adult murine bone marrow. Eur J Immunol. 2001;31(6):1900–9.
99. Fathman JW, Bhattacharya D, Inlay MA, Seita J, Karsunky H, Weissman IL. Identification of the earliest natural killer cell-committed progenitor in murine bone marrow. Blood. 2011;118(20):5439–47.
100. Constantinides MG, Gudjonson H, McDonald BD, Ishizuka IE, Verhoef PA, Dinner AR, et al. PLZF expression maps the early stages of ILC1 lineage development. Proc Natl Acad Sci. 2015;112 (16):5123–8.
101. Hoyler T, Klose CSN, Souabni A, Turqueti-Neves A, Pfeifer D, Rawlins EL, et al. The transcription factor GATA-3 controls cell fate and maintenance of type 2 innate lymphoid cells. Immunity. 2012;37(4):634–48.
102. Yu Y, Tsang JCH, Wang C, Clare S, Wang J, Chen X, et al. Single-cell RNA-seq identifies a PD-1(hi) ILC progenitor and defines its development pathway. Nature. 2016;539(7627):102–6.
103. Seillet C, Mielke LA, Amann-Zalcenstein DB, Su S, Gao J, Almeida FF, et al. Deciphering the innate lymphoid cell transcriptional program. Cell Rep. 2016;17(2):436–47.

104. Chea S, Schmutz S, Berthault C, Perchet T, Petit M, Burlen-Defranoux O, et al. Single-cell gene expression analyses reveal heterogeneous responsiveness of fetal innate lymphoid progenitors to notch signaling. Cell Rep. 2016;14(6):1500–16.
105. Harly C, Kenney D, Ren G, Lai B, Raabe T, Yang Q, et al. The transcription factor TCF-1 enforces commitment to the innate lymphoid cell lineage. Nat Immunol. 2019;20(9):1150–60.
106. Sawa S, Cherrier M, Lochner M, Satoh-Takayama N, Fehling HJ, Langa F, et al. Lineage relationship analysis of ROR$^+$ innate lymphoid cells. Science. 2010;330(6004):665–9.
107. Cherrier M, Eberl G. The development of LTi cells. Curr Opin Immunol. 2012;24(2):178–83.
108. Nagasawa M, Heesters BA, Kradolfer CMA, Krabbendam L, Martinez-Gonzalez I, de Bruijn MJW, et al. KLRG1 and NKp46 discriminate subpopulations of human CD117+CRTH2− ILCs biased toward ILC2 or ILC3. J Exp Med. 2019;216 (8):1762–76.
109. Liu C, Gong Y, Zhang H, Yang H, Zeng Y, Bian Z, et al. Delineating spatiotemporal and hierarchical development of human fetal innate lymphoid cells. Cell Res . 2021
110. Goh W, Huntington ND. Regulation of murine natural killer cell development. Front Immunol. 2017;8:130.
111. Kondo M, Weissman IL, Akashi K. Identification of clonogenic common lymphoid progenitors in mouse bone marrow. Cell. 1997;91(5):661–72.
112. Carotta S, Pang SHM, Nutt SL, Belz GT. Identification of the earliest NK-cell precursor in the mouse BM. Blood. 2011;117(20):5449–52.
113. Seehus CR, Kaye J. The role of TOX in the development of innate lymphoid cells. Mediat Inflamm. 2015;2015:243868.
114. Zhu J. GATA3 regulates the development and functions of innate lymphoid cell subsets at multiple stages. Front Immunol. 2017;8:1571.
115. Cherrier DE, Serafini N, Di Santo JP. Innate lymphoid cell development: a T cell perspective. Immunity. 2018;48(6):1091–103.
116. Califano D, Cho JJ, Uddin MN, Lorentsen KJ, Yang Q, Bhandoola A, et al. Transcription factor Bcl11b controls identity and function of mature type 2 innate lymphoid cells. Immunity. 2015;43(2):354–68.
117. Zhong C, Zhu J. Transcriptional regulatory network for the development of innate lymphoid cells. Mediat Inflamm. 2015;2015:264502.
118. Hosokawa H, Romero-Wolf M, Yang Q, Motomura Y, Levanon D, Groner Y, et al. Cell type–specific actions of Bcl11b in early T-lineage and group 2 innate lymphoid cells. J Exp Med. 2019;217(1)
119. Kamizono S, Duncan GS, Seidel MG, Morimoto A, Hamada K, Grosveld G, et al. Nfil3/E4bp4 is required for the development and maturation of NK cells in vivo. J Exp Med. 2009;206(13):2977–86.
120. Seillet C, Rankin LC, Groom JR, Mielke LA, Tellier J, Chopin M, et al. Nfil3 is required for the development of all innate lymphoid cell subsets. J Exp Med. 2014;211(9):1733–40.
121. Geiger TL, Abt MC, Gasteiger G, Firth MA, O'Connor MH, Geary CD, et al. Nfil3 is crucial for development of innate lymphoid cells and host protection against intestinal pathogens. J Exp Med. 2014;211(9):1723–31.
122. Xu W, Domingues RG, Fonseca-Pereira D, Ferreira M, Ribeiro H, Lopez-Lastra S, et al. NFIL3 orchestrates the emergence of common helper innate lymphoid cell precursors. Cell Rep. 2015;10(12):2043–54.
123. Male V, Nisoli I, Kostrzewski T, Allan DSJ, Carlyle JR, Lord GM, et al. The transcription factor E4bp4/ Nfil3 controls commitment to the NK lineage and directly regulates Eomes and Id2 expression. J Exp Med. 2014;211(4):635–42.
124. Seehus CR, Aliahmad P, de la Torre B, Iliev ID, Spurka L, Funari VA, et al. The development of innate lymphoid cells requires TOX-dependent generation of a common innate lymphoid cell progenitor. Nat Immunol. 2015;16(6):599–608.
125. Kee BL. E and ID proteins branch out. Nat Rev Immunol. 2009;9(3):175–84.
126. Ling F, Kang B, Sun X-H. Chapter five - Id proteins: small molecules, mighty regulators. In: Taneja R, editor. bHLH transcription factors in development and disease. Academic Press; 2014. p. 189–216. Current Topics in Developmental Biology; vol. 110.
127. Zook EC, Li Z-Y, Xu Y, de Pooter RF, Verykokakis M, Beaulieu A, et al. Transcription factor ID2 prevents E proteins from enforcing a naïve T lymphocyte gene program during NK cell development. Sci Immunol. 2018;3(22):eaao2139.
128. Li Z-Y, Morman RE, Hegermiller E, Sun M, Bartom ET, Maienschein-Cline M, et al. The transcriptional repressor ID2 supports natural killer cell maturation by controlling TCF1 amplitude. J Exp Med. 2021;218(6)
129. Berrett H, Qian L, Roman O, Cordova A, Simmons A, Sun X-H, et al. Development of type 2 innate lymphoid cells is selectively inhibited by sustained E protein activity. Immuno Horizons. 2019;3 (12):593–605.
130. Yokota Y, Mansouri A, Mori S, Sugawara S, Adachi S, Nishikawa S, et al. Development of peripheral lymphoid organs and natural killer cells depends on the helix-loop-helix inhibitor Id2. Nature. 1999;397 (6721):702–6.
131. Zhong C, Zheng M, Cui K, Martins AJ, Hu G, Li D, et al. Differential expression of the transcription factor GATA3 specifies lineage and functions of innate lymphoid cells. Immunity. 2020;52(1):83–95e4.
132. Bando JK, Liang H-E, Locksley RM. Identification and distribution of developing innate lymphoid cells in the fetal mouse intestine. Nat Immunol. 2015;16 (2):153–60.

133. Scoville SD, Mundy-Bosse BL, Zhang MH, Chen L, Zhang X, Keller KA, et al. A progenitor cell expressing transcription factor RORγt generates all human innate lymphoid cell subsets. Immunity. 2016;44(5):1140–50.
134. Montaldo E, Teixeira-Alves LG, Glatzer T, Durek P, Stervbo U, Hamann W, et al. Human RORγt(+)CD34 (+) cells are lineage-specified progenitors of group 3 RORγt(+) innate lymphoid cells. Immunity. 2014;41 (6):988–1000.
135. Mielke LA, Groom JR, Rankin LC, Seillet C, Masson F, Putoczki T, et al. TCF-1 controls ILC2 and NKp46 +RORγt+ innate lymphocyte differentiation and protection in intestinal inflammation. J Immunol. 2013;191(8):4383–91.
136. Cortez VS, Fuchs A, Cella M, Gilfillan S, Colonna M. Cutting edge: salivary gland NK cells develop independently of Nfil3 in steady-state. J Immunol. 2014;192(10):4487–91.
137. Sojka DK, Plougastel-Douglas B, Yang L, Pak-Wittel MA, Artyomov MN, Ivanova Y, et al. Tissue-resident natural killer (NK) cells are cell lineages distinct from thymic and conventional splenic NK cells. Elife. 2014;3:e01659.
138. Boulenouar S, Doisne J-M, Sferruzzi-Perri A, Gaynor LM, Kieckbusch J, Balmas E, et al. The residual innate lymphoid cells in NFIL3-deficient mice support suboptimal maternal adaptations to pregnancy. Front Immunol. 2016;7:43.
139. Doisne J-M, Balmas E, Boulenouar S, Gaynor LM, Kieckbusch J, Gardner L, et al. Composition, development, and function of uterine innate lymphoid cells. J Immunol. 2015;195(8):3937–45.
140. Erick TK, Anderson CK, Reilly EC, Wands JR, Brossay L. NFIL3 expression distinguishes tissue-resident NK cells and conventional NK-like cells in the mouse submandibular glands. J Immunol. 2016;197(6):2485–91.
141. Aliahmad P, de la Torre B, Kaye J. Shared dependence on the DNA-binding factor TOX for the development of lymphoid tissue–inducer cell and NK cell lineages. Nat Immunol. 2010;11(10):945–52.
142. Lefrançais E, Ortiz-Muñoz G, Caudrillier A, Mallavia B, Liu F, Sayah DM, et al. The lung is a site of platelet biogenesis and a reservoir for haematopoietic progenitors. Nature. 2017;544 (7648):105–9.
143. Miyazaki M, Miyazaki K, Chen K, Jin Y, Turner J, Moore AJ, et al. The E-id protein Axis specifies adaptive lymphoid cell identity and suppresses Thymic innate lymphoid cell development. Immunity. 2017;46(5):818–834e4.
144. Qian L, Bajana S, Georgescu C, Peng V, Wang H-C, Adrianto I, et al. Suppression of ILC2 differentiation from committed T cell precursors by E protein transcription factors. J Exp Med. 2019;216(4):884–99.
145. Ghaedi M, Shen ZY, Orangi M, Martinez-Gonzalez I, Wei L, Lu X, et al. Single-cell analysis of RORα tracer mouse lung reveals ILC progenitors and effector ILC2 subsets. J Exp Med. 2020;217(3)
146. Bai L, Vienne M, Tang L, Kerdiles Y, Etiennot M, Escalière B, et al. Liver type 1 innate lymphoid cells develop locally via an interferon-γ-dependent loop. Science. 2021;371(6536)
147. Lim AI, Di Santo JP. ILC-poiesis: ensuring tissue ILC differentiation at the right place and time. Eur J Immunol. 2019;49(1):11–8.
148. Massberg S, Schaerli P, Knezevic-Maramica I, Köllnberger M, Tubo N, Moseman EA, et al. Immunosurveillance by hematopoietic progenitor cells trafficking through blood, lymph, and peripheral tissues. Cell. 2007;131(5):994–1008.
149. Cardier JE, Barberá-Guillem E. Extramedullary hematopoiesis in the adult mouse liver is associated with specific hepatic sinusoidal endothelial cells. Hepatology. 1997;26(1):165–75.
150. Wright DE, Wagers AJ, Gulati AP, Johnson FL, Weissman IL. Physiological migration of hematopoietic stem and progenitor cells. Science. 2001;294(5548):1933–6.
151. Saenz SA, Siracusa MC, Monticelli LA, Ziegler CGK, Kim BS, Brestoff JR, et al. IL-25 simultaneously elicits distinct populations of innate lymphoid cells and multipotent progenitor type 2 (MPPtype2) cells. J Exp Med. 2013;210(9):1823–37.
152. Griseri T, McKenzie BS, Schiering C, Powrie F. Dysregulated hematopoietic stem and progenitor cell activity promotes interleukin-23-driven chronic intestinal inflammation. Immunity. 2012;37 (6):1116–29.
153. Siracusa MC, Saenz SA, Wojno EDT, Kim BS, Osborne LC, Ziegler CG, et al. Thymic stromal lymphopoietin-mediated extramedullary hematopoiesis promotes allergic inflammation. Immunity. 2013;39(6):1158–70.
154. Chea S, Possot C, Perchet T, Petit M, Cumano A, Golub R. CXCR6 expression is important for retention and circulation of ILC precursors. Mediat Inflamm. 2015;2015:368427.
155. Stier MT, Zhang J, Goleniewska K, Cephus JY, Rusznak M, Wu L, et al. IL-33 promotes the egress of group 2 innate lymphoid cells from the bone marrow. J Exp Med. 2018;215(1):263–81.
156. Gasteiger G, Fan X, Dikiy S, Lee SY, Rudensky AY. Tissue residency of innate lymphoid cells in lymphoid and nonlymphoid organs. Science. 2015;350 (6263):981–5.
157. Kotas ME, Locksley RM. Why innate lymphoid cells? Immunity. 2018;48(6):1081–90.
158. Koga S, Hozumi K, Hirano K, Yazawa M, Terooatea T, Minoda A, et al. Peripheral PDGFRα+gp38+ mesenchymal cells support the differentiation of fetal

liver-derived ILC2. J Exp Med. 2018;215(6):1609–26.
159. Schneider C, Lee J, Koga S, Ricardo-Gonzalez RR, Nussbaum JC, Smith LK, et al. Tissue-resident group 2 innate lymphoid cells differentiate by layered ontogeny and in situ perinatal priming. Immunity. 2019;50(6):1425–1438e5.
160. Ni F, Sun R, Fu B, Wang F, Guo C, Tian Z, et al. IGF-1 promotes the development and cytotoxic activity of human NK cells. Nat Commun. 2013;4 (1):1479.
161. Dahlgren MW, Jones SW, Cautivo KM, Dubinin A, Ortiz-Carpena JF, Farhat S, et al. Adventitial stromal cells define group 2 innate lymphoid cell tissue niches. Immunity. 2019;50(3):707–722e6.
162. Ricardo-Gonzalez RR, Van Dyken SJ, Schneider C, Lee J, Nussbaum JC, Liang H-E, et al. Tissue signals imprint ILC2 identity with anticipatory function. Nat Immunol. 2018;19(10):1093–9.
163. Kim D-H, Van Dyken SJ. ILC2s in high definition: decoding the logic of tissue-based immunity. Trends Immunol. 2020;41(1):7–16.
164. Simoni Y, Fehlings M, Kløverpris HN, McGovern N, Koo S-L, Loh CY, et al. Human innate lymphoid cell subsets possess tissue-type based heterogeneity in phenotype and frequency. Immunity. 2017;46 (1):148–61.
165. Lehmann FM, von Burg N, Ivanek R, Teufel C, Horvath E, Peter A, et al. Microbiota-induced tissue signals regulate ILC3-mediated antigen presentation. Nat Commun. 2020;11(1):1794.
166. Mucida D, Husain MM, Muroi S, Van Wijk F, Shinnakasu R, Naoe Y, et al. Transcriptional reprogramming of mature CD4+ helper T cells generates distinct MHC class II-restricted cytotoxic T lymphocytes. Nature. 2013;14(3)
167. Ricardo-Gonzalez RR, Schneider C, Liao C, Lee J, Liang H-E, Locksley RM. Tissue-specific pathways extrude activated ILC2s to disseminate type 2 immunity. J Exp Med. 2020;7, 217(4)
168. Robinette ML, Fuchs A, Cortez VS, Lee JS, Wang Y, Durum SK, et al. Transcriptional programs define molecular characteristics of innate lymphoid cell classes and subsets. Nat Immunol. 2015;16(3):306–17.
169. Nussbaum K, Burkhard SH, Ohs I, Mair F, Klose CSN, Arnold SJ, et al. Tissue microenvironment dictates the fate and tumor-suppressive function of type 3 ILCs. J Exp Med. 2017;214(8):2331–47.
170. Gury-BenAri M, Thaiss CA, Serafini N, Winter DR, Giladi A, Lara-Astiaso D, et al. The Spectrum and regulatory landscape of intestinal innate lymphoid cells are shaped by the microbiome. Cell. 2016;166 (5):1231–1246e13.
171. Spencer SP, Wilhelm C, Yang Q, Hall JA, Bouladoux N, Boyd A, et al. Adaptation of innate lymphoid cells to a micronutrient deficiency promotes type 2 barrier immunity. Science. 2014;343(6169):432–7.
172. Van Dyken SJ, Nussbaum JC, Lee J, Molofsky AB, Liang H-E, Pollack JL, et al. A tissue checkpoint regulates type 2 immunity. Nat Immunol. 2016;17 (12):1381–7.

ILC Differentiation in the Thymus

3

Xiao-Hong Sun and Sandra Bajana

Abstract

The thymus provides a microenvironment conducive to the differentiation of innate lymphoid cells (ILCs), supplying IL-7 as well as Notch ligands. Early T cell precursors also express a number of obligatory transcription factors essential for ILC differentiation. Therefore, the thymus could be a powerhouse for ILC production. However, coordinated regulation by transcription factors and T cell receptor signaling events ensure that T cell production is the dominating output of the thymus. One group of the key regulators are the basic helix-loop-helix E protein transcription factors and their inhibitors, Id proteins. When E protein activities are downregulated, T cell development is blocked and massive ILC2 production occurs in the thymus. Normally, the thymus indeed generates a small number of ILCs, mostly group 2 ILCs (ILC2s). It has been shown in vitro that ILC2s can be differentiated from multipotent early T cell progenitors (ETPs) as well as committed T cell precursors. Moreover, thymus-derived ILC precursors have been found in the blood of adult mice. They then home to peripheral tissues and undergo differentiation into distinct ILC subsets. These ILC precursors may replenish tissue ILC pools in steady state or on demand in pathophysiological conditions. Collectively, emerging evidence suggests that the thymus plays an underappreciated role in ILC homeostasis.

Keywords

Thymus · Innate lymphoid cells · ILC2s · E proteins · Id proteins

3.1 Introduction

Innate lymphoid cells (ILCs) originate from hematopoietic stem cells, which reside in fetal liver or adult bone marrow and give rise to all hematopoietic lineages of cells. An array of progenitors with different capacities to produce various subsets of ILCs have been identified in the bone marrow and studied extensively [1–7]. These ILC progenitors are derived from common lymphoid progenitors (CLPs) and lymphoid primed progenitors, which are primarily responsible for generating B cells in the bone marrow [8–10]. However, these lymphoid-biased progenitors also travel through blood circulation to the thymus, where they largely differentiate into T cells [11]. It is known that these thymus-seeding progenitors, though with great T cell-producing capacity, retain multipotency and differentiate into other lineage cells such as dendritic and myeloid cells [12]. Therefore, it is

X.-H. Sun (✉) · S. Bajana
Program in Arthritis and Clinical Immunology, Oklahoma Medical Research Foundation, Oklahoma City, OK, USA
e-mail: sunx@omrf.org; Bajanas@omrf.org

© The Author(s), under exclusive license to Springer Nature Singapore Pte Ltd. 2022

X.-H. Sun (ed.), *Innate Lymphoid Cells*, Advances in Experimental Medicine and Biology 1365,
https://doi.org/10.1007/978-981-16-8387-9_3

plausible that they can also generate ILCs in the thymus. It is however not anticipated that committed T cell precursors can also differentiate into ILCs, at least ILC2s. This chapter intends to summarize available data that illustrate the ability of T cell precursors to differentiate into ILCs and the thymic environment supportive of ILC differentiation, as well as the contribution of the thymus to peripheral ILC pools. The transcriptional regulation of the T cell versus ILC fates has also begun to be elucidated and is best demonstrated in mouse models where the function of E protein transcription factors is specifically ablated in the thymus [13, 14].

3.2 The Requirements for ILC Differentiation

ILC differentiation depends on the cell-extrinsic and -intrinsic properties and conditions. Much of the data accumulated are from studies of ILC progenitors in the bone marrow and fetal liver but they are highly relevant to ILC differentiation in the thymus. ILCs are derived from multipotent lymphoid primed progenitors or common lymphoid progenitors through a series of intermediate progenitors with progressively restricted differentiation potential towards ILC subsets. It is well known that ILC differentiation depends on cytokines like IL-7 and IL-2 [1, 15–18], whose receptors belong to the IL-2 receptor family that utilizes the common γ chain [19, 20]. Consistently, mice lacking the IL-2Rγ gene, *IL2rg*, are deficient of not only B and T lymphocytes but also ILCs [21]. These cytokines are thought to support the proliferation and survival of developing and mature ILCs.

Notch signaling is also important for ILC differentiation even though its impact on different ILC subsets may be divergent [22]. Notch signaling is not necessary for the generation of ILC progenitors such as αLP because ablation of RBPJκ, the DNA-binding partner of Notch intracellular domain (NICD), using *IL-7R-Cre* did not affect the levels of these progenitors [23]. Conversely, constitutive activation of Notch signaling in IL7R-Cre; Rosa26-NICD transgenic mice drives T cell differentiation at the expense of ILC fates [23, 24]. By titrating the strength and duration of Notch signaling, it has been shown that strong Notch signaling favors T cell development over ILCs in in vitro cultures [25]. However, the differentiation of ILC2 and ILC3 depends on intermediate levels of Notch signaling. ILC3 differentiation requires a longer duration of exposure to Notch ligands such as Delta-like 4, relative to ILC2 production. In contrast, the differentiation of NKs and ILC1s is inhibited by Notch signaling when CLPs are cultured on OP9-DL1 stroma [25].

The cell-intrinsic drivers of ILC differentiation include an array of transcription factors which are responsible for either the generation of ILC progenitors or the specification of different ILC subsets. TCF1, encoded by *Tcf7*, is expressed in early ILC progenitors (EILP) and plays a pivotal role in the generation of downstream progenitors for ILC subsets (CHILPs) and NK progenitors (NKP) [26, 27]. In addition, TCF1 promotes ILC2 differentiation by upregulating GATA3, another important transcription factor in ILC development [28, 29]. *Gata3* is expressed at varying levels in ILC progenitors and ILC subsets and activates the transcription of *Il7r*, encoding an obligatory cytokine receptor [30–32]. GATA3 is necessary for the differentiation of ILC1s, ILC2s, and non-LTi ILC3s but not for NKs [33]. In particular, GATA3, highly expressed in ILC2s, is instrumental for the maturation of ILC2s and responsible for the expression of type 2 cytokines including IL-4, IL-5, and IL-13 [32, 34]. It also stimulates the transcription of other genes involved in ILC2 function such as *Il1rl1* and *Il17rb*, which code for receptors of ILC2-activating cytokines, IL-33 and IL-25, respectively.

PLZF, encoded by *Zbtb16*, is expressed in ILC progenitors (ILCPs) which give rise to ILC1 to ILC3 but not NK cells [2]. It is thus thought to promote the differentiation into ILCs but limit the NK and LTi fates. Whether it plays a role in the specification of any ILC subsets is not clear. Additional transcription factors important for ILC progenitors include NFIL3 and TOX whereas T-BET and RORγt help dictate the

ILC1 and ILC3 fates, respectively [28, 35]. Lastly, Id2, an inhibitor of basic helix-loop-helix E protein transcription factors [36], is expressed in ILC progenitors downstream of CLP and is essential for the generation of all ILC subsets [7, 37]. The crucial role of Id2 highlights the importance of this family of transcription factors and their inhibitors, which will be discussed in-depth in subsequent sessions.

3.3 The Ability of the Thymus to Support ILC Differentiation

The primary known function of the thymus is to produce T lymphocytes. Multipotent progenitors from the bone marrow, namely lymphoid-primed multipotent progenitors (LMPPs) and common lymphoid progenitors (CLPs), seed the thymus, where they become early T cell progenitors (ETPs, included in the CD4 and CD8 double-negative stage 1 (DN1) population) [8, 10, 38–40]. Through separate but parallel differentiation programs, ETPs give rise to αβ and γδ T cells, respectively [41]. The developmental cascade for αβ T cells can be characterized by the expression of CD4 and CD8 surface markers. ETPs progress through four DN stages (DN1–4) to the double-positive stage before maturing into CD4 or CD8 single-positive αβ T cells [42, 43]. The massive expansion of DN3 or DN4 T cells makes αβ T cells the major T cell population in the thymus and periphery [44]. γδ T cells are also derived from ETPs and they are largely CD4 and CD8 double negative [45, 46]. The divergence between αβ and γδ T lineages is thought to be governed by either stochastic events or instructive TCR signaling strengths [47, 48]. DN T cells with higher levels of IL-7R and SOX13 are biased towards the γδ T cell fate. Likewise, TCRγδ delivers stronger signals compared to pre-TCR on immature αβ T cells.

Thymic epithelial cells are known to secrete IL-7 and express Notch ligands, Delta-like 4, thus providing a favorable condition for the development of both T cells and innate lymphoid cells [49–51]. Moreover, T cells can produce IL-2, which helps in the survival of ILC2s [52]. In terms of cell-intrinsic factors, early T cell precursors express essential transcription factors for ILC differentiation, including TCF1, GATA3, and Bcl11b [26, 28, 32, 53, 54]. Since ETPs have been shown to retain the multipotent capacity to differentiate into dendritic cells and myeloid cells [12], it is not surprising that they can generate ILCs given the nourishing environment the thymus provides. However, in vivo and in vitro data also illustrate that committed T cell precursors such as DN3 cells can also differentiate into ILC2s [14]. What controls the lineage decision between T cells and ILCs remains to be fully understood but modulating the function of basic helix-loop-helix E protein transcription factors has been shown to play a critical role as detailed below.

3.4 Regulation of T Cell Versus ILC Fates by E Protein Transcription Factors

3.4.1 E Proteins and Their Inhibitors

E proteins are a group of four basic helix-loop-helix transcription factors encoded by the E2A (*Tcf3*), HEB (*Tcf12*), and E2–2 (*Tcf4*) genes [55, 56]. They share extensive sequence homologies in their DNA-binding and transcription activation domains but the expression patterns are different in different cell types and at different developmental stages. They all bind to “E box” sequences and activate transcription of their target genes as homodimers or heterodimers between two E proteins. In addition, there exist four inhibitors of E proteins, called Id1 to Id4, which contain the helix-loop-helix dimerization domain but not the region containing basic amino acids necessary for DNA binding [57]. Id proteins bind all E proteins to form heterodimers which cannot bind to DNA. Therefore, the net E protein activity in a given cell depends on the ratio between E and Id proteins. While expression levels of E protein genes are relatively stable, Id gene transcription is often regulated dynamically, thus controlling the net E protein activity.

3.4.2 E Proteins Are Essential for T Cell Development

E and Id proteins are known to be crucial for T cell development at multiple checkpoints but their roles at early stages are most relevant to ILC differentiation [58–61]. Because both the E2A and HEB genes are expressed in T lineage cells, disrupting either of them did not reveal severe developmental block [58, 62]. However, when the Id1 inhibitor was expressed in early T cell progenitors under the control of the proximal promoter of the *lck* gene, T cell development was completely abrogated in the transgenic mice [63–65]. Likewise, with inducible deletion of both E2A and HEB genes using the *plck-Cre* transgene, which starts to express at the DN3 stage (later than the *plck-Id1* transgene), T cell development, particularly αβ T cells, was dramatically blocked [66]. These data indicate that E proteins are critical for T cell commitment.

Consistent with the function of E proteins, gene expression analyses have shown that E proteins are instrumental for the transcription of *Notch1* [67]. E proteins also activate the transcription of *Ptcra, Rag1, Rag2*, and some of the *Cd3* genes, which encode molecules crucial for the formation of pre-TCRs involved in the β selection of developing T cells that leads to their massive expansion [68]. Interestingly, pre-TCR signaling, and TCR signaling at a later stage, triggers the activation Ras-MAP kinase pathway, which in turn upregulates the Egr transcription factors that stimulate *Id3* transcription [69]. The transient upregulation of Id3 and more sustained Id2 expression are important for T cell selection and NKT cell differentiation [61, 70–75]. In γδ T cells, TCRγδ signaling also leads to high levels of Id3 expression, and thus a downregulation of E protein activities [45, 76].

3.4.3 Suppression of ILC2 Fates by E Proteins

The impact of E proteins on ILC2 differentiation was discovered unexpectedly in Id1 transgenic mice, as well as in E protein-deficient mice [13, 68]. As mentioned in the previous section, ectopic Id1 expression at the ETP stage completely blocks T cell development [63, 65]. However, a large number (about $3x10^5$ per thymus) of lineage-negative cells accumulated in the thymus, which fit the description of ILC2, namely $Lin^-Thy1^+ST2^+$ [13]. In contrast, such cells in wild-type thymus only amount to fewer than 10,000. In addition, these ILC2s exit the thymus, leading to the accumulation of massive amounts of ILC2s in peripheral tissues throughout the body. Functional studies show that these ILC2s made in the thymus elicit type 2 immune responses upon stimulation with papain or helminths [13]. Because the lung and small intestine of Id1 transgenic mice harbor more than ten times ILC2s compared to wild-type mice, they are much more efficient at expulsing worms from the small intestine after infection with *Nippostrongylus brasiliensis*. Besides the cell-intrinsic ability of ETPs to differentiate into ILC2s, this mouse model also illustrated the enormous capacity of the thymus in supporting ILC2 differentiation.

Furthermore, when the E2A and HEB genes were ablated using *plck-Cre*, large numbers of ILC2s were also found in the thymus, as well as in peripheral tissues [13, 14]. Using reporter assays, it has been shown that *plck-Cre* expression begins at the DN3 stage [13], when T cell commitment already takes place. Although *plck-Cre* employs the same promoter as the *plck-Id1* transgene, the different integration sites dictate the different timing of gene expression. The fact that knocking out the E protein genes after T cell commitment can promote ILC2 differentiation suggests that ILC2s can be generated from committed T cell precursors in addition to multipotent progenitors [14]. Moreover, the two E protein genes were also deleted using *Il7r-Cre*, which expresses at the CLP stage in the bone marrow, and increased ILC2 production was also observed [68]. Fetal thymic organ culture studies showed that ILC2s could be generated in the thymus of these mice. Likewise, inducible deletion of the E2A and HEB genes with *Rosa26-CreERT2* after administration of tamoxifen boosted ILC2 production in both the bone marrow and thymus [14]. Conversely, when a gain-of-function mutant of E proteins, called ET2 [77], was expressed after induction with *Il7r-Cre,* ILC2 but not ILC1 or ILC3 differentiation was diminished in the bone marrow [78]. Finally, using the in vitro OP9-DL1 stromal culture system, ILC2 differentiation from CLP, ETP, and DN3 progenitors was all enhanced by 20–40-fold upon ablation of these two E protein genes [14].

Taken together, several lines of evidence strongly suggest that E proteins play crucial roles in suppressing ILC2 differentiation either in the bone marrow or in the thymus. This occurs not only in multipotent progenitors but also in committed T cell precursors.

3.4.4 Molecular Mechanisms Underlying E Protein-Mediated Suppression of ILC2 Differentiation

E proteins are powerful drivers of B and T cell development in the bone marrow and thymus, respectively. One might imagine that the failure of B or T cell differentiation in the absence of E proteins would automatically allow the ILC fate. However, the massive production of ILC2s seen in either Id1 transgenic mice or E protein-deficient mice has not been mirrored in any other mouse models where B or T cell development is arrested at early stages, for example, $Rag1^{-/-}$ or $Rag2^{-/-}$ mice. Comparison of gene expression between ETPs isolated from control and E protein-deficient mice revealed the downregulation of T cell-specific genes and upregulation of ILC2-specific genes [68]. These results could be complicated by the potential differences in the compositions of the ETP fractions in the wild-type and E protein-deficient mice because any contamination of the ETP preparation with ILC2s in the E protein knockout mice would skew the transcription profile towards ILC2. Therefore, the data obtained could not fully explain how E proteins suppress the ILC2 fate.

To identify the root causes of accelerated ILC2 differentiation upon E protein ablation, gene expression changes at the onset of ILC2 differentiation have to be determined. To this end, Qian et al. plated ETP and DN3 cells isolated from $E2AF^{f/f}$;$HEB^{f/f}$;Rosa26-CreERT2;Rosa26-Stop-tdTomato or Rosa26-CreERT2;Rosa26-Stop-tdTomato control mice onto OP9-DL1 stromal cells [14]. Four days later, tamoxifen was added to the cultures and tdTomato-positive cells were collected after 24 or 72 h. At these time points, phenotypic ILC2s had not been made but gene expression already changed. Differentially expressed genes comparing the control and knockout cells at these early stages may be key to the ILC2 differentiation program. Indeed, a network of genes critical for ILC2 differentiation, such as *Gata3, Rora, Lmo4,* and *Zbtb16,* were upregulated within 24 h. This was accompanied with the downregulation of T cell differentiating genes like *Ptcra, Rag1, Rag2, Dntt, Cd3g,* and *Notch1* [14, 79].

Perhaps, a 24-h period is still too long to ascertain if these genes are direct targets of E proteins. The Sun lab next attempted to express an inducible form of E protein, E47ER, in ILC2s isolated from Id1 transgenic mice using retroviral transduction [79]. When tamoxifen was added to the cells, changes in gene expression and chromatin accessibility within 4 and 16 h were examined using RNA sequencing and ATAC (assay for transposase-accessible chromatin)-seq assays. Consistent with the notion that E proteins are transcriptional activators, the majority of the changes in mRNA levels are upregulated. However, ATAC-seq revealed the dynamic changes in chromatin accessibility. At 4 h, induction of E47ER led to increases in accessibilities but by 16 h, a major decrease in open chromatin occurred and the accessible regions lost due to increased E47ER activity contained binding sites for leucine zipper proteins and GATA transcription factors. These results suggested that the suppression of chromatin is a secondary consequence of genes initially upregulated by E47. *Cbfa2t3 and Jdp2,* encoding MTG16 (also called ETO2) and JDP2 transcription repressors, respectively [80–83], were indeed found to be activated by

E47 [79]. Based data from loss of E protein function in ETPs and DN3s and gain of function in ILC2s, a consensus was obtained with regard to genes activated by E proteins, which include *Cbfa2t3, Bach2, Gfi1, Tcf7, Jdp2,* and *Btg2*. Genes directly or indirectly repressed by E proteins include *Tox, Maf, Irf4, Icos, Ikzf2, Klf6, Gata3,* and *Lmo4*, which are known to be involved in ILC2 differentiation [79]. Whether these findings can be verified in vivo using genetic rescuing experiments would depend on how many genes are simultaneously involved in influencing the ILC2 fate and how these genes impact the differentiation of other lineages of cells.

3.5 ILCs Made in the Thymus

Despite the observation of massive ILC2 production in the thymus upon E protein ablation [13, 14], the critical question to be answered is whether the thymus normally contributes to the pools of ILC2 or other ILCs in peripheral tissues. To address this issue, one major challenge is to be able to distinguish thymus-derived ILCs from those generated in the bone marrow or from tissue-resident ILC precursors. Fortunately, several unique properties of thymocytes allow the development of useful genetic tools even though none of them is perfect. These tools include the athymic nude mice, thymus-specific T cell receptor gene rearrangement events that permanently mark the descendant cells, and reporter gene expression induced by a thymus-specifically expressed Cre recombinase [84–88]. Additionally, the single-cell RNA sequencing technology (scRNA-seq) is also a powerful approach, especially when used in combination with the tools mentioned above. These tools have been exploited to vigorously interrogate the impact of the thymus in supplying ILCs to peripheral tissues.

3.5.1 ILCs Within the Thymus

Because of the shared properties between T cells and ILCs, it has been difficult to definitively separate these two cell types with surface markers or transcription factor signatures. Much of the characterization relied on scRNA-seq and inference from in vitro single-cell cultures. Moreover, a majority of the studies were conducted using fetal thymi, which provide advantage in monitoring the developmental progression of ILC production. While analyzing total thymocytes at different embryonic stages using scRNA-seq, Kernfeld et al. found that ILC1-like cells appeared at E12.5–E14.5 whereas ILC2-like cells were predominantly detected at E15.5–E16.5 [89]. In another study, Elsaid et al. used single-cell cultures on OP9 stroma or in fetal organ cultures to show that ETPs from E13.5 thymi had a greater propensity to generate LTi cells but more restricted potential to differentiate into all other ILC subsets [90]. In contrast, the majority of ETPs from E18.5 thymi can give rise to ILCs but have no capacity of generating LTis.

Recently, the McKenzie group has further characterized ILC differentiation in embryonic thymus by using scRNA-seq in conjunction with a five-color polychromILC TF reporter strain of mice, where expression of ID2, BCL11B, GATA-3, RORα, and RORγt was indicated by different fluorescent proteins [91]. ILC2s were found to be more abundant in E19.5 thymi compared to E15.5 thymi. These cells were defined as $ID2^+BCL11B^+GATA3^+ROR\alpha^+$ by flow cytometry and considered ILC2Ps in the thymus. These cells were shown to be able to seed small intestine when the fetal thymi were implanted into recipient mice [91]. Curiously, a subset of cells characterized as $ID2^-BCL11B^+GATA3^+$ also existed in E19.5 thymi but their identity was not clear. $ID2^+BCL11B^-$ cells have the potential to differentiate into NK cells, and thus are designated NKPs. The reporter mice also facilitated the localization of ILC2s to the medulla of the thymus, and the illustration of their close proximity to IL-33-producing thymic epithelial cells. The authors also concluded that ILC differentiation precedes the formation of $CD4^+CD8^+$ T cells, thus indicating a faster kinetics of ILC ontogeny from ETPs compared to T cell development. Mechanistically, the authors concluded that RORα is a critical transcription factor for ILC2 differentiation and suppression of the T cell fate. RORα coordinates binding with GATA3 transcription, which leads to the upregulation of *Bcl11b*, *Id2,* and *Nfil3* [91].

In postnatal thymi, Jones et al. took advantage of the Id2 reporter mice and analyzed the $Lin^{-}IL\text{-}7R^{+}CD3$ (intracellular)$^{-}$ population for the expression of signature transcription factors, T-bet, GATA3, and RORγt, which are enriched in ILC1s, ILC2s, and ILC3s, respectively [92]. Up to the first 7 days after birth, comparable levels of ILC2 and ILC3 were detected but ILC3s disappeared from the thymus by day 14. Of the ILC3s detected on day 7, the $CCR6^{+}$ subset was the predominant form. ILC1s were very rare at all ages. Like in the fetal stage, ILC2 also resides in the medulla.

In human postnatal thymi, ILCs were characterized in the $Lin^{-}CD127^{+}CD161^{+}$ population, within which all three ILC subsets were detected [93]. While $CRTH2^{+}$ cells are considered ILC2s, the $CRTH2^{-}$ cells were further fractionated into ILC1s (c-kit^{-}NP44^{-}) and two subsets of ILC3s (c-kit^{+}NP44^{-} and NP44^{+}). Interestingly, CD5 expression marked immature ILCs in the thymus as well as umbilical cord blood. Upon differentiation of the immature ILCs, CD5 expression is downregulated. In human fetal thymi, all three ILC subsets have been described based on scRNA-seq data [94]. Like in mice, ILC2s appeared to be the predominant subset in the thymus and were more abundant at 12 weeks of gestation compared to 8 weeks. Because of the distinct transcriptomes within each ILC subset, the three ILC subsets were deemed heterogenous and subdivided into multiple groups.

Regardless of the ages of the pre- or postnatal thymi, ILC2s are found to be the prominent group among all ILCs. This is consistent with the thymic microenvironment conducive to ILC2 differentiation. The remaining question concerns their physiological function in the thymus. At early embryonic stages, the production of LTis may facilitate thymic organogenesis [90]. During later embryonic stage and postnatal stages, the presence of ILC2s may contribute to the cytokine milieu [92]. However, whether these functions are essential or redundant remains to be fully understood.

3.5.2 Detection of TCR Rearrangement in ILC2s in the Lung

One of the unique events occurring during T cell development in the thymus is the rearrangement of T cell receptor genes including *Tcrb*, *Tcrg*, *Tcrd,* and *Tcra*. Recombination in *Tcrg* and *Tcrd* takes place at the CD4 and CD8 double-negative (DN2) stage, and leads to the formation of TCRγδ complexes and the development of γδ T cells [95]. Productive rearrangement of *Tcrb* at the DN3 stage allows the assembly of pre-TCR complexes consisting of the TCR β chain paired with the pre-TCRα chain, which is the prerequisite for αβ T cells to progress to the double-positive stage (DP) [44]. It is at the DP stage where *Tcra* rearrangement occurs, which enables the generation of a diverse array of functional αβ TCRs [96]. Because *Tcrd* is imbedded in the *Tcra* locus, *Tcra* rearrangement results in the loss of *Tcrd*.

If ILCs are derived from committed T cell precursors, one would expect that the ILCs in the periphery carry rearranged TCR genes. Indeed, lung ILC2s from wild-type mice but not nude mice harbor rearranged *Tcrb* and *Tcrg* detected by using PCR primer pairs that amplifies Dβ2-Jβ2, Vβ3-DJβ2, and Vγ1-Jγ4 [14]. By comparing to a standard curve made using total thymocytes, these events were estimated to occur in about 10% of mature ILC2s. However, in knockout mice where E proteins are deleted at the DN3 stage, nearly all ILC2s in the lung carried arranged TCRs. By contrast, lung ILC2s from Id1 transgenic mice did not exhibit such rearrangement because these cells were generated at the DN1 stage.

Shin et al. also measured *Tcr* gene recombination in lung ILC2s and detected only rearrangement in the *Tcrg* locus [97]. The failure of detecting *Tcrb* rearrangement may be due to the limited sensitivity of the detection by agarose gel electrophoresis compared to Southern blotting of the PCR products used by Qian et al. *Tcrd* rearrangement was also not found. Additionally, a lower frequency of productive *Tcrg* rearrangement in ILC2s was observed compared to γδ T cells, even though a fraction of the rearranged genes in ILC2s did have open reading frames, which suggests no strong selection against in-frame recombination. Furthermore, the transcripts from the rearranged *Tcrg* genes were not present in ILC2s.

A major challenge in the interpretation of these data is that the studies were carried out with bulk cell preparations. Given the complexity of TCR rearrangement events and the presence of two alleles of each gene, it is difficult to ascertain if the predecessor of individual ILC2s ever had functional TCR genes prior to its conversion to an ILC2. This information would help understand how T cell precursors diverge to the ILC path but it awaits the advancement of the single-cell DNA sequencing technology.

3.5.3 Reporter Expression in ILCs in a Thymus-Dependent Manner

Fate mapping using Cre-mediated recombination is a powerful approach for determining the developmental history of a cell. To monitor thymus-derived ILCs, the Cre recombinase has to be expressed specifically in the thymus and during earlier stages of T cell development. Two such Cre transgenes may be of use. The first one, designated $pT\alpha^{iCre}$, has the Cre gene knocked into *Ptcra* and begins to express at the later part of DN1 stage. It labels about 40% of the DN2/DN3 cells [98]. The second one, called *plck-Cre*, expresses the Cre transgene from the proximal promoter of the *lck* gene [87]. This gene labels few DN1 and DN2 cells but marks about 20% of DN3 cells [13]. When crossed to R26-Stop-tdTomato mice, about 30% of ILC-containing Lin^-Thy1^+ cells in the blood expressed tdTomato [14]. Transplantation of bone marrow cells from plck-Cre:R26-Stop-tdTomato mice into athymic nude mice did not lead to production of $tdTomato^+Lin^-Thy1^+$ cells in the blood, suggesting a thymus dependency. In the peripheral tissues such as lung, small fractions of conventional ILC2s were found to be $tdTomato^+$, which suggested that these cells came from the thymus. In the small intestine, while the $Lin^-Thy1^+KLRG1^+$ ILC2s expressed little tdTomato, their $KLRG1^-$ counterparts had high frequencies of $tdTomato^+$ cells. These cells have recently been shown to contain immature forms of ILCs [99].

Intriguingly, the frequencies of $tdTomato^+$ mature ILC2s in the lung or small intestine are not as high as those seen in the immature forms or in the blood [14]. Since the immature forms of ILCs are permanently labeled by tdTomato, these cells should have been detected after they mature. Although this might mean that thymus-derived ILCs do not contribute significantly to the pools, one possible explanation is that the expression of tdTomato causes some disadvantage in cell growth or survival such that the small fraction of $tdTomato^+$ cells cannot compete with the majority of $tdTomato^-$ cells. We indeed found this to be the case in OP9 stromal cultures where $tdTomato^+$ progenitors generate two- to threefold fewer ILC2s compared to cells not expressing tdTomato (Qian and Sun, unpublished). Therefore, this drawback combined with a low efficiency of tdTomato expression before the DN3 stage makes plck-Cre;R26-Stop-tdTomato mice less than ideal reporters for tracing thymus-derived ILCs.

3.5.4 In Vitro Differentiation of ILC2s from T Cell Precursors

To directly assess the potentials of wild-type thymocytes to differentiate into ILC2s, Qian et al. compared such abilities of purified DN1 (ETP) and DN3 thymocytes to those of CLPs from the bone marrow by using in vitro cultures on OP9-DL1 stromal cells in the presence of IL-2, IL-7, and stem cell factor [14]. CLPs and DN1 cells displayed similar kinetics and proliferative potential during differentiation. By day 17, they yielded 4–7 × 10^4 ILC2s per progenitor. This result indicates that earliest T cell progenitors have similar capacity in producing ILC2s as the lymphoid-biased bone marrow progenitors, CLPs.

In contrast, DN3 cells differentiated into ILC2s faster and ILC2s were detectable on day 7. However, the cell number increased at a slower pace and plateaued by day 14, thus yielding about ten times fewer ILC2s on a per cell basis. This may be due to the lower intrinsic proliferative ability of the committed T cell precursors compared to multipotent progenitors such as CLPs and DN1s. Nevertheless, this finding is of great significance because it suggests that committed T cell precursors can differentiate into ILC2s. Considering the vast excess in the number of DN3s over DN1s, differentiation from DN3 cells could amount a significant output of ILCs.

3.5.5 Identification of Thymus-Derived ILC Precursors in the Blood

Given the limitations of using reporter mice and TCR arrangement events as handles to examine thymus-derived ILCs, the Sun laboratory has taken advantage of the scRNA-seq technology and the availability of different mouse strains to examine the ILC populations in the blood of 2-month-old mice and assess the contribution of the thymus to circulating ILCs [99]. By comparing the abundance of different subsets in wild-type and athymic nude mice, we found that several clusters, collectively called td-ILC, accounted for about 50% of the $Lin^{-}Thy1^{+}$ population in the blood and about 0.025% of the total peripheral blood mononuclear cells (PBMCs). They were suspected to be derived from the thymus because the athymic nude mice had dramatically reduced frequencies of these cells. On the contrary, mice with two of the E protein genes specifically deleted at the DN3 stage possess over 1000 times more of the td-ILCss. Transcriptome analyses of td-ILCs showed expression of genes commonly found in ILCs such as *Il7r* and *Tcf7*. They also expressed *Sell* and *S1pr1,* which encode proteins necessary for cells to exit the thymus. Genes coding for different CD3 chains were also detected but other T cell-specific genes were not, which suggested that td-ILCs may be generated from committed T cell precursors. Based on the scRNA-seq data, a flow cytometric staining scheme ($Lin^{-}CD127^{+}CD62L^{+}$ intracellular (ic) $CD3\varepsilon^{+}$) was designed to detect td-ILCs in the blood of wild-type mice but not nude mice. CD127 and CD62L are encoded by *Il7r* and *Sell,* respectively [99].

However, signature genes for transcription factors dictating ILC1, ILC2, and ILC3 fates, *Tbx21*, *Gata3*, and *Rorc*, were not found in td-ILCs, which raised the possibility that these cells are the precursors of ILCs. To further investigate the nature of td-ILCs, the scRNA-seq data was interrogated by performing trajectory inference of WT $Lin^{-}Thy1^{+}$ cells using the pseudotime progression method in the *Monocle2* program. This study placed clusters containing td-ILCs in between immature T cells and differentiated ILC subsets which may or may not be made in the thymus. Analyses of pseudotime-dependent genes along the trajectory confirmed such an assessment in that T cell-specific genes are highly expressed at the beginning of the trajectory whereas ILC-specific genes are enriched at the other end of the spectrum [99].

The observation that substantial fractions of $Lin^{-}Thy1^{+}$ population in mouse blood represent ILC precursors is reminiscent of that found in human blood, where $Lin^{-}CD127^{+}CRTH2^{-}c\text{-}kit^{+}$ cells have been shown to be ILC precursors [99, 100]. These human ILCs were further demonstrated to be heterogeneous with distinct biases towards different ILC subsets [100, 101]. Likewise, the mouse ILC precursors are likely to have diverse differential potentials and tissue preferences. It would be interesting to ascertain if any of the human ILCPs at the adult stage are made in the thymus.

3.5.6 Identification of Immature Thymus-Derived ILCs in Peripheral Tissues and Their Differentiation

To trace the thymus-derived ILC precursors in the blood to peripheral tissues, intracellular CD3ε (icCD3ε) was used as a marker [99]. This marker was first validated by analyzing the lung and small intestine of wild-type, nude, and E protein-deficient mice. The numbers of $Lin^-CD127^+ST2^-icCD3\varepsilon^+$ cells in the lung and $Lin^-CD127^+KLRG1^-icCD3\varepsilon^+$ cells in the small intestine were dramatically reduced in the athymic nude mice and greatly increased in E protein-deficient mice, thus suggesting that icCD3ε marks thymus-derived immature ILCs, which are likely from the blood. Furthermore, these cells could only be labeled by intracellular but not surface staining for CD3ε, which excludes the possibility that these cells are T cell contaminants [99].

Since the $icCD3\varepsilon^+$ cells were mainly found in the immature forms in the lung and small intestine, their ability to differentiate into different ILC subsets in these peripheral tissues was examined by analyzing the expression of T-BET, GATA3, and RORγt transcription factors [99]. Interestingly, an inverse relationship between the levels of the transcription factors and icCD3ε was found, which suggests that as the ILC precursors differentiate into different ILC subsets, they downregulate CD3ε expression. Additionally, the $icCD3\varepsilon^+$ ILC precursors bias towards different ILC subsets in different tissues. For example, they differentiated primarily into ILC2s in the lung whereas in the small intestine, the majority of the cells expressed T-BET, suggesting an NK/ILC1 cell fate. There are also $icCD3\varepsilon^+$ mature ILC2s and ILC3s but they constitute very small fractions, which is consistent with the notion that CD3 expression is downregulated as ILCs mature.

When mice were infected with helminths, *Nippostrongylus brasiliensis*, there was an initial increase of $Lin^-CD127^+ST2^-icCD3\varepsilon^+$ cells in the lung, which could be due to recruitment of ILC precursors from the blood or the local expansion of these cells [99]. This was then followed by marked elevation of mature ILC2 counts and secretion of type 2 cytokines, IL-5 and IL-13. Similarly, IL-5 and IL-13 production can also be induced by helminth infection in the mesenteric lymph nodes. These results suggest that a known stimulus of ILC2s promotes the differentiation of the thymus-derived ILC precursors. Considering that these $icCD3\varepsilon^+$ ILC precursors are constantly circulating in the blood, they may replenish ILC pools in tissues, particularly under nonphysiological conditions such as parasite infections.

3.6 Conclusions and Future Directions

The thymus provides a nurturing environment for ILC differentiation. Thymocytes, either multipotent progenitors (e.g., ETPs/DN1 cells) or committed T cell precursors (e.g., DN3 cells), have the potential to differentiate into ILCs, at least ILC2s (Fig. 3.1). A key question is whether the thymus is one of the major sources of ILC pools prenatally or postnatally. Recent data from our laboratory showed that in the blood of 2-month-old wild-type mice, a substantial fraction of PBMCs represents thymus-derived ILC precursors (Fig. 3.2). These cells were shown to home to peripheral tissues and differentiate into diverse ILC subsets. Infection with helminth parasites initially increased the level of immature ILCs in the lung and their subsequent maturation into functional ILC2s, which suggests that thymus-derived ILC precursors in the blood could replenish tissue ILCs on demand. Further studies are necessary to understand the participation of circulating ILC precursors in a variety of immunological responses involving all three ILC subsets.

The recognition of the thymus as a source of ILCs helps establish a new paradigm in ILC ontogeny (Fig. 3.1). The fact that committed T cell precursors can differentiate into ILC2s suggests that not all ILC differentiations have to go through the route involving specialized ILC

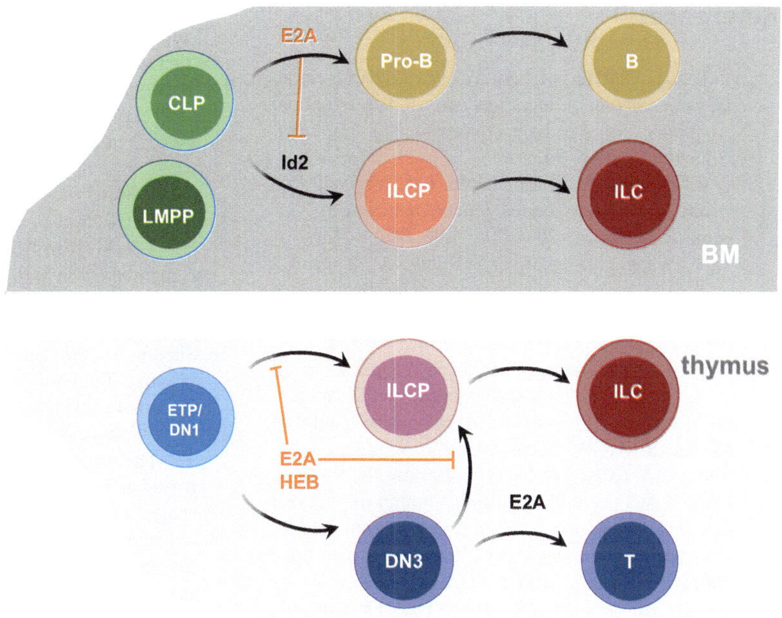

Fig. 3.1 ILC differentiation program taking place in the thymus, in parallel to the program occurring in the bone marrow

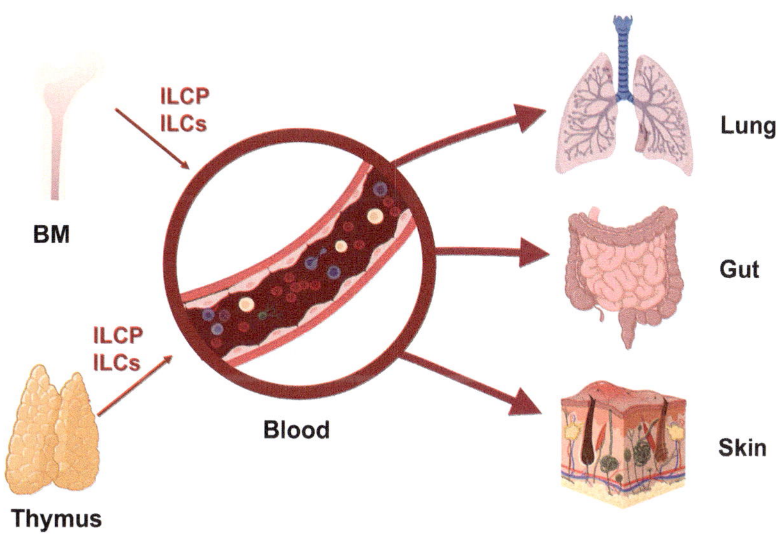

Fig. 3.2 Contribution of ILC precursors and mature ILCs by the bone marrow and thymus to tissue ILC pools through blood circulation

progenitors as defined in the bone marrow. This then leads to interesting questions as to how the T and ILC cell fates are determined and what molecular switches are involved. Considering the dynamics of thymus maintenance through lifetime, the contribution of thymus in ILC homeostasis is of great significance in age-related immunological functions in physiological and pathophysiological conditions.

Acknowledgements We thank Aneta Pankow for sharing unpublished data. Work in the Sun laboratory was supported by a grant from the NIH (R01AI126851).

References

1. Yang Q, Li F, Harly C, Xing S, Ye L, Xia X, et al. TCF-1 upregulation identifies early innate lymphoid progenitors in the bone marrow. Nat Immunol. 2015;16(10):1044–50.
2. Constantinides MG, McDonald BD, Verhoef PA, Bendelac A. A committed precursor to innate lymphoid cells. Nature. 2014;508(7496):397–401.
3. Yu X, Wang Y, Deng M, Li Y, Ruhn KA, Zhang CC, et al. The basic leucine zipper transcription factor NFIL3 directs the development of a common innate lymphoid cell precursor. Elife. 2014;3 e04406.
4. Yu Y, Tsang JC, Wang C, Clare S, Wang J, Chen X, et al. Single-cell RNA-seq identifies a PD-1hi ILC progenitor and defines its development pathway. Nature. 2016;539(7627):102–6.
5. Walker JA, Clark PA, Crisp A, Barlow JL, Szeto A, Ferreira ACF, et al. Polychromic reporter mice reveal unappreciated innate lymphoid cell progenitor heterogeneity and elusive ILC3 progenitors in bone marrow. Immunity. 2019;51(1):104–18 e7.
6. Wong SH, Walker JA, Jolin HE, Drynan LF, Hams E, Camelo A, et al. Transcription factor RORalpha is critical for nuocyte development. Nat Immunol. 2012;13(3):229–36.
7. Klose CS, Flach M, Mohle L, Rogell L, Hoyler T, Ebert K, et al. Differentiation of type 1 ILCs from a common progenitor to all helper-like innate lymphoid cell lineages. Cell. 2014;157(2):340–56.
8. Kondo M, Weissman IL, Akashi K. Identification of clonogenic common lymphoid progenitors in mouse bone marrow. Cell. 1997;91(5):661–72.
9. Adolfsson J, Borge OJ, Bryder D, Theilgaard-Monch K, Astrand-Grundstrom I, Sitnicka E, et al. Upregulation of Flt3 expression within the bone marrow Lin(−)Sca1(+)c-kit(+) stem cell compartment is accompanied by loss of self-renewal capacity. Immunity. 2001;15(4):659–69.
10. Lai AY, Kondo M. Asymmetrical lymphoid and myeloid lineage commitment in multipotent hematopoietic progenitors. J Exp Med. 2006;203(8): 1867–73.
11. Bhandoola A, Sambandam A. From stem cell to T cell: one route or many? Nat Rev Immunol. 2006;6(2):117–26.
12. Bell JJ, Bhandoola A. The earliest thymic progenitors for T cells possess myeloid lineage potential. Nature. 2008;452(7188):764–7.
13. Wang HC, Qian L, Zhao Y, Mengarelli J, Adrianto I, Montgomery CG, et al. Downregulation of E protein activity augments an ILC2 differentiation program in the thymus. J Immunol. 2017;198(8):3149–56.
14. Qian L, Bajana S, Georgescu C, Peng V, Wang HC, Adrianto I, et al. Suppression of ILC2 differentiation from committed T cell precursors by E protein transcription factors. J Exp Med. 2019;216(4):884–99.
15. Van Gool F, Molofsky AB, Morar MM, Rosenzwajg M, Liang HE, Klatzmann D, et al. Interleukin-5-producing group 2 innate lymphoid cells control eosinophilia induced by interleukin-2-therapy. Blood. 2014;124(24):3572–6.
16. Roediger B, Kyle R, Tay SS, Mitchell AJ, Bolton HA, Guy TV, et al. IL-2 is a critical regulator of group 2 innate lymphoid cell function during pulmonary inflammation. J Allergy Clin Immunol. 2015;136(6):1653–63 e7.
17. Martin CE, Spasova DS, Frimpong-Boateng K, Kim HO, Lee M, Kim KS, et al. Interleukin-7 availability is maintained by a hematopoietic cytokine sink comprising innate lymphoid cells and T cells. Immunity. 2017;47(1):171–82 e4.
18. Sheikh A, Abraham N. Interleukin-7 receptor alpha in innate lymphoid cells: more than a marker. Front Immunol. 2019;10:2897.
19. He YW, Nakajima H, Leonard WJ, Adkins B, Malek TR. The common gamma-chain of cytokine receptors regulates intrathymic T cell development at multiple stages. J Immunol. 1997;158(6):2592–9.
20. Noguchi M, Nakamura Y, Russell SM, Ziegler SF, Tsang M, Cao X, et al. Interleukin-2 receptor gamma chain: a functional component of the interleukin-7 receptor. Science. 1993;262(5141):1877–80.
21. Moro K, Yamada T, Tanabe M, Takeuchi T, Ikawa T, Kawamoto H, et al. Innate production of T(H) 2 cytokines by adipose tissue-associated c-Kit(+) Sca-1(+) lymphoid cells. Nature. 2010;463(7280): 540–4.
22. Golub R. The notch signaling pathway involvement in innate lymphoid cell biology. Biom J. 2021;44(2): 133–43.
23. Chea S, Schmutz S, Berthault C, Perchet T, Petit M, Burlen-Defranoux O, et al. Single-cell gene expression analyses reveal heterogeneous responsiveness of fetal innate lymphoid progenitors to notch signaling. Cell Rep. 2016;14(6):1500–16.
24. Gentek R, Munneke JM, Helbig C, Blom B, Hazenberg MD, Spits H, et al. Modulation of signal strength switches notch from an inducer of T cells to an inducer of ILC2. Front Immunol. 2013;4:334.
25. Koga S, Hozumi K, Hirano KI, Yazawa M, Terooatea T, Minoda A, et al. Peripheral PDGFRalpha(+)gp38(+) mesenchymal cells support the differentiation of fetal liver-derived ILC2. J Exp Med. 2018;215(6):1609–26.
26. Yang Q, Monticelli LA, Saenz SA, Chi AW, Sonnenberg GF, Tang J, et al. T cell factor 1 is required for group 2 innate lymphoid cell generation. Immunity. 2013;38(4):694–704.
27. Mielke LA, Groom JR, Rankin LC, Seillet C, Masson F, Putoczki T, et al. TCF-1 controls ILC2 and NKp46+RORgammat+ innate lymphocyte differentiation and protection in intestinal inflammation. J Immunol. 2013;191(8):4383–91.
28. De Obaldia ME, Bhandoola A. Transcriptional regulation of innate and adaptive lymphocyte lineages. Annu Rev Immunol. 2015;33:607–42.

29. Weber BN, Chi AW, Chavez A, Yashiro-Ohtani Y, Yang Q, Shestova O, et al. A critical role for TCF-1 in T-lineage specification and differentiation. Nature. 2011;476(7358):63–8.
30. Zhong C, Zheng M, Cui K, Martins AJ, Hu G, Li D, et al. Differential expression of the transcription factor GATA3 specifies lineage and functions of innate lymphoid cells. Immunity. 2020;52(1):83–95 e4.
31. Yagi R, Zhong C, Northrup DL, Yu F, Bouladoux N, Spencer S, et al. The transcription factor GATA3 is critical for the development of all IL-7Ralpha-expressing innate lymphoid cells. Immunity. 2014;40(3):378–88.
32. Zhu J. GATA3 regulates the development and functions of innate lymphoid cell subsets at multiple stages. Front Immunol. 2017;8:1571.
33. Serafini N, Klein Wolterink RG, Satoh-Takayama N, Xu W, Vosshenrich CA, Hendriks RW, et al. Gata3 drives development of RORgammat+ group 3 innate lymphoid cells. J Exp Med. 2014;211(2):199–208.
34. Wei G, Wei L, Zhu J, Zang C, Hu-Li J, Yao Z, et al. Global mapping of H3K4me3 and H3K27me3 reveals specificity and plasticity in lineage fate determination of differentiating CD4+ T cells. Immunity. 2009;30(1):155–67.
35. Serafini N, Vosshenrich CA, Di Santo JP. Transcriptional regulation of innate lymphoid cell fate. Nat Rev Immunol. 2015;15(7):415–28.
36. Sun X-H, Copeland NG, Jenkins NA, Baltimore D. Id proteins, Id1 and Id2, selectively inhibit DNA binding by one class of helix-loop-helix proteins. Mol Cell Biol. 1991;11:5603–11.
37. Seillet C, Mielke LA, Amann-Zalcenstein DB, Su S, Gao J, Almeida FF, et al. Deciphering the innate lymphoid cell transcriptional program. Cell Rep. 2016;17(2):436–47.
38. Allman D, Sambandam A, Kim S, Miller JP, Pagan A, Well D, et al. Thymopoiesis independent of common lymphoid progenitors. Nat Immunol. 2003;4(2):168–74.
39. Adolfsson J, Mansson R, Buza-Vidas N, Hultquist A, Liuba K, Jensen CT, et al. Identification of Flt3+ lympho-myeloid stem cells lacking erythro-megakaryocytic potential a revised road map for adult blood lineage commitment. Cell. 2005;121(2): 295–306.
40. Porritt HE, Rumfelt LL, Tabrizifard S, Schmitt TM, Zuniga-Pflucker JC, Petrie HT. Heterogeneity among DN1 prothymocytes reveals multiple progenitors with different capacities to generate T cell and non-T cell lineages. Immunity. 2004;20(6):735–45.
41. Balciunaite G, Ceredig R, Fehling HJ, Zuniga-Pflucker JC, Rolink AG. The role of Notch and IL-7 signaling in early thymocyte proliferation and differentiation. Eur J Immunol. 2005;35(4):1292–300.
42. Godfrey DI, Kennedy J, Suda T, Zlotnik A. A developmental pathway involving four phenotypically and functionally distinct subsets of CD3-CD4-CD8- triple-negative adult mouse thymocytes defined by CD44 and CD25 expression. J Immunol. 1993;150 (10):4244–52.
43. Kisielow P, von Boehmer H. Development and selection of T cells: facts and puzzles. Adv Immunol. 1995;58:87–209.
44. Fehling HJ, Krotkova A, Saint-Ruf C, von Boehmer H. Crucial role of the pre-T-cell receptor alpha gene in development of alpha beta but not gamma delta T cells. Nature. 1995;375(6534):795–8.
45. Ciofani M, Knowles GC, Wiest DL, von Boehmer H, Zuniga-Pflucker JC. Stage-specific and differential notch dependency at the alpha-beta and gamma-delta T lineage bifurcation. Immunity. 2006;25(1): 105–16.
46. Kreslavsky T, Garbe AI, Krueger A, von Boehmer H. T cell receptor-instructed alpha-beta versus gamma-delta lineage commitment revealed by single-cell analysis. J Exp Med. 2008;205(5): 1173–86.
47. Xiong N, Raulet DH. Development and selection of gamma-delta T cells. Immunol Rev. 2007;215:15–31.
48. Munoz-Ruiz M, Sumaria N, Pennington DJ, Silva-Santos B. Thymic determinants of gammadelta T cell differentiation. Trends Immunol. 2017;38(5):336–44.
49. Koch U, Fiorini E, Benedito R, Besseyrias V, Schuster-Gossler K, Pierres M, et al. Delta-like 4 is the essential, nonredundant ligand for Notch1 during thymic T cell lineage commitment. J Exp Med. 2008;205(11):2515–23.
50. Alves NL, Richard-Le Goff O, Huntington ND, Sousa AP, Ribeiro VS, Bordack A, et al. Characterization of the thymic IL-7 niche in vivo. Proc Natl Acad Sci. 2009;106(5):1512–7.
51. Abramson J, Anderson G. Thymic epithelial cells. Annu Rev Immunol. 2017;35:85–118.
52. Liao W, Lin JX, Leonard WJ. IL-2 family cytokines: new insights into the complex roles of IL-2 as a broad regulator of T helper cell differentiation. Curr Opin Immunol. 2011;23(5):598–604.
53. Hosokawa H, Romero-Wolf M, Yang Q, Motomura Y, Levanon D, Groner Y, et al. Cell type-specific actions of Bcl11b in early T-lineage and group 2 innate lymphoid cells. J Exp Med. 2020;217(1) e20190972.
54. Okamura RM, Sigvardsson M, Galceran J, Verbeek S, Clevers H, Grosschedl R. Redundant regulation of T cell differentiation and TCRalpha gene expression by the transcription factors LEF-1 and TCF-1. Immunity. 1998;8(1):11–20.
55. Murre C. Helix-loop-helix proteins and lymphocyte development. Nat Immunol. 2005;6(11):1079–86.
56. Sun XH. Multitasking of helix-loop-helix proteins in lymphopoiesis. Adv Immunol. 2004;84:43–77.
57. Ling F, Kang B, Sun XH. Id proteins: small molecules, mighty regulators. Curr Top Dev Biol. 2014;110:189–216.
58. Bain G, Engel I, Robanus Maandag EC, te Riele HP, Voland JR, Sharp LL, et al. E2A deficiency leads to

abnormalities in alphabeta T-cell development and to rapid development of T-cell lymphomas. Mol Cell Biol. 1997;17(8):4782–91.
59. Engel I, Johns C, Bain G, Rivera RR, Murre C. Early thymocyte development is regulated by modulation of E2A protein activity. J Exp Med. 2001;194(6): 733–45.
60. Zhuang Y, Soriano P, Weintraub H. The helix-loop-helix gene E2A is required for B cell differentiation. Cell. 1994;79:875–84.
61. Rivera RR, Johns CP, Quan J, Johnson RS, Murre C. Thymocyte selection is regulated by the helix-loop-helix inhibitor protein, Id3. Immunity. 2000;12:17–26.
62. Yan W, Young AZ, Soares VC, Kelley R, Benezra R, Zhuang Y. High incidence of T-cell tumors in E2A-null mice and E2A/Id1 double-knockout mice. Mol Cell Biol. 1997;17:7317–27.
63. Kim D, Peng XC, Sun XH. Massive apoptosis of thymocytes in T-cell-deficient Id1 transgenic mice. Mol Cell Biol. 1999;19(12):8240–53.
64. Kim D, Xu M, Nie L, Peng XC, Jimi E, Voll RE, et al. Helix-loop-helix proteins regulate pre-TCR and TCR signaling through modulation of Rel/NF-kappaB activities. Immunity. 2002;16(1):9–21.
65. Wang HC, Perry SS, Sun XH. Id1 attenuates Notch signaling and impairs T-cell commitment by elevating Deltex1 expression. Mol Cell Biol. 2009;29(17):4640–52.
66. Wojciechowski J, Lai A, Kondo M, Zhuang Y. E2A and HEB are required to block thymocyte proliferation prior to pre-TCR expression. J Immunol. 2007;178(9):5717–26.
67. Yashiro-Ohtani Y, He Y, Ohtani T, Jones ME, Shestova O, Xu L, et al. Pre-TCR signaling inactivates Notch1 transcription by antagonizing E2A. Genes Dev. 2009;23(14):1665–76.
68. Miyazaki M, Miyazaki K, Chen K, Jin Y, Turner J, Moore AJ, et al. The E-Id protein axis specifies adaptive lymphoid cell identity and suppresses thymic innate lymphoid cell development. Immunity. 2017;46(5):818–34.
69. Bain G, Cravatt CB, Loomans C, Alberola-Ila J, Hedrick SM, Murre C. Regulation of the helix-loop-helix proteins, E2A and Id3, by the Ras-ERK MAPK cascade. Nat Immunol. 2001;2(2):165–71.
70. Yang CY, Best JA, Knell J, Yang E, Sheridan AD, Jesionek AK, et al. The transcriptional regulators Id2 and Id3 control the formation of distinct memory CD8+ T cell subsets. Nat Immunol. 2011;12(12): 1221–9.
71. Hu T, Wang H, Simmons A, Bajana S, Zhao Y, Kovats S, et al. Increased level of E protein activity during invariant NKT development promotes differentiation of invariant NKT2 and invariant NKT17 subsets. J Immunol. 2013;191(10):5065–73.
72. Constantinides MG, Bendelac A. Transcriptional regulation of the NKT cell lineage. Curr Opin Immunol. 2013;25(2):161–7.
73. Verykokakis M, Krishnamoorthy V, Iavarone A, Lasorella A, Sigvardsson M, Kee BL. Essential functions for ID proteins at multiple checkpoints in invariant NKT cell development. J Immunol. 2013;191(12):5973–83.
74. Li J, Wu D, Jiang N, Zhuang Y. Combined deletion of Id2 and Id3 genes reveals multiple roles for E proteins in invariant NKT cell development and expansion. J Immunol. 2013;191(10):5052–64.
75. D'Cruz LM, Stradner MH, Yang CY, Goldrath AW. E and Id proteins influence invariant NKT cell sublineage differentiation and proliferation. J Immunol. 2014;192(5):2227–36.
76. Lauritsen JP, Wong GW, Lee SY, Lefebvre JM, Ciofani M, Rhodes M, et al. Marked induction of the helix-loop-helix protein Id3 promotes the gammadelta T cell fate and renders their functional maturation Notch independent. Immunity. 2009;31 (4):565–75.
77. Park ST, Sun XH. The Tal1 oncoprotein inhibits E47-mediated transcription. Mechanism of inhibition. J Biol Chem. 1998;273(12):7030–7.
78. Berrett H, Qian L, Roman O, Cordova A, Simmons A, Sun XH, et al. Development of Type 2 innate lymphoid cells is selectively inhibited by sustained E protein activity. Immunohorizons. 2019;3(12):593–605.
79. Peng V, Georgescu C, Bakowska A, Pankow A, Qian L, Wren JD, et al. E proteins orchestrate dynamic transcriptional cascades implicated in the suppression of the differentiation of group 2 innate lymphoid cells. J Biol Chem. 2020;295(44): 14866–77.
80. Chyla BJ, Moreno-Miralles I, Steapleton MA, Thompson MA, Bhaskara S, Engel M, et al. Deletion of Mtg16, a target of t(16,21), alters hematopoietic progenitor cell proliferation and lineage allocation. Mol Cell Biol. 2008;28(20):6234–47.
81. Hunt A, Fischer M, Engel ME, Hiebert SW. Mtg16/Eto2 contributes to murine T-cell development. Mol Cell Biol. 2011;31(13):2544–51.
82. Jin C, Li H, Murata T, Sun K, Horikoshi M, Chiu R, et al. JDP2, a repressor of AP-1, recruits a histone deacetylase 3 complex to inhibit the retinoic acid-induced differentiation of F9 cells. Mol Cell Biol. 2002;22(13):4815–26.
83. Jonak CR, Lainez NM, Roybal LL, Williamson AD, Coss D. c-JUN dimerization protein 2 (JDP2) is a transcriptional repressor of follicle-stimulating hormone beta (FSHbeta) and is required for preventing premature reproductive senescence in female mice. J Biol Chem. 2017;292(7):2646–59.
84. Cordier AC, Haumont SM. Development of thymus, parathyroids, and ultimo-branchial bodies in NMRI and nude mice. Am J Anat. 1980;157(3):227–63.
85. Jenkinson EJ, Van Ewijk W, Owen JJ. Major histocompatibility complex antigen expression on the epithelium of the developing thymus in normal and nude mice. J Exp Med. 1981;153(2):280–92.
86. Mallick CA, Dudley EC, Viney JL, Owen MJ, Hayday AC. Rearrangement and diversity of T cell receptor beta chain genes in thymocytes: a critical

role for the beta chain in development. Cell. 1993;73 (3):513–9.
87. Hennet T, Hagen FK, Tabak LA, Marth JD. T-cell-specific deletion of a polypeptide N-acetylgalactosaminyl-transferase gene by site-directed recombination. Proc Natl Acad Sci. 1995;92(26):12070–4.
88. Dudley EC, Petrie HT, Shah LM, Owen MJ, Hayday AC. T cell receptor beta chain gene rearrangement and selection during thymocyte development in adult mice. Immunity. 1994;1(2):83–93.
89. Kernfeld EM, Genga RMJ, Neherin K, Magaletta ME, Xu P, Maehr R. A single-cell transcriptomic atlas of thymus organogenesis resolves cell types and developmental maturation. Immunity. 2018;48 (6):1258–70.
90. Elsaid R, Meunier S, Burlen-Defranoux O, Soares-da-Silva F, Perchet T, Iturri L, et al. A wave of bipotent T/ILC-restricted progenitors shapes the embryonic thymus microenvironment in a time-dependent manner. Blood. 2021;137(8):1024–36.
91. Ferreira ACF, Szeto ACH, Heycock MWD, Clark PA, Walker JA, Crisp A, et al. RORalpha is a critical checkpoint for T cell and ILC2 commitment in the embryonic thymus. Nat Immunol. 2021;22(2): 166–78.
92. Jones R, Cosway EJ, Willis C, White AJ, Jenkinson WE, Fehling HJ, et al. Dynamic changes in intrathymic ILC populations during murine neonatal development. Eur J Immunol. 2018; 48(9):1481-91.
93. Nagasawa M, Germar K, Blom B, Spits H. Human CD5(+) innate lymphoid cells are functionally immature and their development from CD34(+) progenitor cells is regulated by Id2. Front Immunol. 2017;8: 1047.
94. Liu C, Gong Y, Zhang H, Yang H, Zeng Y, Bian Z, et al. Delineating spatiotemporal and hierarchical development of human fetal innate lymphoid cells. Cell Res. 2021; 31:1106–22.
95. Kreslavsky T, Gleimer M, Garbe AI, von Boehmer H. Alphabeta versus gammadelta fate choice: counting the T-cell lineages at the branch point. Immunol Rev. 2010;238(1):169–81.
96. Krangel MS. Mechanics of T cell receptor gene rearrangement. Curr Opin Immunol. 2009;21(2):133–9.
97. Shin SB, Lo BC, Ghaedi M, Scott RW, Li Y, Messing M, et al. Abortive gammadeltaTCR rearrangements suggest ILC2s are derived from T-cell precursors. Blood Adv. 2020;4(21):5362–72.
98. Luche H, Nageswara Rao T, Kumar S, Tasdogan A, Beckel F, Blum C, et al. In vivo fate mapping identifies pre-TCRalpha expression as an intra- and extrathymic, but not prethymic, marker of T lymphopoiesis. J Exp Med. 2013;210(4):699–714.
99. Bajana S, Pankow A, Liu K, Michniowska M, Urban JF Jr, Chen WR, et al. Correlation between circulating innate lymphoid cell precursors and thymic function. iScience. 2022;25(2):103732.
100. Lim AI, Li Y, Lopez-Lastra S, Stadhouders R, Paul F, Casrouge A, et al. Systemic human ILC precursors provide a substrate for tissue ILC differentiation. Cell. 2017;168(6):1086–100.
101. Nagasawa M, Heesters BA, Kradolfer CMA, Krabbendam L, Martinez-Gonzalez I, de Bruijn MJW, et al. KLRG1 and NKp46 discriminate subpopulations of human CD117(+)CRTH2(−) ILCs biased toward ILC2 or ILC3. J Exp Med. 2019;216(8):1762–76.

NK Cells and ILC1s in Cancer Biology

4

Anna Sliz and Wayne M. Yokoyama

Abstract

Natural killer (NK) cells are cytotoxic innate lymphocytes that can kill tumor cells. While a majority of the early studies on the role of NK cells in cancer focused on hematopoietic tumors, there has been a growing interest in the role of NK cells in solid tumors. NK cells are grouped with innate lymphoid cells (ILCs) that include ILC1, a closely related but distinct cell whose role in antitumor immunity is incompletely understood. In this review we focus primarily on the role of NK cells in solid tumors and review the limited data available on the role of ILC1s in cancer.

Keywords

NK cells · ILC1s · Solid tumor · Cancer

Abbreviations used in this chapter:

cDC1	Conventional type I dendritic cells
CLP	Common lymphoid progenitor
cNK cell	Conventional natural killer cell
DC	Dendritic cell
Eomes	Eomesodermin
FBP1	Fructose-1,6-bisphosphatase
Flt3-L	Fms-like tyrosine kinase 3-ligand
GM-CSF	Granulocyte-macrophage-colony-stimulating factor
IBD	Inflammatory bowel disease
ILC1s	Innate lymphoid cells group 1
Lag3	Lymphocyte activation gene 3
MCA	Methylchoanthrene
MCMV	Murine cytomegalovirus
MHC	Major histocompatibility complex
NETs	Neutrophil extracellular traps
NK cell	Natural killer cell
PD1	Programmed death protein 1
PMA	Phorbol 12-myristate 13-acetate
SMAD4	Smad family member 4
T-bet	T-box transcription factor
Tim3	T cell immunoglobulin domain and mucin domain 3
TGCA	The Cancer Genome Atlas
TGFβ	Transforming growth factor β
TIGIT	T cell immunoreceptor with Ig and ITIM domains
trNK	Tissue-resident NK cells

4.1 Introduction

Natural killer (NK) cells and innate lymphoid cells group 1 (ILC1s) are closely related innate lymphocytes that have been described in mice and humans under homeostatic conditions and in cancer [1–3]. The antitumor functions of NK cells were recognized at the time of their discovery

A. Sliz · W. M. Yokoyama (✉)
Rheumatology Division, Washington University School of Medicine, St. Louis, MO, USA
e-mail: yokoyama@wustl.edu

© The Author(s), under exclusive license to Springer Nature Singapore Pte Ltd. 2022

X.-H. Sun (ed.), *Innate Lymphoid Cells*, Advances in Experimental Medicine and Biology 1365,
https://doi.org/10.1007/978-981-16-8387-9_4

[3]. In mouse models, NK cell depletion and perturbations of NK cell effector functions such as cytotoxicity and cytokine production result in significantly increased tumor burdens [4–8]. In humans, a large, long-term study found that decreased cytotoxic potential of NK cells was associated with increased risk of cancer development [9], highlighting the critical role NK cells play in antitumor immunity. With the more recent identification of ILC1s, there has been an intense focus on understanding how these two cell types can be differentiated and function in the development or control of cancer. In this review we focus primarily on NK cells in solid non-hematological tumors, as the role of NK cells in hematological malignancies has been extensively reviewed [3, 10–12], and discuss the limited data regarding the role of ILC1s in antitumor immunity.

4.2 NK Cells and ILC1s

Most early discoveries regarding NK cell function in both mice and humans were derived primarily from what we now know to be conventional NK cells (cNK) that circulate in the blood vasculature and are readily found in the mouse spleen and human peripheral blood. In this review, we will refer to cNK cells whenever NK cells are discussed, unless otherwise stated. However, ILC1s are now known to express many markers originally thought to be selectively expressed on NK cells, such as NK1.1 and NKp46 in mice [13, 14]. One of the early challenges following the identification of ILC1s was finding markers that could consistently differentiate the two, such as CD49a and CD49b that are selectively found on NK cells and ILC1s in the naïve mouse, respectively [13]. Yet CD49a expression can be induced on NK cells during infection and CD49b can be downregulated on activated NK cells [15, 16]. Similar issues were uncovered with other markers and molecules, such that it remains challenging to determine if functions attributed to one cell type could be due to the other, and vice versa, or due to both innate lymphocyte populations.

While NK cells and ILC1s develop from the common lymphoid progenitor (CLP), there are key differences in their developmental requirements that help discriminate the two cell types [2, 17]. NK cells differentiate from the CLP at a point proximal to when ILC commitment occurs [18]. In addition, NK cells require the T-box transcription factor Eomesodermin (Eomes) for their development while ILC1s can develop in the absence of Eomes but are critically dependent on a related T-box transcription factor (T-bet) for their development [19–21]. By contrast, NK cells only require T-bet for their maturation [22]. There is significant overlap in the tissue distribution between cNK cells and ILC1s; both can be found in the liver, kidney, uterus, salivary glands, skin, and lungs which also contain tissue-resident NK (trNK) cells that are often thought to be ILC1s, in part because ILC1s are primarily tissue resident and generally do not circulate and trNK cells also require T-bet [13, 23–26]. However, during inflammation, ILC progenitors can circulate and seed other organs [27]. Moreover, during *Toxoplasma gondii* infection, cNK cells appear to be converted to ILC1-like cells that circulate [28]. Thus, while NK cells and ILC1s can be discriminated through their developmental programs, recent discovery of their plasticity has further made it difficult to distinguish them unequivocally, particularly during immune responses.

4.3 NK Cell Functions

NK cells express a wide variety of germline-encoded receptors that allow NK cells to recognize self-MHC class I, pathogen-derived, and stress-induced/damage molecules that have been extensively reviewed [29, 30] and will be briefly discussed here. NK cells classically use inhibitory receptors to ignore cells with normal MHC class I (MHC-I) expression, whereas NK cells kill targets lacking MHC-I expression, as in "missing-self." Tumors may express molecules that are recognized by NK cell activation receptors, but when MHC-I is normally expressed, NK cells are generally inhibited. Under certain circumstances, tumors can activate NK cells via NKG2D which recognizes stress-induced ligands, even in the presence of normal MHC-I [31, 32]. Once activated, NK cells kill targets via the release of cytolytic granules containing perforin and granzymes that induce target apoptosis [5, 33, 34] though activated NK cells can also kill targets via death receptor pathways [35–37]. However, chronic stimulation of NK cells through their activation receptors can lead to NK cell dysfunction [38], akin to T cell anergy. Nonetheless, in addition to direct cytotoxicity, NK cells can shape the immune response to the tumors (and other inflammatory situations) via cytokine and chemokine production that can be triggered via their activation receptors or through direct pro-inflammatory cytokine stimulation [39–43]. This in turn can drive activation and infiltration of other immune cells into the tumor and antagonize the development of an immunosuppressive tumor microenvironment. Thus, NK cells can modulate the immune response to targets either directly through cytotoxicity or indirectly via cytokine or chemokine production.

4.4 NK Cells and Other Tumor-Infiltrating Immune Cells

The tumor microenvironment is a complex structure composed of malignant cells as well as nonmalignant stromal and immune cells [44, 45], all of which can impact tumor growth and immune response. In addition, soluble factors in the tumor microenvironment can affect tumor-infiltrating NK cells, like other infiltrating immune cells [46]. Recently published studies have highlighted how these complex relationships can affect NK cells and tip the balance between tumor progression and tumor control (Fig. 4.1).

A well-recognized relationship between NK cells and dendritic cells (DCs) in cancer was established by early studies in mice indicating that DC-derived cytokines such as type I interferons and IL-2 were critical for NK cell activation and enhanced NK cell-mediated antitumor immunity [47–49]. Other studies revealed that NK cell immunoediting contributes to improved antitumor immunity through NK cell killing of immature DCs while sparing mature DCs which prevents the induction of a tolerogenic tumor microenvironment and supports a strong $CD8^+$ T cell response to the tumor [50, 51]. Moreover, NK cell-derived CCL5 and XCL1 are critical for the recruitment of conventional type I DCs (cDC1s) into the tumor microenvironment [52]. More importantly, disruption of this signaling axis led to a cascade effect of poor tumor infiltration by cDC1s, resulting in a weakened $CD8^+$ T cell response and a significantly higher tumor burden. In addition to supporting the adaptive response to the tumor, cDC1 migration and infiltration into the tumor provide support to NK cells through cell-cell contact and production of cytokines such as IL-12, driving NK cell activation and promoting continued NK cell infiltration into the tumor [51, 53, 54]. Data from The Cancer Genome Atlas (TCGA) provide evidence for a similar NK cell/cDC1 signaling axis in human melanoma, breast cancer, lung cancer, and head and neck squamous cell carcinoma that is associated with improved patient survival, demonstrating that this signaling axis is relevant not only in mice but in humans as well [52]. Other studies reported similar findings but identified NK cells as a critical source of fms-like tyrosine kinase 3-ligand (Flt3-L), the formative cytokine for cDC1s [55]. Moreover, an in-depth analysis of human patient samples revealed increased NK cell and cDC1 infiltration to be associated with elevated Flt3-L expression and improved melanoma patient survival [56]. Though some of the details differ which could be tumor specific, these findings nonetheless highlight a signaling axis between NK cells and cDC1s that potentially could be targeted for new NK cell immunotherapies or improving the existing ones.

While the relationship between NK cells and DCs has been well established, emerging studies also indicate that NK cells are affected by and in turn influence a wide variety of other tumor-infiltrating immune cells including ILC2s, eosinophils [57], neutrophils [58], T cells, and myeloid cells (Fig. 4.1, [59]). With respect to

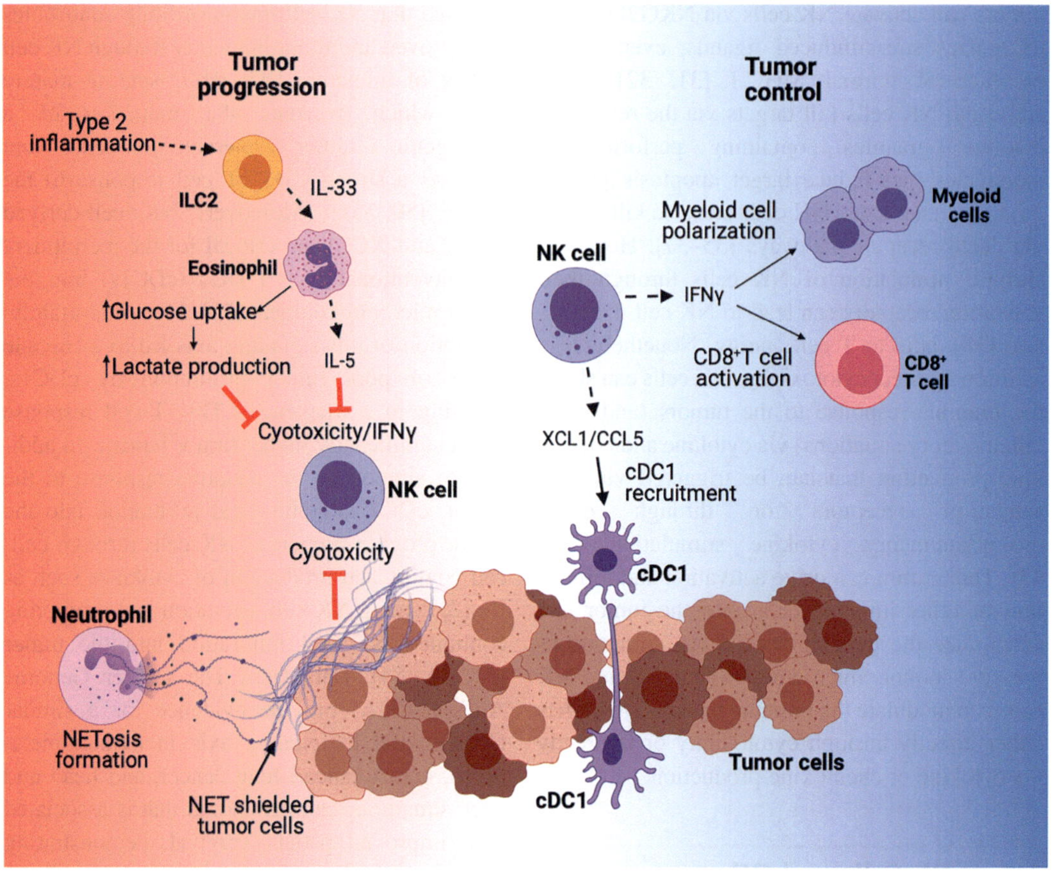

Fig. 4.1 In the tumor microenvironment NK cells engage in complex interactions with other tumor-infiltrating immune cells which have the potential to tip the balance between tumor progression and tumor control. *cDC1* conventional type I dendritic cell, *IFNγ* interferon γ, *NK cell* natural killer cell, *ILC2* innate lymphoid cell 2. Generated with BioRender.com

ILC2s and eosinophils, pretreatment with Aspergillus protease allergen to induce type 2 inflammation in the airways prior to tumor challenge resulted in significantly higher tumor burdens compared to untreated controls [57]. More specifically, ILC2-derived IL-33 triggered an influx of IL-5-producing eosinophils into the lungs that led to metabolic changes in the lung microenvironment. The altered lung tissue microenvironment resulted in decreased NK cell interferon-γ (IFNγ) production and cytotoxicity, leading to increased tumor metastasis. Moreover, inhibition of this signaling axis with blocking antibodies against IL-5 and IL-33 was sufficient to restore NK cell function, highlighting potential novel targets affecting NK cells, ILC2s, and eosinophils for immunotherapy [57].

Regarding neutrophils and NK cells, high levels of tumor-infiltrating neutrophils are associated with poor patient outcomes [60]. One way that neutrophils contribute to host defense is through NETosis, a unique form of cell death specific to neutrophils occurring when their DNA is extruded and released into the surrounding tissues, resulting in structures, the so-called neutrophil extracellular traps (NETs). These structures trap and neutralize bacteria, fungi, parasites, and other pathogens; however if this process is dysregulated they can contribute to disease pathology and progression [61]. While

NETs have been identified in cancer, in many cases the contribution of NETs to disease outcome remains elusive. Recently published studies, using an in vitro co-culture model, revealed that the formation of NETs around tumor cells significantly inhibited NK cell- and $CD8^+$ T cell-mediated cytotoxicity against the tumor cells [58]. Interestingly, pretreatment of tumor cell cultures with DNaseI and NETosis inhibitors was sufficient to restore NK cell and $CD8^+$ T cell cytotoxicity. Moreover, the combination of DNase I and checkpoint blockade synergized, resulting in significantly smaller tumors in a preclinical mouse model of breast cancer [58]. These findings illustrate a unique mechanism by which tumors evade NK cell killing but it is unclear if evasion is mediated primarily by preventing the NK cells from physically interacting with the tumor cells or if there are additional suppressive effects on NK cell function following exposure to tumors coated in NETs. Nonetheless, neutrophils appear to be able to create a shield around the tumor cells, at least physically and perhaps functionally, protecting them from NK cell and $CD8^+$ T cell cytotoxicity, suggesting that resistance to checkpoint blockade may be related to this barrier between the cytotoxic cells and the tumor cells [58].

In terms of T cells and myeloid cells interacting with NK cells, the temporal relationship of NK cell to control of tumors is relevant. Depletion of NK cells at the time of tumor challenge significantly impacted tumor outgrowth whereas NK cell depletion after tumors had been established made no difference for tumor rejection, highlighting how the NK cell response early after tumor challenge is essential for controlling tumor growth [59]. Moreover, disruption of NK cell responsiveness shortly after tumor challenge significantly impaired the resulting $CD8^+$ T cell response [52, 59]. Interestingly, stimulation of NK cells via NKG2D ligands on tumor cells can prime tumor-specific $CD8^+$ CTL responses that are protective [32]. NK cell-derived IFNγ is critical for tumor-associated macrophage polarization to a phenotype conducive to adaptive immunity and tumor control. If NK cells are unable to produce IFNγ, this reprogramming of the myeloid compartment does not occur [59], leading to the development of an immunosuppressive microenvironment and uncontrolled tumor growth. These findings emphasize how myeloid cells can influence both NK cell cytotoxicity and cytokine production that work in concert to coordinate the immune response, including T cells, and create a tumor microenvironment more conducive to a strong antitumor response and tumor control. Thus, a greater understanding of how NK cells interact with other tumor-infiltrating immune cells is essential for a deeper understanding of the role of NK cells in solid tumors.

4.5 NK Cell Exhaustion in Cancer

NK cell exposure to the tumor microenvironment can induce significant dysfunction, characterized by decreased IFNγ production, cytotoxicity, and proliferative capacity [62]. Studies have examined if this dysfunction is related to "exhaustion," a broad term used in immunology to describe a progressive and hierarchal loss of effector functions that develops due to persistent exposure to antigen or chronic inflammation [63, 64]. In extensive studies of T cells, the hallmark characteristics of exhaustion include loss of key effector functions (cytotoxicity, cytokine production, and proliferation) and sustained upregulation of multiple inhibitory receptors. However, it remains unknown if the NK cell dysfunction in the tumor microenvironment is equivalent to T cell exhaustion because definitive states of NK cell dysfunction (anergy, exhaustion, or senescence) have not been as extensively studied and characterized [65].

Nonetheless, chronic stimulation of mouse NK cells through their activation receptors, such as Ly49H and NKG2D, can lead to NK cells displaying an anergy-like state [38, 66, 67]. This phenomenon has been better described for NKG2D whose ligands are rapidly upregulated on proliferating infected or tumor cells but are not normally expressed [68, 69]. Chronic stimulation of NK cells via NKG2D ligands in vitro resulted in significantly decreased NK cell cytotoxicity and impaired missing self-rejection

[70]. Moreover, similar NK cell defects were obtained through chronic NKG2D stimulation in vivo using transgenic mice expressing its ligands [66, 67]. These defects in turn resulted in increased tumor burden in Rae1-transgenic mice [67]. In humans, it was observed that many tumors express NKG2D ligands which are then cleaved and can be found in blood [71]. The soluble NKG2D ligands induced NKG2D downregulation and receptor dysfunction, touted as a mechanism for tumor evasion of NK cells. While these studies highlight the negative impact of chronic NK cell activation receptor stimulation on effector functions essential for NK cell-mediated antitumor immunity, it remains unclear if this represents a state of exhaustion equivalent to T cell anergy or exhaustion.

Regardless, similar observations have been made for other activating NK cell receptors including NKp30 in humans, suggesting that the expression of NK cell ligands is a conserved strategy used by the tumor to evade NK cell activation and killing [72, 73]. In a recent extensive phenotypic characterization, mouse NK cells chronically activated by a wide variety of activating stimuli {cytokines, receptor activation, and chronic murine cytomegalovirus (MCMV) infection} resulted in significantly impaired cytotoxicity, proliferation, and cytokine secretion [74]. Moreover, following chronic stimulation, human NK cells have impaired function and upregulation of classic T cell exhaustion markers and show signs of epigenetic changes as well, the latter an additional characteristic of exhausted T cells that has yet to be extensively examined in NK cells [75].

Downregulation of MHC-I by tumor cells is also relevant to NK cell dysfunction, as it appears to be a ploy to avoid recognition and killing by cytotoxic $CD8^+$ T cells. Although MHC-I downregulation should make the tumors vulnerable to NK cell killing, MHC-I-deficient tumors can persist in human patients and mouse models, suggesting that these tumors are able to avoid killing by NK cells as well [76]. Recognition of self-MHC-I molecules is integral for fully functional NK cells in a process termed licensing or education that ironically requires an inhibitory MHC-I-specific receptor that binds self-MHC-I [77]. NK cells that develop in MHC-I-deficient hosts are hyporesponsive and cannot reject MHC-I-deficient cells [78, 79]. Moreover, NK cells that develop in an MHC-I-sufficient environment and are subsequently exposed to an MHC-I-deficient environment take on a hyporesponsive or dysfunctional phenotype [80–83]. In addition, NK cells isolated from RMA (MHC-I-sufficient) tumors were more responsive to activating stimuli than those isolated from RMA-S (MHC-I-deficient) tumors [62]. This effect appeared to be restricted to NK cells isolated from the tumors as compared to NK cells isolated from non-tumor draining lymph nodes or spleen. However, it remains unclear if these NK cells are exhausted or display a phenotype more related to that associated with chronic activation receptor stimulation, and how well this phenotype applies to MHC-I-deficient, non-hematopoietic tumors that have been less well studied.

Nevertheless, cytokine therapy can reverse NK cell hyporesponsiveness in MHC-I-deficient hosts, and also tumors [77]. In a study of RMA (MHC-I-sufficient) or RMA-S (MHC-I-deficient) tumor-bearing mice treated with IL-12/IL-18 or H9 (IL-2 mutant superkine), RMA tumor-bearing mice were refractory to treatment while RMA-S tumor-bearing mice showed significantly improved survival compared to untreated controls [62]. Moreover, cytokine treatment of RMA-S tumor-bearing mice was able to significantly improve tumor-infiltrating NK cell responsiveness to activating stimuli compared to untreated controls. These results indicate that dysfunction induced by MHC-I deficiency in the tumor microenvironment is reversible. However, one caveat of this study was the focus on tumors of hematopoietic origin leaving open the question of the efficacy of this therapy in solid tumors. In a more recent study, IL-21 was also found to reverse MHC-I deficiency-induced NK cell dysfunction in mice and in humans [84]. Specifically, treatment of tumor-infiltrating NK cells in vitro or in vivo with IL-21 was sufficient to increase NK cell cytotoxicity and IFNγ secretion. Importantly, this treatment was effective in improving NK cell-mediated antitumor immunity against multiple solid tumors including TC1 (lung cancer), CT26 (colon cancer), and MC38 (colon cancer). Taken together, these findings indicate that in both hematological malignancies and solid tumors NK cell dysfunction induced by MHC-I deficiency is reversible by cytokine therapy.

Tumor-infiltrating NK cells can express inhibitory receptors associated with T cell exhaustion such as programmed death protein 1 (PD1, [84, 85]), lymphocyte activation gene 3 (Lag3, [75, 86]), T cell immunoglobulin domain and mucin domain 3 (Tim3, 86), and T cell immunoreceptor with Ig and ITIM domains (TIGIT) [86]. However, there are little data on the impact of these receptors on NK cell intracellular signaling and the downstream effects on NK cell effector functions in tumors. The expression of PD1 on NK cells remains controversial in both mice and humans. Two studies found high levels of PD1 expression on tumor-infiltrating NK cells in multiple murine tumors such as RMA/RMA-S (lymphoma), CT26 (colon cancer), B16 (melanoma), and MC38 (colon cancer) [84, 85]. A functional analysis of $PD1^+$ versus $PD1^-$ NK cells isolated from MC38 tumors by one group revealed that $PD1^+$ NK cells showed a significantly lower level of IFNγ production, degranulation, and expression of granzyme B and perforin in response to restimulation with anti-NK1.1, suggestive of an "exhausted" phenotype. They also found Tim3, 2B4 (CD244), and CD160 to be co-expressed with PD1, another potential hallmark of exhaustion [84]. Conversely, the other study indicated that $PD1^+$ NK cells isolated from RMA-S and CT26 tumors were more functional compared to their $PD1^-$ counterparts following restimulation ex vivo with phorbol 12-myristate 13-acetate (PMA) and ionomycin, suggesting that PD1 expression on tumor-infiltrating NK cells is associated with more NK cell activity [85]. Interestingly, both groups indicated that prolonged exposure to MHC-I-deficient tumors (RMA-S- and MHC-I-deficient MC38 tumors, respectively) potently induced PD1 expression on tumor-infiltrating NK cells. In direct contrast to both studies, a recent paper found no PD1 expression on tumor-infiltrating NK cells in any of the tumors analyzed, including CT26, 4T1 (breast cancer), B16, and methylcholanthrene (MCA)-induced sarcomas. However, there were high levels of CD96, Lag3, Tim3, and TIGIT on tumor-infiltrating NK cells in all tumors, consistent with some of the aforementioned findings [86]. While it is still unclear why the expression of PD1 on NK cells varies so greatly between studies and experimental models, these studies highlight the need for more work to be done to fully understand the role of PD1 as well as other receptors and markers associated with T cell exhaustion on NK cell signaling and effector function. The findings highlighted here emphasize that NK cells can take on a phenotype that can mirror T cell exhaustion, but many aspects of this phenotype in NK cells remain to be fully understood and characterized.

4.6 NK Cell Metabolism and Cancer

In addition to NK cell changes in the tumor microenvironment from chronic stimulation, a growing body of literature indicates that changes in cellular metabolism are tightly linked to key NK cell effector functions [87]. Classic NK cell functions such as cytokine secretion and cytotoxicity rely on NK cell metabolic fitness and ability to meet increased energy demands under inflammatory conditions [88–91]. One hallmark of cancer is the dysregulation of cellular proliferation, accompanied by significantly dysregulated cellular metabolism, that in turn can affect tumor-infiltrating immune cells, including NK cells [44]. Indeed, emerging data from multiple groups have shown that NK cells are sensitive to metabolic stresses in the tumor microenvironment.

In a recent study of lung cancer, it was discovered that dysregulation of a metabolic enzyme, fructose-1,6-bisphosphatase (FBP1), in NK cells significantly impaired their function [92]. NK cells isolated from tumor-bearing lungs at multiple stages of disease had decreased cytolytic capacity and IFNγ production compared to wild-type controls. Moreover, the glycolytic capacity of NK cells isolated from tumor-bearing lungs was significantly impaired. Inhibition of FBP1 early during disease development was sufficient to reverse the defects in NK cell glycolysis, cytotoxicity, and cytokine secretion, leading to improved control of tumor growth. These findings highlight the importance of glycolysis for NK cell-mediated antitumor function.

In other studies, impaired metabolic fitness of NK cells caused by obesity resulted in impaired NK cell cytotoxicity, cytokine production, and tumor control [93]. This appeared to be a direct result of lipid accumulation within the NK cells which significantly impaired their ability to undergo degranulation after forming a synapse with the tumor target. Moreover, culturing NK cells from lean mice with lipids was sufficient to induce defects in cytotoxicity, cytokine secretion, and glycolytic capacity. Finally, these functional impairments induced by obesity resulted in significantly larger tumors in mice fed a high-fat diet [93]. Whether these disturbances in lipid metabolism are related to NK cell defects in the tumor microenvironment remains to be investigated.

In addition, multiple groups showed that the accumulation of lactic acid leading to low pH in the tumor microenvironment has an inhibitory effect on NK cell effector functions [94, 95]. It was been widely held that tumor cells preferentially consume glucose and produce lactate; however, a recent study found tumor-infiltrating myeloid cells, followed by T cells, to be the primary consumers of intra-tumoral glucose compared to tumor cells [96]. Interestingly, tumor cells preferentially utilized large quantities of glutamine to subsidize their proliferation. These findings are consistent with the abovementioned study on eosinophil effects on NK cells in that eosinophils disrupt NK cell metabolic fitness through glucose deprivation and increased lactate levels in the lung, leading to dysfunctional NK cells and increased tumor burdens [57]. While the latter study did not specifically rule out the tumor cells as contributing to these results, their findings highlight the potential negative effect of other tumor-infiltrating immune cells on NK cell metabolic function.

Thus, these findings suggest the sensitivity of NK cells to different types of metabolic stress encountered in the tumor microenvironment, and the therapeutic potential of restoring NK cell metabolic fitness may be a novel NK cell-based immunotherapy strategy for solid tumors. Moreover, these findings also indicate that altered nutrient uptake in the tumor microenvironment is not dictated solely by the tumor cells as previously thought. Instead, it appears that nutrient utilization within the tumor microenvironment can be altered by the tumor-infiltrating immune cells with the potential to negatively impact tumor-infiltrating NK cells.

4.7 ILC1s and Cancer

The role of ILC1s in cancer is controversial in that a recent study showed that early tumor control in mice is dependent on CD49a$^+$ NK1.1$^+$ cells (ILC1 phenotype) while other studies have demonstrated that the ILC1 and ILC1-like cells were unable to protect the host from tumor growth [97, 98]. Similarly in human patients, some studies suggest a protective role of ILC1s in cancers while others suggest a pathogenic role [99, 100]. Inflammatory bowel disease (IBD) patients who have elevated numbers of ILC1s have increased pathology associated with IBD and increased rates of colorectal carcinoma [101, 102]. However, a definite link between increased numbers of ILC1s, autoimmunity, and cancer incidence has yet to be demonstrated.

Given that ILC1s are tissue resident, they are ideally located to respond to tissue-specific cues; however in many cases the impact of tissue-specific cues on ILC1 function and their role in tumor immunosurveillance are incompletely understood [103, 104]. As previously mentioned, solid tumors are infiltrated by many different types of immune cells, but for the most part, the interactions between tumor-associated ILC1s and other tumor-infiltrating immune cells remain uninvestigated. Such studies will be crucial for understanding the role of ILC1s in solid tumors because ILC1s, as tissue-resident, cytokine-producing helper cells, may potentially act early during the tumor immune response and significantly impact disease outcome. To again emphasize a point made earlier, one of the caveats to many of the studies reviewed here is a lack of differentiation between cNK cells and ILC1s. While this is understandable given the difficulty in distinguishing cNK cells from ILC1s, especially during inflammation and immune responses, it is quite possible that a role for ILC1s may have been overlooked in many of these studies that instead attributed tumor responses to cNK cells.

4.8 NK Cell Plasticity in Cancer

One reason that ILCs in cancer may have been overlooked is that ILCs are heterogenous, plastic populations that can undergo phenotypic and functional changes in response to alterations in their microenvironment. Indeed, ILCs have the capacity to transdifferentiate into other ILC subsets during diseases, including cancer [105]. Until recently, it was thought that there was only minimal plasticity between NK cells and ILC1s that was limited to a short period during development. However, it was recently shown by two groups that transforming growth factor β (TGFβ) signaling facilitates plasticity between NK cells and ILC1s (Fig. 4.2). TGFβ is an immunosuppressive cytokine associated with the development of a tolerogenic tumor microenvironment that is not conducive to antitumor immunity [106]. Moreover, TGFβ can directly inhibit the antitumor activity of NK cells through direct suppression of NK cell cytotoxicity, as well as the frequency and anti-metastatic function of tumor-infiltrating NK cells [107, 108]. While the impact of TGFβ signaling on ILC1s is incompletely understood, TGFβ signaling was shown to be critical for the development and maintenance of salivary gland ILC1s; however the exact function of these cells remains poorly understood as they have features of both ILC1s and NK cells [109]. While TGFβ signaling was sufficient to convert NK cells into ILC1-like cells in the salivary gland, a noncanonical pathway is involved. SMAD family member 4 (SMAD4) is a critical negative regulator of noncanonical TGFβ signaling in NK cells [110]. SMAD4-deficient NK cells were readily converted into ILC1s whereas SMAD4 expression was sufficient to prevent NK cell conversion into ILC1-like cells and SMAD4-deficient NK cells had decreased cytotoxicity and IFNγ production compared to wild-type controls [110]. This is biologically relevant as the SMAD4-deficient mice are significantly more susceptible to viral infection and tumor outgrowth but the mechanism is incompletely understood [110].

In studies of an MCA-induced fibrosarcoma, TGFβ signaling drove conversion of tumor-infiltrating NK cells into an ILC1-like characterized by a decrease in Eomes expression and upregulation of CD49a (and other markers), both hallmarks of ILC1s [98]. NK cells outside the tumor did not undergo this conversion. As demonstrated with a series of TGFβ receptor transgenic mice, TGFβ signaling was required for the conversion of NK cells into ILC1 intermediates in vitro and in the tumor microenvironment. Moreover, in mice whose NK cells were unresponsive to TGFβ signaling, the ILC1-like intermediate population did not develop in the tumors, demonstrating a TGFβ-dependent process. Mice whose NK cells were unresponsive to TGFβ signaling were significantly protected from MCA tumor outgrowth and B16 melanoma metastases as compared to mice that constitutively expressed the TGFβ receptor on NK cells. While NK cells typically produce large amounts of IFNγ in the tumor microenvironment which is critical for a strong antitumor response [6, 111], converted NK cells showed a significant decrease in IFNγ production and an increase in granulocyte-macrophage-colony-stimulating factor (GM-CSF) production [98]. While it has been shown by many groups that NK cells and ILC1s both produce GM-CSF, it is unclear as to what effect GM-CSF has on tumor outgrowth as the function of GM-CSF signaling in cancer remains controversial [112, 113].

Taken together these findings indicated that there is potential for conversion between NK cells and ILC1s in the tumor microenvironment and that this process is detrimental to antitumor immunity. However, the studies highlighted here did not use the same markers to differentiate NK cells and ILC1s, thus making it difficult to ascertain the similarity between the subsets of the two

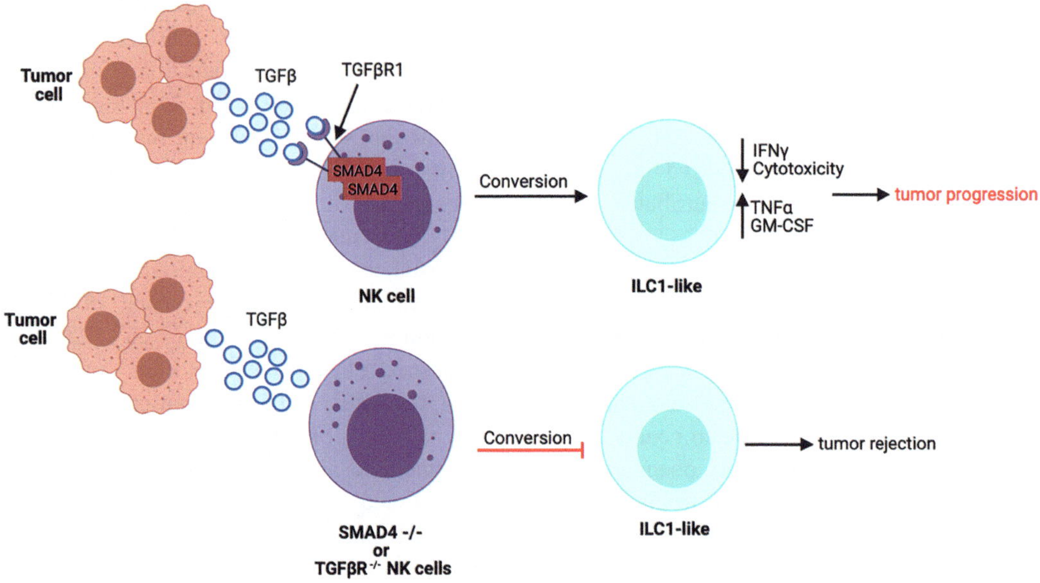

Fig. 4.2 NK cell plasticity in cancer. NK cells can be converted into ILC1-like cells in solid tumors by TGFβ signaling. The conversion of NK cells into ILC1-like cells is accompanied by decreased cytotoxicity and altered cytokine secretion profiles and impaired control of tumor growth. *ILC1* innate lymphoid cell 1, *TGFβ transforming growth factor β*, *NK cell* natural killer cell. Generated with BioRender.com

groups studied. Despite these differences, these studies suggest that conversion between NK cells and ILC1s may be limited to specific inflammatory stimuli, including conditions found in the tumor microenvironment. While these papers identified a novel mechanism by which tumors can evade NK cell-mediated killing, it is unclear if this phenomenon is specific to TGFβ-rich tumors or if this can be mediated by other soluble mediators produced in the tumor microenvironment. NK cell conversion to ILC1s has been observed in *Toxoplasma gondii* infection, indicating that conversion of NK cells can be mediated in other inflammatory settings, independent of the tumor microenvironment [28]. However, other studies indicate that NK cells do not undergo conversion into ILC1s during MCMV infection, suggesting that NK cell conversion into ILC1-like cells does not occur uniformly in infections. Interestingly, NK cells in the livers of MCMV-infected mice did upregulate CD49a, a marker typically associated with ILC1s, but maintained Eomes expression and other phenotypic hallmarks of NK cells [114]. These findings highlight the need for continued study of the impact inflammation within the tumor microenvironment has on shaping NK cells and ILC1 functionality.

4.9 Concluding Remarks

In both mice and humans, NK cells are important for initiating and coordinating antitumor immunity. In solid tumors, NK cells engage in complex interactions with other tumor-infiltrating immune cells, which can impact NK cell function and tumor growth. NK cells are also affected by other elements found in the tumor microenvironment such as NK cell activation receptor ligands, immunosuppressive cytokines, and nutrient deprivation. The role of ILC1s in tumors remains to be completely elucidated as they are difficult to distinguish from NK cells, and NK cells can be converted to ILC1-like cells. As the field moves forward and continues to unravel the role of NK cells and ILC1s in antitumor immunity, we will gain improved understanding of how NK cells and ILC1s contribute to antitumor immunity that will be critical for the development of novel immunotherapies targeting these innate immune cells.

Acknowledgments A.S. wrote the manuscript. W.Y. edited the manuscript. The Yokoyama Lab appreciates funding from the National Institutes of Health (R01-AI128845, R01-AI129545, R01-AI131680, and R01-AI140397) to study NK cells.

References

1. Morvan MG, Lanier LL. NK cells and cancer: you can teach innate cells new tricks. Nat Rev Cancer. 2016;16(1):7–19.
2. Vivier E, Artis D, Colonna M, Diefenbach A, Di Santo JP, Eberl G, et al. Innate lymphoid cells: 10 years on. Cell. 2018;174(5):1054–66.
3. Chiossone L, Dumas PY, Vienne M, Vivier E. Natural killer cells and other innate lymphoid cells in cancer. Nat Rev Immunol. 2018;18(11): 671–88.
4. Seaman WE, Sleisenger M, Eriksson E, Koo GC. Depletion of natural killer cells in mice by monoclonal antibody to NK-1. 1. Reduction in host defense against malignancy without loss of cellular or humoral immunity. J Immunol. 1987;138(12): 4539–44.
5. Smyth MJ, Thia KY, Cretney E, Kelly JM, Snook MB, Forbes CA, et al. Perforin is a major contributor to NK cell control of tumor metastasis. J Immunol. 1999;162(11):6658–62.
6. Street SEA, Cretney E, Smyth MJ. Perforin and interferon-γ activities independently control tumor initiation, growth, and metastasis. Blood. 2001;97(1): 192–7.
7. Takeda K, Nakayama M, Sakaki M, Hayakawa Y, Imawari M, Ogasawara K, et al. IFN-γ production by lung NK cells is critical for the natural resistance to pulmonary metastasis of B16 melanoma in mice. J Leukoc Biol. 2011;90(4):777–85.
8. Sliz A, Locker KCS, Lampe K, Godarova A, Plas DR, Janssen EM, et al. Gab3 is required for IL-2– and IL-15–induced NK cell expansion and limits trophoblast invasion during pregnancy. Sci Immunol. 2019;4(38):eaav3866.
9. Imai K, Matsuyama S, Miyake S, Suga K, Nakachi K. Natural cytotoxic activity of peripheral-blood lymphocytes and cancer incidence: an 11-year follow-up study of a general population. Lancet. 2000;356(9244):1795–9.
10. Velardi A, Ruggeri L, Mancusi A, Aversa F, Christiansen FT. Natural killer cell allorecognition of missing self in allogeneic hematopoietic transplantation: a tool for immunotherapy of leukemia. Curr Opin Immunol. 2009;21(5):525–30.
11. Fehniger TA, Cooper MA. Harnessing NK cell memory for cancer immunotherapy. Trends Immunol. 2016;37(12):877–88.
12. Myers JA, Miller JS. Exploring the NK cell platform for cancer immunotherapy. Nat Rev Clin Oncol. 2021;18(2):85–100.
13. Peng H, Jiang X, Chen Y, Sojka DK, Wei H, Gao X, et al. Liver-resident NK cells confer adaptive immunity in skin-contact inflammation. J Clin Invest. 2013;123(4):1444–56.
14. Gasteiger G, Fan X, Dikiy S, Lee SY, Rudensky AY. Tissue residency of innate lymphoid cells in lymphoid and nonlymphoid organs. Science (80-). 2015;350(6263):981–5.
15. Bezman NA, Kim CCC, Sun JC, Min-Oo G, Hendricks DW, Kamimura Y, et al. Molecular definition of the identity and activation of natural killer cells. Nat Immunol. 2012;13(10):1000–8.
16. Arase H, Saito T, Phillips JH, Lanier LL. Cutting edge: the mouse NK cell-associated antigen recognized by DX5 Moncoclonal antibody is CD49b (α 2 integrin, very late Antigen-2). J Immunol. 2001;167(3):1141–4.
17. Riggan L, Freud AG, O'Sullivan TE. True detective: unraveling group 1 innate lymphocyte heterogeneity. Trends Immunol. 2019;40(10):909–21.
18. Klose CSN, Flach M, Möhle L, Rogell L, Hoyler T, Ebert K, et al. Differentiation of type 1 ILCs from a common progenitor to all helper-like innate lymphoid cell lineages. Cell. 2014;157(2):340–56.
19. Gordon SM, Chaix J, Rupp LJ, Wu J, Madera S, Sun JC, et al. The transcription factors T-bet and Eomes control key checkpoints of natural killer cell maturation. Immunity. 2012;36(1):55–67.
20. Daussy C, Faure F, Mayol K, Viel S, Gasteiger G, Charrier E, et al. T-bet and Eomes instruct the development of two distinct natural killer cell lineages in the liver and in the bone marrow. J Exp Med. 2014;211(3):563–77.
21. Pikovskaya O, Chaix J, Rothman NJ, Collins A, Chen Y-H, Scipioni AM, et al. Cutting edge: Eomesodermin is sufficient to direct type 1 innate lymphocyte development into the conventional NK lineage. J Immunol. 2016;196(4):1449–54.
22. Townsend MJ, Weinmann AS, Matsuda JL, Salomon R, Farnham PJ, Biron CA, et al. T-bet regulates the terminal maturation and homeostasis of NK and Vα14i NKT cells. Immunity. 2004;20 (4):477–94.
23. Victorino F, Sojka DK, Brodsky KS, McNamee EN, Masterson JC, Homann D, et al. Tissue-resident NK cells mediate ischemic kidney injury and are not depleted by anti–asialo-GM1 antibody. J Immunol. 2015;195(10):4973–85.
24. Sojka DK, Yang L, Plougastel-Douglas B, Higuchi DA, Croy BA, Yokoyama WM. Cutting edge: local proliferation of uterine tissue-resident NK cells

during decidualization in mice. J Immunol. 2018;201 (9):2551–6.
25. Cortez VS, Fuchs A, Cella M, Gilfillan S, Colonna M. Cutting edge: salivary gland NK cells develop independently of Nfil3 in steady-state. J Immunol. 2014;192(10):4487–91.
26. Sojka DK, Plougastel-Douglas B, Yang L, Pak-Wittel MA, Artyomov MN, Ivanova Y, et al. Tissue-resident natural killer (NK) cells are cell lineages distinct from thymic and conventional splenic NK cells. elife. 2014;3:1–21.
27. Huang Y, Mao K, Chen X, Sun MA, Kawabe T, Li W, et al. S1P-dependent interorgan trafficking of group 2 innate lymphoid cells supports host defense. Science (80-). 2018;359(6371):114–9.
28. Park E, Patel S, Wang Q, Andhey P, Zaitsev K, Porter S, et al. Toxoplasma gondii infection drives conversion of NK cells into ILC1-like cells. elife. 2019;8:1–25.
29. Vivier E, Tomasello E, Baratin M, Walzer T, Ugolini S. Functions of natural killer cells. Nat Immunol. 2008;9(5):503–10.
30. Lanier LL. Up on the tightrope: natural killer cell activation and inhibition. Nat Immunol. 2008;9(5): 495–502.
31. Cerwenka A, Baron JL, Lanier LL. Ectopic expression of retinoic acid early inducible-1 gene (RAE-1) permits natural killer cell-mediated rejection of a MHC class I-bearing tumor in vivo. Proc Natl Acad Sci. 2001;98(20):11521–6.
32. Diefenbach A, Jensen ER, Jamieson AM, Raulet DH. Rae1 and H60 ligands of the NKG2D receptor stimulate tumour immunity. Nature. 2001;413:165–71.
33. van den Broek MF, Kägi D, Zinkernagel RM, Hengartner H. Perforin dependence of natural killer cell-mediated tumor control in vivo. Eur J Immunol. 1995;25(12):3514–6.
34. Orange JS. Formation and function of the lytic NK-cell immunological synapse. Nat Rev Immunol. 2008;8(9):713–25.
35. Kayagaki N, Yamaguchi N, Nakayama M, Takeda K, Akiba H, Tsutsui H, et al. Expression and function of TNF-related apoptosis-inducing ligand on murine activated NK cells. J Immunol. 1999;163(4): 1906–13.
36. Screpanti V, Wallin RPA, Ljunggren H-G, Grandien A. A central role for death receptor-mediated apoptosis in the rejection of tumors by NK cells. J Immunol. 2001;167(4):2068–73.
37. Cretney E, Takeda K, Yagita H, Glaccum M, Peschon JJ, Smyth MJ. Increased susceptibility to tumor initiation and metastasis in TNF-related apoptosis-inducing ligand-deficient mice. J Immunol. 2002;168(3):1356–61.
38. Tripathy SK, Keyel PA, Yang L, Pingel JT, Cheng TP, Schneeberger A, et al. Continuous engagement of a self-specific activation receptor induces NK cell tolerance. J Exp Med. 2008;205(8):1829–41.
39. Martín-Fontecha A, Thomsen LL, Brett S, Gerard C, Lipp M, Lanzavecchia A, et al. Induced recruitment of NK cells to lymph nodes provides IFN-γ for TH1 priming. Nat Immunol. 2004;5(12):1260–5.
40. Wherry JC, Schreiber RD, Unanue ER. Regulation of gamma interferon production by natural killer cells in scid mice: roles of tumor necrosis factor and bacterial stimuli. Infect Immun. 1991;59(5):1709–15.
41. Gazzinelli RT, Hieny S, Wynn TA, Wolf S, Sher A. Interleukin 12 is required for the T-lymphocyte independent induction of interferon gamma by an intracellular parasite and induces resistance in T-cell-deficient hosts. Proc Natl Acad Sci. 1993;90 (July):6115–9.
42. Scharton TM, Scott P. Natural killer cells are a source of interferon γ that drives differentiation of CD4+ T cell subsets and induces early resistance to leishmania major in mice. J Exp Med. 1993;178(2):567–78.
43. Js O, Wang B, Terhorst C, Ca B. Requirement for natural killer cell-produced interferon γ in defense against murine cytomegalovirus infection and enhancement of this defense pathway by interleukin 12 administration. J Exp Med. 1995;182(4):1045–56.
44. Hanahan D, Weinberg RA. Hallmarks of cancer: the next generation. Cell. 2011;144(5):646–74.
45. Palucka AK, Coussens LM. The basis of oncoimmunology. Cell. 2016;164(6):1233–47.
46. Chambers AM, Lupo KB, Matosevic S. Tumor microenvironment-induced immunometabolic reprogramming of natural killer cells. Front Immunol. 2018;9:1–10.
47. Fernandez NC, Lozier A, Flament C, Ricciardi-Castagnoli P, Bellet D, Suter M, et al. Dendritic cells directly trigger NK cell functions: cross-talk relevant in innate anti-tumor immune responses in vivo. Nat Med. 1999;5(4):405–11.
48. Adam C, King S, Allgeier T, Braumüller H, Lüking C, Mysliwietz J, et al. DC-NK cell cross talk as a novel CD4+ T-cell-independent pathway for antitumor CTL induction. Blood. 2005;106(1): 338–44.
49. Mocikat R, Braumüller H, Gumy A, Egeter O, Ziegler H, Reusch U, et al. Natural killer cells activated by MHC class I low targets prime dendritic cells to induce protective CD8 T cell responses. Immunity. 2003;19(4):561–9.
50. Alexandre YO, Ghilas S, Sanchez C, Le Bon A, Crozat K, Dalod M. XCR1+ dendritic cells promote memory CD8+ T cell recall upon secondary infections with listeria monocytogenes or certain viruses. J Exp Med. 2016;213(1):75–92.
51. Mittal D, Vijayan D, Putz EM, Aguilera AR, Markey KA, Straube J, et al. Interleukin-12 from CD103+ Batf3-dependent dendritic cells required for NK-cell suppression of metastasis. Cancer Immunol Res. 2017;5(12):1098–108.
52. Böttcher JP, Bonavita E, Chakravarty P, Blees H, Cabeza-Cabrerizo M, Sammicheli S, et al. NK cells stimulate recruitment of cDC1 into the tumor

microenvironment promoting cancer immune control. Cell. 2018;172(5):1022–1037.e14.
53. Walzer T, Dalod M, Robbins SH, Zitvogel L, Vivier E. Natural-killer cells and dendritic cells: " l ' union fait la force ". October. 2005;106(7):2252–8.
54. Degli-Esposti MA, Smyth MJ. Close encounters of different kinds: dendritic cells and NK cells take Centre stage. Nat Rev Immunol. 2005;5(2):112–24.
55. Karsunky H, Merad M, Cozzio A, Weissman IL, Manz MG. Flt3 ligand regulates dendritic cell development from Flt3+ lymphoid and myeloid-committed progenitors to Flt3+ dendritic cells in vivo. J Exp Med. 2003;198(2):305–13.
56. Barry KC, Hsu J, Broz ML, Cueto FJ, Binnewies M, Combes AJ, et al. A natural killer–dendritic cell axis defines checkpoint therapy–responsive tumor microenvironments. Nat Med. 2018;24(8):1178–91.
57. Schuijs MJ, Png S, Richard AC, Tsyben A, Hamm G, Stockis J, et al. ILC2-driven innate immune checkpoint mechanism antagonizes NK cell antimetastatic function in the lung. Nat Immunol. 2020;21(9): 998–1009.
58. Teijeira Á, Garasa S, Gato M, Alfaro C, Migueliz I, Cirella A, et al. CXCR1 and CXCR2 chemokine receptor agonists produced by tumors induce neutrophil extracellular traps that interfere with immune cytotoxicity. Immunity. 2020;52(5):856–71.
59. Bonavita E, Bromley CP, Jonsson G, Pelly VS, Sahoo S, Walwyn-Brown K, et al. Antagonistic inflammatory phenotypes dictate tumor fate and response to immune checkpoint blockade. Immunity. 2020;53(6):1215–1229.e8.
60. Ireland AS, Oliver TG. Neutrophils create an impeNETrable shield between tumor and cytotoxic immune cells. Immunity. 2020;52(5):729–31.
61. Papayannopoulos V. Neutrophil extracellular traps in immunity and disease. Nat Rev Immunol. 2018;18 (2):134–47.
62. Ardolino M, Garcia KC, Raulet DH, Ardolino M, Azimi CS, Iannello A, et al. Cytokine therapy reverses NK cell anergy in MHC-deficient tumors. J Clin Invest. 2014;124(11):4781–94.
63. Wherry EJ. T cell exhaustion. Nat Immunol. 2011;12 (6):492–9.
64. Wherry EJ, Kurachi M. Molecular and cellular insights into T cell exhaustion. Nat Rev Immunol. 2015;15(8):486–99.
65. Judge SJ, Murphy WJ, Canter RJ. Characterizing the dysfunctional NK cell: assessing the clinical relevance of exhaustion, anergy, and senescence. Front Cell Infect Microbiol. 2020;10(February):49.
66. Wiemann K, Mittrücker H-W, Feger U, Welte SA, Yokoyama WM, Spies T, et al. Systemic NKG2D down-regulation impairs NK and CD8 T cell responses in vivo. J Immunol. 2005;175(2):720–9.
67. Oppenheim DE, Roberts SJ, Clarke SL, Filler R, Lewis JM, Tigelaar RE, et al. Sustained localized expression of ligand for the activating NKG2D receptor impairs natural cytotoxicity in vivo and reduces tumor immunosurveillance. Nat Immunol. 2005;6(9):928–37.
68. Lanier LL. NKG2D receptor and its ligands in host defense. Cancer Immunol Res. 2015;3(6):575–82.
69. Raulet DH, Gasser S, Gowen BG, Deng W, Jung H, Mistry AR, et al. Regulation of ligands for the activating receptor NKG2D. Annu Rev Immunol. 2013;31(4):413–41.
70. Coudert JD, Scarpellino L, Gros F, Vivier E, Held W. Sustained NKG2D engagement induces cross-tolerance of multiple distinct NK cell activation pathways. Blood. 2008;111(7):3571–8.
71. Groh V, Wu J, Yee C, Spies T. Tumour-derived soluble MIC ligands impair expression of NKG2D and T-cell activation. Nature. 2002;419:734–8.
72. Pazina T, Shemesh A, Brusilovsky M, Porgador A, Campbell KS. Regulation of the functions of natural cytotoxicity receptors by interactions with diverse ligands and alterations in splice variant expression. Front Immunol. 2017;8
73. Textor S, Bossler F, Henrich KO, Gartlgruber M, Pollmann J, Fiegler N, et al. The proto-oncogene Myc drives expression of the NK cell-activating NKp30 ligand B7-H6 in tumor cells. Onco Targets Ther. 2016;5(7):1–12.
74. Alvarez M, Simonetta F, Baker J, Pierini A, Wenokur AS, Morrison AR, et al. Regulation of murine NK cell exhaustion through the activation of the DNA damage repair pathway. JCI Insight. 2019;4(14): 1–17.
75. Merino A, Zhang B, Dougherty P, Luo X, Wang J, Blazar BR, et al. Chronic stimulation drives human NK cell dysfunction and epigenetic reprograming. J Clin Invest. 2019;129(9):3770–85.
76. Garcia-Lora A, Algarra I, Garrido F. MHC class I antigens, immune surveillance, and tumor immune escape. J Cell Physiol. 2003;195(3):346–55.
77. Kim S, Poursine-Laurent J, Truscott SM, Lybarger L, Song Y-J, Yang L, et al. Licensing of natural killer cells by self-major histocompatibility complex class I. Nature. 2005;436(4):709–13.
78. Bix M, Liao NS, Zijlstra M, Loring J, Jaenisch R, Raulet D. Rejection of class I MHC-deficient haemopoietic cells by irradiated MHC-matched mice. Nature. 1991;353:412–4.
79. Liao NS, Bix M, Zijlstra M, Jaenisch R, Raulet D. MHC class I deficiency: susceptibility to natural killer (NK) cells and impaired NK activity. Science (80-). 1991;253(5016):199–202.
80. Elliott JM, Wahle JA, Yokoyama WM. MHC class I-deficient natural killer cells acquire a licensed phenotype after transfer into an MHC class I-sufficient environment. J Exp Med. 2010;207(10):2073–9.
81. Joncker NT, Shifrin N, Delebecque F, Raulet DH. Mature natural killer cells reset their responsiveness when exposed to an altered MHC environment. J Exp Med. 2010;207(10):2065–72.

82. Ebihara T. Jonsson a H, Yokoyama WM. Natural killer cell licensing in mice with inducible expression of MHC class I. Proc Natl Acad Sci. 2013;110:1–6.
83. Bern MD, Parikh BA, Yang L, Beckman DL, Poursine-Laurent J, Yokoyama WM. Inducible down-regulation of MHC class I results in natural killer cell tolerance. J Exp Med. 2019;216(1):99–116.
84. Seo H, Jeon I, Kim BS, Park M, Bae EA, Song B, et al. IL-21-mediated reversal of NK cell exhaustion facilitates anti-tumour immunity in MHC class I-deficient tumours. Nat Commun. 2017;8:1–14.
85. Hsu J, Hodgins JJ, Marathe M, Nicolai CJ, Bourgeois-Daigneault MC, Trevino TN, et al. Contribution of NK cells to immunotherapy mediated by PD-1/PD-L1 blockade. J Clin Invest. 2018;128(10): 4654–68.
86. Zhang Q, Bi J, Zheng X, Chen Y, Wang H, Wu W, et al. Blockade of the checkpoint receptor TIGIT prevents NK cell exhaustion and elicits potent anti-tumor immunity. Nat Immunol. 2018;19(7):723–32.
87. O'Brien KL, Finlay DK. Immunometabolism and natural killer cell responses. Nat Rev Immunol. 2019;19(5):282–90.
88. Assmann N, O'Brien KL, Donnelly RP, Dyck L, Zaiatz-Bittencourt V, Loftus RM, et al. Srebp-controlled glucose metabolism is essential for NK cell functional responses. Nat Immunol. 2017;18 (11):1197–206.
89. Loftus RM, Assmann N, Kedia-Mehta N, O'Brien KL, Garcia A, Gillespie C, et al. Amino acid-dependent cMyc expression is essential for NK cell metabolic and functional responses in mice. Nat Commun. 2018;9(1):152–60.
90. Donnelly RP, Loftus RM, Keating SE, Liou KT, Biron CA, Gardiner CM, et al. mTORC1-dependent metabolic reprogramming ss a prerequisite for NK cell effector function. J Immunol. 2014;193(9): 4477–84.
91. Mah AY, Rashidi A, Keppel MP, Saucier N, Moore EK, Alinger JB, et al. Glycolytic requirement for NK cell cytotoxicity and cytomegalovirus control. JCI Insight. 2017;2(23)
92. Cong J, Wang X, Zheng X, Wang D, Fu B, Sun R, et al. Dysfunction of natural killer cells by FBP1-induced inhibition of glycolysis during lung cancer progression. Cell Metab. 2018;28(2):243–255. e.5.
93. Michelet X, Dyck L, Hogan A, Loftus RM, Duquette D, Wei K, et al. Metabolic reprogramming of natural killer cells in obesity limits antitumor responses. Nat Immunol. 2018;19(12):1330–40.
94. Brand A, Singer K, Koehl GE, Kolitzus M, Schoenhammer G, Thiel A, et al. LDHA-associated lactic acid production blunts tumor immunosurveillance by T and NK cells. Cell Metab. 2016;24 (5):657–71.
95. Harmon C, Robinson MW, Hand F, Almuaili D, Mentor K, Houlihan DD, et al. Lactate-mediated acidification of tumor microenvironment induces apoptosis of liver-resident NK cells in colorectal liver metastasis. Cancer Immunol Res. 2019;7(2): 335–46.
96. Reinfeld BI, Madden MZ, Wolf MM, Chytil A, Bader JE, Patterson AR, et al. Cell-programmed nutrient partitioning in the tumour microenvironment. Nature. 2021;2020:1–30.
97. Dadi S, Chhangawala S, Whitlock BM, Franklin RA, Luo CT, Oh SA, et al. Cancer immunosurveillance by tissue-resident innate lymphoid cells and innate-like T cells. Cell. 2016;164(3):365–77.
98. Gao Y, Souza-Fonseca-Guimaraes F, Bald T, Ng SS, Young A, Ngiow SF, et al. Tumor immunoevasion by the conversion of effector NK cells into type 1 innate lymphoid cells. Nat Immunol. 2017;18(9):1004–15.
99. Eckl J, Buchner A, Prinz PU, Riesenberg R, Siegert SI, Kammerer R, et al. Transcript signature predicts tissue NK cell content and defines renal cell carcinoma subgroups independent of TNM staging. J Mol Med. 2012 Jan;90(1):55–66.
100. Rusakiewicz S, Perier A, Semeraro M, Pitt JM, von Strandmann EP, Reiners KS, et al. NKp30 isoforms and NKp30 ligands are predictive biomarkers of response to imatinib mesylate in metastatic GIST patients. Onco Targets Ther. 2017;6(1):1–13.
101. Fuchs A, Vermi W, Lee JS, Lonardi S, Gilfillan S, Newberry RD, et al. Intraepithelial type 1 innate lymphoid cells are a unique subset of Il-12- and Il-15-responsive IFN-γ-producing cells. Immunity. 2013;38(4):769–81.
102. Karvellas CJ, Fedorak RN, Hanson J, Wong CKW. Increased risk of colorectal cancer in ulcerative colitis patients diagnosed after 40 years of age. Can J Gastroenterol. 2007;21(7):443–6.
103. Spits H, Bernink JH, Lanier L. NK cells and type 1 innate lymphoid cells: partners in host defense. Nat Immunol. 2016;17(7):758–64.
104. Stamatiades EG, Li MO. Tissue-resident cytotoxic innate lymphoid cells in tumor immunosurveillance. Semin Immunol. 2019;41(March):1–7.
105. Bal SM, Golebski K, Spits H. Plasticity of innate lymphoid cell subsets. Nat Rev Immunol. 2020;20 (9):552–65.
106. Colak S, ten Dijke P. Targeting TGF-β signaling in cancer. Trends Cancer. 2017;3(1):56–71.
107. Laouar Y, Sutterwala FS, Gorelik L, Flavell RA. Transforming growth factor-β controls T helper type 1 cell development through regulation of natural killer cell interferon-γ. Nat Immunol. 2005;6(6): 600–7.
108. Viel S, Marçais A, Guimaraes FSF, Loftus R, Rabilloud J, Grau M, et al. TGF-β inhibits the activation and functions of NK cells by repressing the mTOR pathway. Sci Signal. 2016;9(415)
109. Cortez VS, Cervantes-Barragan L, Robinette ML, Bando JK, Wang Y, Geiger TL, et al. Transforming growth factor-β signaling guides the differentiation of innate lymphoid cells in salivary glands. Immunity. 2016;44(5):1127–39.

110. Cortez VS, Ulland TK, Cervantes-Barragan L, Bando JK, Robinette ML, Wang Q, et al. SMAD4 impedes the conversion of NK cells into ILC1-like cells by curtailing non-canonical TGF-β signaling. Nat Immunol. 2017;18(9):995–1003.
111. Smyth BMJ, Cretney E, Takeda K, Wiltrout RH, Sedger LM, Kayagaki N. Tumor necrosis factor–related apoptosis-inducing ligand (TRAIL) contributes to interferon γ - dependent natural killer cell protection from tumor metastasis natural killer cell protection from tumor metastasis. J Exp Med. 2001;193(6):661–70.
112. Becher B, Tugues S, Greter M. GM-CSF: from growth factor to central mediator of tissue inflammation. Immunity. 2016;45(5):963–73.
113. Zhan Y, Lew AM, Chopin M. The pleiotropic effects of the GM-CSF rheostat on myeloid cell differentiation and function: more than a numbers game. Front Immunol. 2019;10(November):1–15.
114. El Weizman O, Adams NM, Schuster IS, Krishna C, Pritykin Y, Lau C, et al. ILC1 confer early host protection at initial sites of viral infection. Cell. 2017;171(4):795–808.

Finding a Niche: Tissue Immunity and Innate Lymphoid Cells

5

Haerin Jung, Do-Hyun Kim, Yilin Wang, and Steven J. Van Dyken

Abstract

The immune system plays essential roles in maintaining homeostasis in mammalian tissues that extend beyond pathogen clearance and host defense. Recently, several homeostatic circuits comprised of paired hematopoietic and non-hematopoietic cells have been described to influence tissue composition and turnover in development and after perturbation. Crucial circuit components include innate lymphoid cells (ILCs), which seed developing organs and shape their resident tissues by influencing progenitor fate decisions, microbial interactions, and neuronal activity. As they develop in tissues, ILCs undergo transcriptional imprinting that encodes receptivity to corresponding signals derived from their resident tissues but ILCs can also shift their transcriptional profiles to adapt to specific types of tissue perturbation. Thus, ILC functions are embedded within their resident tissues, where they constitute key regulators of homeostatic responses that can lead to both beneficial and pathogenic outcomes. Here, we examine the interactions between ILCs and various non-hematopoietic tissue cells, and discuss how specific ILC-tissue cell circuits form essential elements of tissue immunity.

Keywords

Innate lymphoid cells · Cytokines · Transcriptomics · Tissue-resident lymphocytes

5.1 Introduction

Innate lymphoid cells (ILCs) are a diverse group of lymphocytes that resemble T cells and mainly reside in peripheral organs. ILCs are broadly grouped by signature transcription factor (TF) and cytokine expression, mirroring their adaptive T cell counterparts, such that ILC1s, ILC2s, and ILC3s correspond to CD4+ T helper (Th)1, Th2, and Th17 cell subsets, respectively, while NK cells closely resemble cytotoxic CD8+ T cells [1]. In contrast to T cells, however, ILCs do not rely on specific antigen receptors for priming or effector function, but rather integrate signals emanating from surrounding cells to sense and respond to various tissue perturbations. This tissue responsiveness is a defining feature of ILCs, and is established initially during embryogenesis, as ILC precursors develop prenatally and seed the peripheral organs in which they reside and mature in situ throughout adulthood. Some ILCs, known as lymphoid tissue-inducer (LTi) cells, also directly contribute to the formation of

Haerin Jung, Do-Hyun Kim and Yilin Wang contributed equally with all other contributors.

H. Jung · D.-H. Kim · Y. Wang · S. J. Van Dyken (✉)
Department of Pathology and Immunology, Washington University School of Medicine, St. Louis, MO, USA
e-mail: svandyken@wustl.edu

© The Author(s), under exclusive license to Springer Nature Singapore Pte Ltd. 2022

X.-H. Sun (ed.), *Innate Lymphoid Cells*, Advances in Experimental Medicine and Biology 1365,
https://doi.org/10.1007/978-981-16-8387-9_5

secondary lymphoid tissues through close communication with stromal mesenchymal cells. In adult tissues, most ILCs are long-lived, exhibiting relatively low turnover rates and minimal replacement by circulating or tissue-resident precursors in the steady state.

Such ontogeny intimates that ILC development and function are integrated with the surrounding tissue microenvironment. Indeed, ILCs exhibit tissue-specific transcriptional imprinting that enables networking with neighboring epithelial, neural, and stromal cells in addition to other hematopoietic cells [2–7]. Correspondingly, tissue cells also exhibit organ-specific epigenomic profiles that encode immune signal responsiveness, providing a framework for immune-structural cell communication [8, 9]. In response to tissue perturbation or inflammation, ILCs engage various ligand–receptor pairs to proliferate, differentiate, and redistribute to adapt to local stimuli. This sentinel function of ILCs has emerged as a key feature of tissue-based immunity that is also comprised of tissue-resident memory T cells (Trm) and innate-like T cells that similarly mature and reside in mucosal tissues while relying on tissue signals for their function. In this chapter, we discuss recent studies that have connected tissue signals and cell types comprising peripheral organs with the maturation and function of ILCs. We focus primarily on recent characterizations of ILC2-tissue cell circuits in mucosal barriers and other tissues involved in mammalian metabolic homeostasis. These connections fuel ILC form and function in the steady state and in response to tissue perturbations, subsequently shaping organ-specific immune composition.

5.2 Respiratory Tract

The lungs perform vital gas exchange functions and undergo constant mechanical stress and environmental exposures that shape the resident ILC populations (Fig. 5.1). In mice, ILC2s populate lung tissue prenatally, and then expand robustly during the first week after birth, coinciding with alveolar development and mechanical distension associated with ventilation [10, 11]. Similar ILC2 expansion can be recapitulated in adult mouse lungs after surgical lung lobe removal (pneumonectomy), which triggers compensatory growth in remaining lobes and rapid epithelial proliferation to generate new alveoli [12]. These studies indicate that ILC2 homeostasis is synchronized with signals emanating from cells comprising the developing and regenerating lung tissue. Key lung cells in this regard include alveolar type II epithelial cells (ATII), along with several other lung epithelial and perivascular stromal cell types that comprise the developing lung and respond to perturbation. These cell types are sources of interleukin (IL)-33, a widely expressed nuclear factor and IL-1 family member that is a major activating signal for lung ILC2s [3, 13–16]. IL-33 influences basal ILC2 cytokine expression during lung development [3, 15], and also functions as an alarmin released by damaged lung epithelial cells to further activate ILC2s in response to inhaled environmental stimuli such as viruses, allergens, toxicants, and particulate matter [17–20]. IL-33 drives ILC2 lung responses in synergy with additional signals like leukotrienes [21], neuropeptides [22, 23], and other cytokines, particularly thymic stromal lymphopoietin (TSLP), a signal transducer and activator of transcription 5 (STAT5)-activating cytokine associated with allergic lung inflammation [14, 20, 24, 25]. TSLP and IL-33 are co-expressed by platelet-derived growth factor receptor (PDGFR)α + Sca-1+ adventitial stromal cells comprising collagen-rich areas surrounding airways and vasculature, where ILC2s are preferentially localized [10, 16]. ILC2s are also strategically positioned near airway branch points and rare pulmonary neuroendocrine cells (PNECs), which coordinate ILC2 responses to environmental cues via neuropeptide and neurotransmitter signaling [23].

Lung epithelial cells can also provide signals that influence ILC2 trafficking in response to environmental exposures. For example, after house dust mite (HDM) allergen inhalation, mice lacking lung epithelial transforming growth factor β (TGF-β1) exhibit impaired ILC2 migration into airways and reduced allergic airway

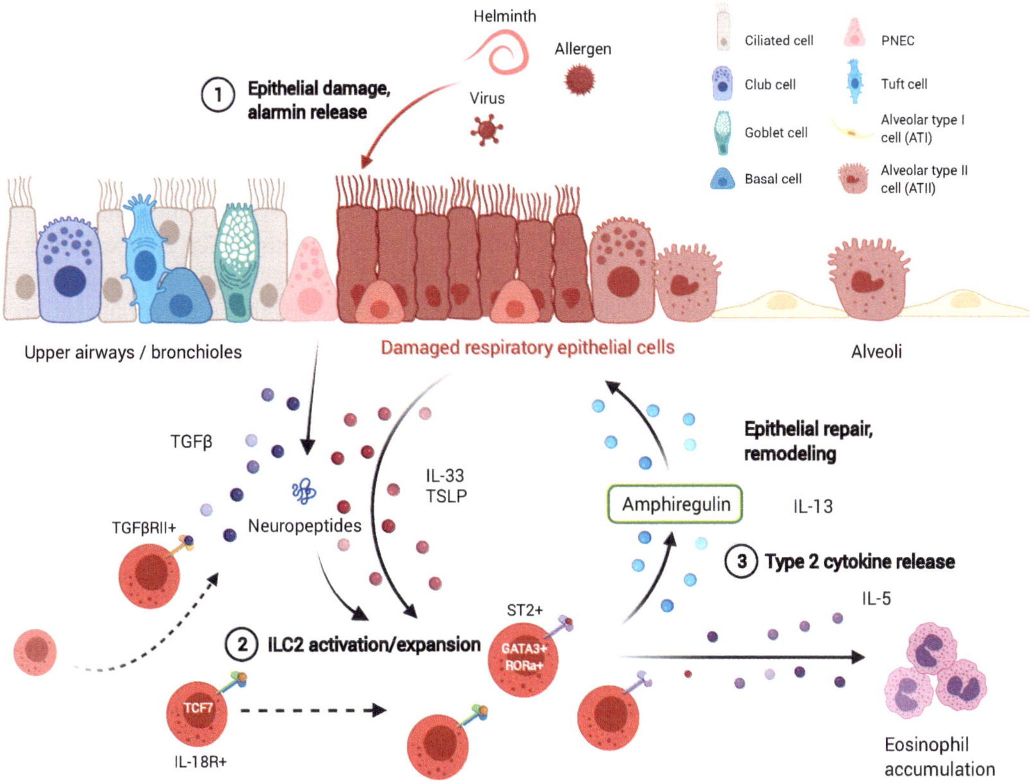

Fig. 5.1 ILC2-tissue cell interactions in the lung. Lung epithelial stress or damage leads to release of IL-33, which activates resident ILC2s to produce type 2 cytokines and amphiregulin that contribute to type 2 inflammation as well as epithelial repair and remodeling responses. Epithelial cells also provide TGFβ that recruits ILC2s into the airways, while pulmonary neuroendocrine cells (PNECs) influence ILC2s via neuropeptide and neurotransmitter signaling. Most mature lung ILC2s express the IL-33 receptor (IL-33R; ST2), but a smaller progenitor-like subset expresses IL-18R and can give rise to multiple ILC lineages, including the mature ST2-expressing ILC2 population

disease while retaining normal T cell and NK cell responses to HDM [26]. Epithelial TGF-β1-deficient mice also show normal HDM-induced alarmin production and ILC2 proliferation, implicating TGF-β1 in regulating the specific migration of TGFβRII-expressing ILC2s into inflamed airways, likely by heightening responsiveness to other chemotactic stimuli [26]. Such stimuli may include prostaglandin D_2 (PGD2), which mediates ILC2 chemotaxis in vitro and lung migration in vivo via chemoattractant receptor-homologous molecule expressed on Th2 cells (CRTH2), a cell surface receptor for PGD2 expressed on human and mouse ILC2s [27–29]. Interestingly, PGD2 increases the expression of the IL-33 receptor ST2 on ILC2s [28], suggesting that ILC2s acquire enhanced receptiveness to other lung tissue-derived signals as they transit into the tissue. Under some circumstances, ILC2s can also migrate out of tissues; for example, in response to helminth-induced type 2 immune activation and tissue damage, ILC2s are extruded from infected lung and intestinal tissues into the circulation [30, 31], thereby disseminating type 2 immune responses systemically.

ILC2 responses in the lung are also regulated by neuroimmune interactions. For example, cholinergic neurons express neuromedin U (NMU), a neuropeptide that activates NMU receptor

(NMUR1)-expressing ILC2s to proliferate and produce type 2 cytokines, particularly in the presence of co-stimulatory IL-25; mice deficient in NMUR1 displayed reduced ILC2 frequency and cytokine production after intranasal HDM challenge [22]. Adrenergic neurons, on the other hand, produce norepinephrine, a neurotransmitter that binds to β2 adrenergic receptor (β2AR) expressed on ILC2s to inhibit allergic inflammation. Correspondingly, mice lacking β2AR show exaggerated lung ILC2 responses after helminth infection or in response to *Alternaria* extract [32]. Another neuropeptide, calcitonin gene-related peptide (CGRP), can similarly antagonize ILC2 function in the lung via its receptor components Calcr/Ramp1 expressed on ILC2s [33]; however, PNEC-derived CGRP has also been reported to promote allergic inflammation [23], suggesting that the effects of this neuropeptide are context and microenvironment dependent. Adding to this complexity, both CGRP and acetylcholine can be produced by ILC2s after allergic stimulation and exposure to IL-25 and IL-33 [33–35], implying that neuronal derived and ILC2-intrinsic factors cross-regulate each other. The presence of other systemically distributed signals, such as sex hormones, can further modulate lung ILC2 activity. For example, testosterone and androgen receptor signaling suppresses ILC2 responses to allergen and IL-33 challenges [36, 37], by a mechanism that perhaps also underlies sex-based disparities in ILC2s in human allergic diseases such as asthma [37].

In response to local and systemic activating signals, ILC2s produce cytokines and growth factors that in turn influence surrounding lung tissue cells. A defining feature of ILC2 activation, for example, is production of the canonical type 2 cytokine IL-13, which can act directly on respiratory epithelial and smooth muscle cells to influence their differentiation and function in the context of allergic airway inflammation and rhinitis [38–41]. Activated ILC2s also increase their expression of amphiregulin (Areg), a member of the epidermal growth factor family that mediates remodeling and repair following epithelial injury, which occurs after severe respiratory viral infection. In response to influenza A (IAV) infection, for instance, ILC2s contribute to acute airway hyperresponsiveness but also contribute to Areg-mediated epithelial repair during the later recovery phase [17, 42]. During the initial stages of infection, ILC2 repair functions are tempered by vigorous antiviral responses characterized by interferons and the IL-12 cytokine family member IL-27, which suppress ILC2s and restrain type 2 immune-mediated epithelial repair and remodeling processes that develop later in post-viral lung lesions [43–45]. Intriguingly, these chronic lesions contain stemlike precursor and epithelial cell types not typically found in distal alveolar regions of the lung, including tuft cell-like chemosensory cells that are normally distributed throughout upper airway epithelia [46]. Such chemosensory cells closely resemble small intestinal tuft cells that produce IL-25 to activate ILC2s and mediate intestinal responses to helminth infection (discussed below). In the nasal and tracheal epithelium, tuft-like chemosensory cells regulate breathing, mucociliary clearance, and responses to airborne allergens and bacteria [47–50], highlighting the tissue context-dependent manner in which specialized epithelial cells coordinate responses to environmental stimuli.

Correspondingly, the ILC landscape in the lung is shaped by distinct environmental stimuli. Although prominent in acute respiratory viral infection, interferons and IL-27 also counteract ILC2 function in the context of allergic inflammation. In particular, interferon-γ (IFN-γ) dampens IL-33-stimulated ILC2 responses, and mice genetically lacking receptors for IFN-γ or IL-27 displayed exacerbated ILC2 responses after type 2 immune stimulation and infection with the migratory helminth *Nippostrongylus brasiliensis* [51–53], suggesting that an ability to respond quickly to life-threatening viral or bacterial infections is paramount to type 2 immune-driven tissue remodeling and regenerative mechanisms. After helminth infection, subsets of multifunctional ILCs also emerge that enable responses to separate infectious agents. Specifically, IL-25 drives the expansion of GATA-3/Rorγt-expressing ILC2s that produce IL-17 and IL-13 upon stimulation with TGFβ and IL-6, and also contribute to lower morbidity and fungal burden after adoptive transfer into mice infected with *Candida albicans* [54]. In addition, chronic cigarette smoke inhalation drives lung ILC2s to acquire characteristics of ILC1 and ILC3-like cells, particularly expression of IL12Rβ2, IL18Rα, and reduced GATA3 [55]. These toxicant-induced ILC2s also shift their production of IL-5 and IL-13 to IFN-γ, suggesting that ILCs can exhibit multifunctional cytokine capacity like their effector T cell counterparts. Recently, a similar subset of ILC1-like ILC2s expressing IL18Rα and low levels of GATA3, ST2, and IL-5 has been described to promote protection against *Mycobacterium tuberculosis* lung infection in mice [56], contrasting with a detrimental role attributed to similar cells in mice infected with influenza virus [55]. ILC2 subsets thus mediate distinct effects that may be adapted to different types of infections.

Recently, single-cell RNA-sequencing (scRNA-seq) and lineage-tracing studies have revealed that rare, systemically distributed ILC progenitor populations may underlie heterogeneity and adaptability observed among lung tissue ILC populations. For instance, although most mouse lung ILC2s express the IL-33 receptor, ST2, a smaller subset bearing resemblance to ILC precursors in the bone marrow (BM), was identified to express IL-18Rα and produce type 2 cytokines in response to IL-18 stimulation, independently of TSLP, IL-33, and IL-25 signaling [3], similar to subsets of human ILC2s [57]. Subsequent studies confirmed that mouse lung IL-18Rα+ ILC2s are enriched in progenitor markers, particularly the Notch target *Tcf7*, and can give rise to multiple ILC lineages in vitro and in vivo, including more mature ST2+ ILC2 populations [58, 59]. The presence of similar precursor-like ILCs in human blood and lung tissue [60] supports a model by which ILCs at various stages of maturation are systemically distributed and can differentially contribute to the total ILC population in each tissue. Although the role of IL-18 in directing distinct ILC fates in the lung and other tissues remains unresolved, these findings suggest that in situ "ILC-poiesis" underlies phenotypic shifts in resident ILC populations, enabling local renewal of the ILC niche and dynamic adaptation to specific perturbations.

5.3 Gastrointestinal Tract

The gastrointestinal tract is densely innervated and populated with immune cells that integrate signals and metabolites derived from diet, commensal microbes, and pathogens. ILC precursors seed the developing mouse intestinal tract prenatally [61], and subsequently adapt to dietary and microbial cues that shape the maturing ILC populations within the small intestinal lamina propria [62–68]. Many of these cues also influence intestinal epithelial cells, which comprise the tissue interface between the luminal environment and underlying ILC populations. In contrast with relatively quiescent mucosal tissues like the lung, the intestinal epithelium is highly active, undergoing rapid and continuous renewal fueled by a dynamic stem cell niche that regenerates the epithelial barrier every 3–5 days in mice [69]. This enables an enhanced ability to promptly shift epithelial cell composition, thereby modulating absorptive, sensory, and secretory capacity to adapt to constantly changing luminal contents. These dynamics are closely linked with ILC functions, which establish bidirectional circuits with epithelial cells to influence gut physiology in response to the various dietary and microbial constituents.

ILC2-epithelial cell communication underlies the characteristic "weep-and-sweep" type 2 immune responses to parasitic helminth infection. This coordinated tissue reaction to gastrointestinal parasites like *N. brasiliensis* enables expulsion by boosting intestinal mucus production, fluid secretion, inflammatory cell recruitment, and smooth muscle contraction. Crucial to this response is IL-13, which is mainly produced by ILC2s and Th2 cells in the small intestine in response to IL-25 and other tissue-derived factors. Initial studies on IL-25 connected its function with ILC2s, which were observed to robustly increase IL-13 expression after IL-25 administration, together with tissue eosinophilia, mucous cell expansion, and epithelial hyperplasia, all in the absence of adaptive immunity [70–72]. Later studies more thoroughly characterized ILC2 function [73–75], and the source of IL-25 was identified to be tuft cells, rare chemosensory cells that increase in frequency along with mucus-producing goblet cells in the intestinal epithelium after helminth or protist infection [76–78]. Tuft cells, like ILC2s, exhibit tissue-specific transcriptional differences that may underlie functional divergence in various organs; for instance, in addition to IL-25, subpopulations of tuft cells express ILC2-activating factors such as cysteinyl leukotrienes and TSLP [67, 79], possibly reflecting tissue niche-specific ILC2-tuft cell relationships. In this regard, ILC2s in the small intestine highly express the IL-25 receptor IL17RB, contrasting with other tissue ILC2s that predominantly express the IL-33 receptor ST2; this divergence forms the basis of tissue-specific differences in basal ILC2 activation that is influenced by IL-33 in lung and adipose tissue and IL-25 in small intestine [3, 15, 51, 65]. In response to gastrointestinal helminths or protists, intestinal tuft cells upregulate IL-25, which triggers resident ILC2s to produce IL-13, which in turn promotes increased tuft and goblet cell differentiation from epithelial progenitors [76–78], thereby forming an ILC2-tuft cell feedback circuit (Fig. 5.2).

Although ILC2s populate tissues and are basally activated in the absence of commensal microbiota [3], alterations to the intestinal microbiome also disrupt ILCs and can trigger the tuft cell–ILC2 circuit. For example, the intestinal protozoan *Tritrichomonas* ferments dietary fiber to generate the metabolite succinate, which is sensed by tuft cells expressing the succinate receptor GPR91 (Sucnr1) to produce IL-25 that activates ILC2s [65–67]. Perturbing the mouse microbiome with antibiotics alters transcriptional and epigenetic states in small intestinal ILC populations [64], and induces Sucnr1-dependent tuft cell expansion [66], implying that commensal bacteria normally suppress succinate-mediated tuft cell triggering. Sustained activation of the tuft cell-ILC2 circuit leads to small intestinal lengthening, an adaptation that retains absorptive enterocyte capacity and increases secretory epithelial cell frequency to enable systemic energy balance and immunity to helminths [65]. These dual effects reinforce a model in which tissue and metabolic homeostasis are interconnected with ILC2 function, hinting at broader physiologic roles for mammalian type 2 immunity. At the same time, certain pathogens exploit the tuft cell-ILC2 axis to invade host tissues, as in the case of murine norovirus, which gains entry via the CD300lf receptor expressed on tuft cells [80]. In the stomach, *Helicobacter pylori* infection induces ILC2 activation and IL-5 production, which reciprocally contributes to B cell and IgA responses that curtail infection [81]. Similar stomach ILC2 activation occurs in response to acute chemical injury, which drives IL-33-mediated ILC2 activation that influences subsequent epithelial reprogramming and tuft cell hyperplasia [82]. This type of tuft cell expansion can be detrimental in the context of carcinogenesis, particularly gastric cancer, in which aberrant tuft cell-neuron interactions contribute to tumor progression [83]. Accordingly, tuft-ILC2 circuit firing appears to be tightly controlled under normal circumstances. In the small intestine, for example, ILC2-intrinsic expression of the ubiquitin-editing enzyme A20 constrains IL-25-mediated signaling and ILC2 activation, thereby preventing spontaneous induction of type 2 immune responses and intestinal remodeling [65].

In addition to the microbiome, dietary constituents also play a major role in shaping epithelial and ILC function in the small intestine. For example, several ILC subsets express the ligand-inducible aryl hydrocarbon receptor (Ahr) TF which directs gene transcription in response to exogenous ligands, including dietary metabolites. In the small intestine, Ahr ligation induces divergent cell-intrinsic responses in ILC2s and ILC3s to modulate immune output; in ILC2s, Ahr signaling interferes with Gfi1-mediated ST2 expression to suppress ILC2 and anti-helminth responses, whereas in ILC3s Ahr expression enhances immune responses to *Citrobacter rodentium* infection [68]. In this regard, ILC3 function, particularly production of IL-22, is critical to forming feedback loops with intestinal

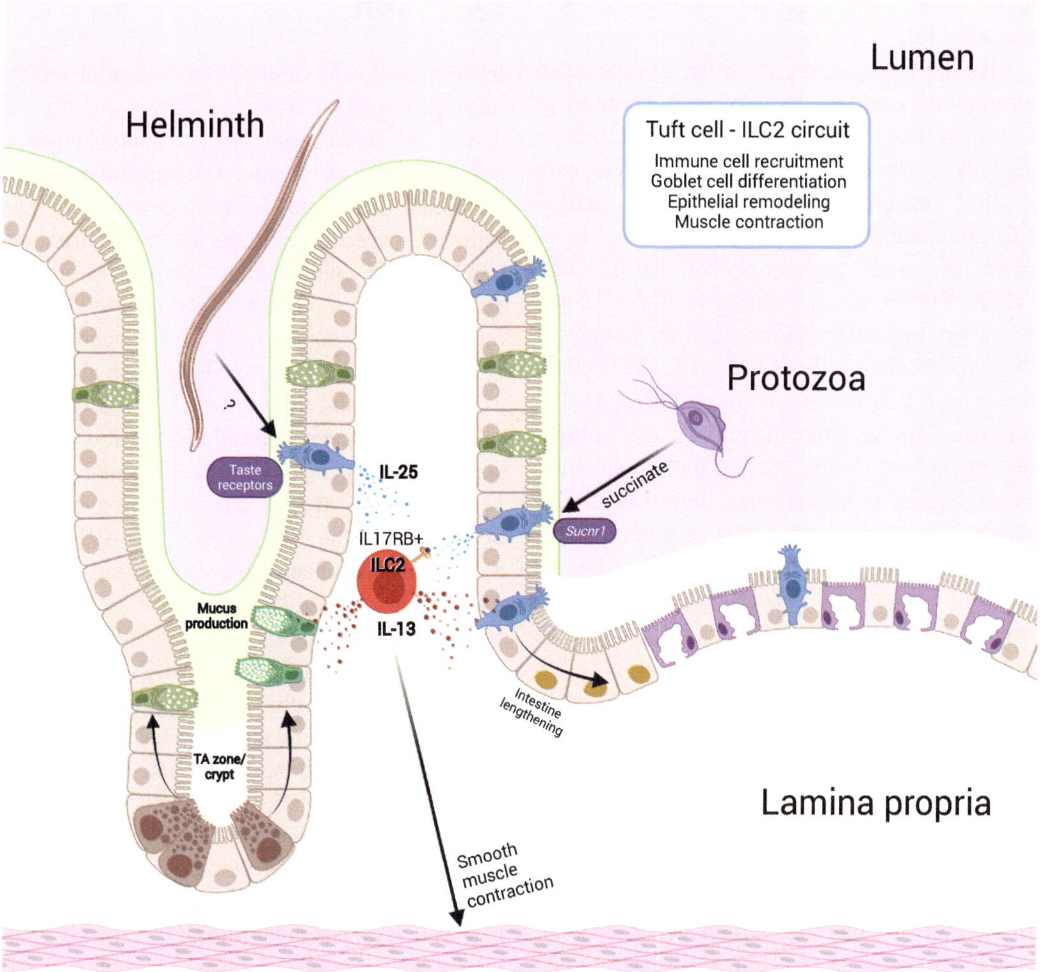

Fig. 5.2 Tuft cell-ILC2 circuit in the small intestine. In response to triggering by helminth infection or increased luminal succinate derived from intestinal protozoans, epithelial tuft cells release IL-25, which activates IL-17RB-expressing ILC2s in the lamina propria. Activated ILC2s produce IL-13, which promotes smooth muscle hypercontractility, immune cell recruitment, and increased goblet and tuft cell frequencies from epithelial progenitors, thereby comprising a feed-forward circuit

epithelial cells to maintain regenerative capacity, mucous barrier composition, and interactions with microbial constituents [84–86]. Epithelial cells correspondingly adapt their ability to digest and absorb macronutrients, such as carbohydrates, a response that can be modulated by innate-like lymphocytes, particularly γδ T cells, that suppress IL-22 production from ILC3s [87]. Gut ILC3s are also sensitive to dietary micronutrient availability; for instance, vitamin A deficiency leads to a loss of small intestinal ILC3s and a compensatory increase in ILC2s that enhances anti-helminth immunity, perhaps representing an adaptation that preserves host survival when food is scarce [63]. In human intestinal tissue, ILC spatial distribution, subset composition, and transcriptomic profiles are heterogeneous and vary with age and BMI, suggesting that dietary and environmental factors also shape the human intestinal ILC landscape [5].

Like in the respiratory tract, ILCs in the small intestine network extensively with neurons to coordinate tissue-wide responses to food intake and other external cues. ILCs co-localize with enteric neurons that express several ILC-modulating neuropeptides, including vasoactive intestinal peptide (VIP), NMU, and CGRP. VIP receptors expressed on ILC2s and ILC3s are engaged to induce cytokine production [10, 88], although VIP suppresses IL-22 production from ILC3s in some contexts [89], possibly reflecting the differential impact of multiple microbiota and homeostatic inputs, including light-entrained circadian cues. Intestinal ILC2s also highly express the NMU receptor Nmur1 and respond to NMU, which promotes nuclear factor of activated T cell (NFAT) activity and synergizes with IL-25 to drive ILC2 cytokine production and type 2 inflammation [22, 90, 91]. NMU also triggers IL-10 production from intestinal ILC2s, which comprise a major source of IL-10 in mouse intestine [92]; however, the contribution of ILC2-derived IL-10 in physiologic settings remains unclear. Cholinergic enteric neurons also express CGRP, which dampens intestinal ILC2 proliferation but promotes IL-5 expression; genetic absence of CGRP receptor leads to greater tuft cell frequencies and an enhanced ability to expel helminths from infected mice [33, 93], highlighting the dynamic interplay between epithelial cells, neurons, and ILC2s that coordinates intestinal responses to infection. As in the lung, ILC2s induce ChAT expression in response to alarmins and helminth infection [34, 35], and enteric neurons also express cytokines, particularly IL-18, which is required for antimicrobial peptide production from intestinal goblet cells [94] and may further shape intestinal ILC differentiation and function. Thus, future studies employing cell-specific deletion strategies are needed to clarify the precise mechanistic details of how these neuroimmune interactions control in vivo gastrointestinal function.

5.4 Skin

In contrast to the single-layer intestinal epithelium, skin consists of multiple layers (epidermis, dermis, subcutis), providing a substantial physical barrier that resists external stressors and microbial incursion. In mammals, hair grows from hair follicles, specialized regenerative structures consisting of epithelial and stem cells that are also reservoirs of oily sebum secreted by contiguous sebaceous glands (Fig. 5.3). Mechanosensory neurons innervate hair follicles, which also serve as immune hubs that attract T cells, dendritic cells, macrophages, and ILCs. In this regard, skin ILCs rely on CCR6-CCL20 signaling as well as CCR10 expression for migration and positioning along the hair follicles [95–97], which additionally produce cytokines required for lymphocyte growth and maintenance [98]. These cytokines further differentially influence the microanatomic distribution of ILCs in the skin; whereas subcutaneous ILC2s require IL-7 for development, epidermal and dermal ILCs are partially retained in the absence of IL-7 in a TSLP-dependent manner [97, 99].

Skin ILCs are marked by distinct transcriptomic profiles that diverge from other organs and among ILCs distributed in different layers of skin. Subcutaneous ILCs, for example, resemble other fat-associated ILC2s that highly express *Il1rl1* (encoding the IL-33 receptor ST2), *Il5*, *Il1rl1*, *Gata3*, and *Klrg1*, whereas dermal and epidermal ILCs exhibit hybrid profiles enriched in both ILC3/LTi-related (*Rorc* and *Lta*) and ILC2 signature genes (*Il13*), as well as *Tcf7* and *Il18r1*, that mark progenitor-like ILCs in other tissues [3, 97]. Consistent with this profile, the turnover rate of skin ILC2s is relatively higher compared with other tissues [11], and cytokine production from IL18Rα-expressing skin ILC2s is influenced by IL-18 in the steady state and in response to allergic inflammatory stimuli [3], likely in synergy with other activating cytokines including TSLP and IL-33 [100, 101]. Similar to the inflammatory ILC2s that emerge after helminth infection, skin ILC2s adapt their transcriptional trajectories to acquire ILC3-like

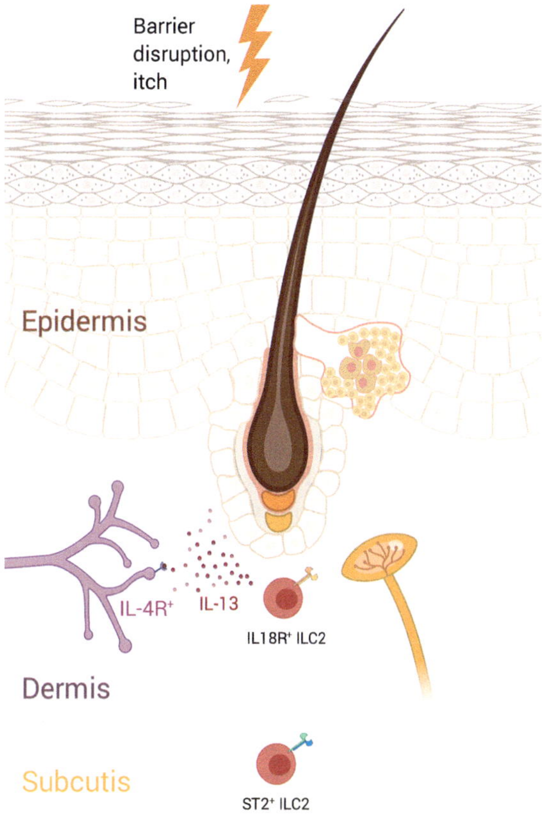

Fig. 5.3 ILC2-neuron interactions in the skin. ILC2s in skin respond to barrier disruption, which promotes release of tissue signals such as IL-18, TSLP, and IL-33 that initiate allergic inflammation. Whereas subcutaneous ILC2s resemble IL-33R/ST2-expressing adipose tissue ILC2s, epidermal and dermal ILC2s express IL-18R, and are closely associated with hair follicles. Sensory neurons in the skin express IL-4Rα and directly respond to type 2 cytokines IL-4 and IL-13, which sensitize them to pruritogens like histamine that instigate and perpetuate the itch-scratch cycle

characteristics including Rorγt and IL-17 expression after exposure to IL-1β or IL-23, in models of psoriasis, and in human dermal samples stimulated with *C. albicans* [102, 103].

Skin barrier function is modulated by ILCs in the steady state and after mechanical perturbation. Loss of skin ILCs in Rag2/common γ chain DKO or IL7/TSLP DKO mice decreases tumor necrosis factor (TNF) and lymphotoxin (LT) signals that suppress sebocyte growth and sebaceous gland size, leading to overproduction of antimicrobial lipids and a corresponding shift in the composition of the commensal skin microbiome in these mice [102]. After cutaneous injury, healing responses including reepithelialization and wound closure are impaired in mice that lack ILCs [104], suggesting similarities with the respiratory tract, where ILCs produce factors such as amphiregulin that promote epithelial repair and regeneration but may also contribute to persistent pathogenic inflammation. In the skin, triggering mechanisms for ILC2 activation after barrier breach often involve mechanical disruption. Keratinocytes, for example, express the cell adhesion glycoprotein E-cadherin, which is rapidly downregulated in response to epithelial perturbation in several skin diseases including atopic dermatitis (AD), characterized by allergic inflammation, itch, and impaired skin barrier function. Skin ILC2s express killer cell lectin-like receptor subfamily G member 1 (KLRG1), which acts as an inhibitory receptor when bound to E-cadherin; this interaction is lost upon physical disruption, resulting in increased ILC2 proliferation and cytokine production [100], providing one mechanism for how ILC2s in AD skin may be persistently triggered. Mechanical disruption of skin also induces keratinocyte release of the alarmin IL-33 that is sufficient to systemically trigger ILC2s in distal tissues such as the small intestine [105]. Such dissemination of local perturbations may underlie the "atopic march," a frequently observed temporal progression of skin, respiratory tissue, and intestinal involvement in human allergic disease.

Type 2 immune-mediated activation of neural circuits in the skin influences both inflammatory responses and complex mammalian behaviors related to skin disease. For example, allergic skin inflammation in AD is often accompanied by recurrent bouts of the itch-scratch cycle, a reflex-like response triggered by the sensation of itch followed by self-perpetuating scratching behavior. In this context, type 2 cytokines IL-4 and IL-13 can act directly on sensory neurons expressing both IL-4Rα and its downstream signaling component JAK1 in dorsal root ganglia, sensitizing them to pruritogens like histamine and IL-31; deletion of neuronal IL-4Rα attenuates both itch and skin inflammation, highlighting the dual role for type 2 cytokine signaling in mediating immune and behavioral responses [105]. This type of coordinated response is also promoted by TSLP produced by keratinocytes, which not only activates ILC2s, but can also act directly on TSLPR-expressing sensory neurons in the skin to promote itch behavior [106]. Such neuroimmune effects reflect the broader integration of ILC2s and type 2 immune signaling within the peripheral and central nervous systems, the effects of which have been linked to learning and memory formation [107], age-related cognitive decline [108], autoimmune disease [109], and responses to traumatic brain injury [110, 111]. Thus, ILC2-neuronal circuits coordinate both local skin responses and system-wide effects to influence tissue immunity.

5.5 Metabolically Active Tissues

Reciprocal interactions between ILC2s and resident tissue cells also underlie immune maintenance of metabolic homeostasis. Adipose tissue, for instance, is a dynamic endocrine organ that serves as an energy depot integrating both local and system-wide effects to mediate metabolic homeostasis. Whereas excessive accumulation of fatty acids in white adipose tissue (WAT) reflects overnutrition linked to obesity, brown adipose tissue (BAT) promotes healthy metabolic responses by converting dietary nutrients into heat via adaptive thermogenesis mediated by mitochondrial uncoupling protein 1 (UCP1) activation. Resident ILC2s are present in both mouse and human adipose tissue, and can produce IL-5 and IL-13 to recruit eosinophils that enhance glucose tolerance and promote a lean body state [112, 113]. Activated adipose ILC2s also produce methionine-enkephalin peptides, which induce UCP1 activation and "beiging" of adipocytes in WAT that is associated with a metabolically healthy state [114]. Cytokine production from adipose tissue ILC2s is primarily mediated by IL-33, which can be derived from adipose stromal cells and, along with IL-33-responsive T regulatory cells (Tregs), is also correlated with metabolic health [3, 51, 114] (Fig. 5.4). In particular, stem and progenitor stromal cells as well as PDGFRα+Sca-1 + podoplanin+ stromal cells in the WAT express IL-33, which is decreased in response to high-fat diet [16, 115–117]. In the lean state, PDGFRα+ multipotent stromal cells promote ICAM-1-mediated proliferation and activation of LFA-1-expressing ILC2s, which in turn induce eotaxin secretion from stromal cells via IL-13 signaling to enhance subsequent eosinophil

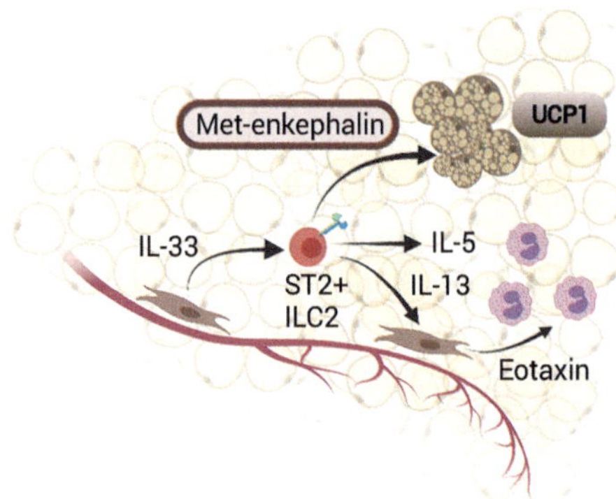

Fig. 5.4 Metabolic functions of ILC2s in the adipose tissue. In adipose tissue, resident ILC2s respond to stromal IL-33 by producing type 2 cytokines that increase eotaxin expression and recruit eosinophils that are associated with improved glucose homeostasis and a lean body state. Adipose tissue ILC2s are also activated by IL-33 to produce methionine-enkephalin (Met-enkephalin) peptides that activate UCP1-mediated "beiging" of adipocytes and adaptive thermogenesis to promote metabolic health

recruitment [116]. The positive correlation between IL-33-expressing stromal cells and ILC2s/Treg cells is lost during high-fat-induced obesity, however, indicating that metabolic dysregulation overrides the basal immune feedback circuit. This dysregulation may be further enhanced by increased expression of the soluble form of the IL-33 receptor, sST2, which is induced in adipocytes during high-fat feeding and exacerbates obesity-induced insulin resistance by dampening the effects of IL-33-driven ILC2/Treg function in WAT [118].

Additional metabolically active tissues such as liver and skeletal muscle engage similar ILC2-stromal circuits in response to homeostatic cues. For example, humans with muscular dystrophy accumulate eosinophils in muscle tissue, and a corresponding mouse model indicated that increased IL-33 production from fibro/adipogenic progenitor cells activated ILC2s to recruit eosinophils and drive the expression of genes associated with muscle fibrosis [119]. ILC2 activation in muscle tissue also contributes to muscle-intrinsic metabolic adaptations to increased whole-body activity, as IL-13 conditions skeletal muscle during endurance exercise; mice with muscle cell-intrinsic deletion of IL13Rα1 show a reduction in endurance running capacity and muscle fatty acid oxidation [120], highlighting the broad metabolic impact of type 2 cytokine signaling in different tissues. Additional elegant studies have parsed the role of type 2 cytokine signaling on diverse cellular targets within the same organ, ascribing specific roles to IL-13-responsive cell lineages in different aspects of organ pathology. In mouse models of liver fibrosis, for example, fibrosis is due to direct IL-13 signaling in PDGFRβ+ fibroblasts, whereas IL-4Rα-mediated signaling in hepatocytes and biliary cells controls the ductal reaction and microvesicular steatosis, thereby dissociating the pro-fibrotic effects of type 2 cytokine signaling from epithelial cell changes associated with regeneration [121]. Although the role of ILC2s in these processes remains to be determined, as with other tissue circuits, type 2 cytokine signaling in liver fibrosis contributes to repair and regeneration of damaged tissue but can contribute to pathologic outcomes if unresolved.

5.6 Conclusions

ILCs are integrated with tissue function throughout development and adult life, forming circuits with tissue cells to coordinate homeostatic responses to perturbation. This is evident from the earliest stages of ILC development, where ILC precursors in the fetal liver rely initially on local IL-7 and Notch signaling, and then PDGFRα+ mesenchymal cell support to acquire tissue responsiveness [122]. As they mature, ILCs become imprinted with corresponding tissue signatures by mechanisms not yet fully resolved, but which establish receptiveness to signals produced later in life from differentiated tissue cells. The tuft cell-ILC2 axis in the small intestine exemplifies this type of interaction, which is fueled primarily by tuft cell-derived IL-25 and ILC2-derived IL-13, but the circuit can be amplified and modulated by additional inputs from a host of other neighboring cells that converge to generate tissue-wide effects. Key to further decoding the cell types and signals in this regard will be advanced genomic, proteomic, and computational approaches that enable high-resolution and systems-level analysis of in vivo perturbations. In addition, new and highly specific conditional deletion strategies will help to clarify the precise roles of neurons, muscle cells, and additional epithelial and stromal progenitor cells that reciprocally shape ILC responses and differentiation. For type 2 immune tissue responses, a recurring pattern in which IL-13 is the primary means to direct resident tissue cell responses has emerged; future work may further discern the identity and roles of diverse immune-responsive tissue cells, enabling tailored approaches that promote beneficial versus detrimental outcomes in various organ niches.

On a broader scale, ILC-tissue interactions represent a paradigm for understanding how immunity is embedded or "trained" within tissues, and may be particularly relevant to tissue T cells that correspond to tissue ILC subsets. For example, tissue Th2 cells share epigenetic and transcriptomic profiles with mature ILC2s in the same tissue, and rely on similar tissue-derived factors such as IL-25, IL-33, and TSLP for their terminal effector function [123–125]. After viral infection, antigen-experienced Trm cells acquire ILC-like characteristics and persist in tissues for over a year, poised for rapid effector function even in the absence of T cell receptor signaling, continuously accruing with other immune cells in peripheral tissues, which act as expandable immune reservoirs [126]. These ILC-like features of tissue T cells reinforce a model in which adaptive immune cells compete with, or are layered upon, innate cells in tissues that respond to common niche factors, and similarly shape the local tissue composition. Future studies aimed at understanding how innate and adaptive immunity is integrated within tissue development and specification will enable tissue-specific therapeutic approaches based on specific immune-tissue cell circuits.

Acknowledgements We thank members of the Van Dyken laboratory for their insightful comments and discussion. Figures were created with BioRender.com. This work was supported by the US National Institutes of Health (R01 HL148033) and the Department of Pathology and Immunology at Washington University School of Medicine.

References

1. Vivier E, Artis D, Colonna M, Diefenbach A, Di Santo JP, Eberl G, et al. Innate lymphoid cells: 10 years on. Cell. 2018;174(5):1054–66.
2. Robinette ML, Fuchs A, Cortez VS, Lee JS, Wang Y, Durum SK, et al. Transcriptional programs define molecular characteristics of innate lymphoid cell classes and subsets. Nat Immunol. 2015;16(3): 306–17.
3. Ricardo-Gonzalez RR, Van Dyken SJ, Schneider C, Lee J, Nussbaum JC, Liang H-E, et al. Tissue signals imprint ILC2 identity with anticipatory function. Nat Immunol. 2018;19(10):1093–9.
4. Crinier A, Milpied P, Escalière B, Piperoglou C, Galluso J, Balsamo A, et al. High-dimensional single-cell analysis identifies organ-specific signatures and conserved NK cell subsets in humans and mice. Immunity. 2018;49(5):971–86.
5. Yudanin NA, Schmitz F, Flamar A-L, Thome JJC, Tait Wojno E, Moeller JB, et al. Spatial and temporal mapping of human innate lymphoid cells reveals elements of tissue specificity. Immunity. 2019;50 (2):505–19.
6. McFarland AP, Yalin A, Wang S-Y, Cortez VS, Landsberger T, Sudan R, et al. Multi-tissue single-cell analysis deconstructs the complex programs of mouse natural killer and type 1 innate lymphoid cells in tissues and circulation. Immunity. 2021;54(6): 1320–37.
7. Mazzurana L, Czarnewski P, Jonsson V, Wigge L, Ringnér M, Williams TC, et al. Tissue-specific transcriptional imprinting and heterogeneity in human innate lymphoid cells revealed by full-length single-cell RNA-sequencing. Cell Res. 2021;31(5):554–68.
8. Oetjen LK, Mack MR, Feng J, Whelan TM, Niu H, Guo CJ, et al. Sensory neurons co-opt classical immune signaling pathways to mediate chronic itch. Cell. 2017;171(1):217–28.
9. Krausgruber T, Fortelny N, Fife-Gernedl V, Senekowitsch M, Schuster LC, Lercher A, et al. Structural cells are key regulators of organ-specific immune responses. Nature. 2020;583(7815): 296–302.
10. Nussbaum JC, Van Dyken SJ, von Moltke J, Cheng LE, Mohapatra A, Molofsky AB, et al. Type 2 innate lymphoid cells control eosinophil homeostasis. Nature. 2013;502(7470):245–8.
11. Schneider C, Lee J, Koga S, Ricardo-Gonzalez RR, Nussbaum JC, Smith LK, et al. Tissue-resident group 2 innate lymphoid cells differentiate by layered ontogeny and in situ perinatal priming. Immunity. 2019;50(6):1425–38.
12. Lechner AJ, Driver IH, Lee J, Conroy CM, Nagle A, Locksley RM, et al. Recruited monocytes and type 2 immunity promote lung regeneration following pneumonectomy. Cell Stem Cell. 2017;21(1): 120–34.
13. Pichery M, Mirey E, Mercier P, Lefrancais E, Dujardin A, Ortega N, et al. Endogenous IL-33 is highly expressed in mouse epithelial barrier tissues, lymphoid organs, brain, embryos, and inflamed tissues: in situ analysis using a novel Il-33-LacZ gene trap reporter strain. J Immunol. 2012;188(7): 3488–95.
14. Mohapatra A, Van Dyken SJ, Schneider C, Nussbaum JC, Liang H-E, Locksley RM. Group 2 innate lymphoid cells utilize the IRF4-IL-9 module to coordinate epithelial cell maintenance of lung homeostasis. Mucosal Immunol. 2016;9(1):275–86.
15. Saluzzo S, Gorki A-D, Rana BMJ, Martins R, Scanlon S, Starkl P, et al. First-breath-induced type 2 pathways shape the lung immune environment. Cell Rep. 2017;18(8):1893–905.
16. Dahlgren MW, Jones SW, Cautivo KM, Dubinin A, Ortiz-Carpena JF, Farhat S, et al. Adventitial stromal cells define group 2 innate lymphoid cell tissue niches. Immunity. 2019;50(3):707–22.
17. Monticelli LA, Sonnenberg GF, Abt MC, Alenghat T, Ziegler CGK, Doering TA, et al. Innate

lymphoid cells promote lung-tissue homeostasis after infection with influenza virus. Nat Immunol. 2011;12 (11):1045–54.
18. Yasuda K, Muto T, Kawagoe T, Matsumoto M, Sasaki Y, Matsushita K, et al. Contribution of IL-33-activated type II innate lymphoid cells to pulmonary eosinophilia in intestinal nematode-infected mice. Proc Natl Acad Sci USA. 2012;109(9):3451–6.
19. Halim TYF, Steer CA, Mathä L, Gold MJ, Martinez-Gonzalez I, McNagny KM, et al. Group 2 innate lymphoid cells are critical for the initiation of adaptive T helper 2 cell-mediated allergic lung inflammation. Immunity. 2014;40(3):425–35.
20. Van Dyken SJ, Mohapatra A, Nussbaum JC, Molofsky AB, Thornton EE, Ziegler SF, et al. Chitin activates parallel immune modules that direct distinct inflammatory responses via innate lymphoid type 2 and γδ T cells. Immunity. 2014;40(3):414–24.
21. von Moltke J, O'Leary CE, Barrett NA, Kanaoka Y, Austen KF, Locksley RM. Leukotrienes provide an NFAT-dependent signal that synergizes with IL-33 to activate ILC2s. J Exp Med. 2017;214(1):27–37.
22. Wallrapp A, Riesenfeld SJ, Burkett PR, Abdulnour R-EE, Nyman J, Dionne D, et al. The neuropeptide NMU amplifies ILC2-driven allergic lung inflammation. Nature. 2017;549(7672):351–6.
23. Sui P, Wiesner DL, Xu J, Zhang Y, Lee J, Van Dyken S, et al. Pulmonary neuroendocrine cells amplify allergic asthma responses. Science. 2018;360(6393):8546.
24. Zhou B, Comeau MR, De Smedt T, Liggitt HD, Dahl ME, Lewis DB, et al. Thymic stromal lymphopoietin as a key initiator of allergic airway inflammation in mice. Nat Immunol. 2005;6(10):1047–53.
25. Al-Shami A, Spolski R, Kelly J, Keane-Myers A, Leonard WJ. A role for TSLP in the development of inflammation in an asthma model. J Exp Med. 2005;202(6):829–39.
26. Denney L, Byrne AJ, Shea TJ, Buckley JS, Pease JE, Herledan GMF, et al. Pulmonary epithelial cell-derived cytokine TGF-β1 is a critical cofactor for enhanced innate lymphoid cell function. Immunity. 2015;43(5):945–58.
27. Mjösberg JM, Trifari S, Crellin NK, Peters CP, van Drunen CM, Piet B, et al. Human IL-25- and IL-33-responsive type 2 innate lymphoid cells are defined by expression of CRTH2 and CD161. Nat Immunol. 2011;12(11):1055–62.
28. Xue L, Salimi M, Panse I, Mjösberg JM, McKenzie ANJ, Spits H, et al. Prostaglandin D2 activates group 2 innate lymphoid cells through chemoattractant receptor-homologous molecule expressed on TH2 cells. J Allergy Clin Immunol. 2014;133(4):1184–94.
29. Wojno EDT, Tait Wojno ED, Monticelli LA, Tran SV, Alenghat T, Osborne LC, et al. The prostaglandin D2 receptor CRTH2 regulates accumulation of group 2 innate lymphoid cells in the inflamed lung. Mucosal Immunol. 2015;8(6):1313–23.
30. Huang Y, Mao K, Chen X, Sun M-A, Kawabe T, Li W, et al. S1P-dependent interorgan trafficking of group 2 innate lymphoid cells supports host defense. Science. 2018;359(6371):114–9.
31. Ricardo-Gonzalez RR, Schneider C, Liao C, Lee J, Liang H-E, Locksley RM. Tissue-specific pathways extrude activated ILC2s to disseminate type 2 immunity. J Exp Med. 2020;217(4):e20191172.
32. Moriyama S, Brestoff JR, Flamar A-L, Moeller JB, Klose CSN, Rankin LC, et al. β2-adrenergic receptor-mediated negative regulation of group 2 innate lymphoid cell responses. Science. 2018;359(6379): 1056–61.
33. Nagashima H, Mahlakõiv T, Shih H-Y, Davis FP, Meylan F, Huang Y, et al. Neuropeptide CGRP limits group 2 innate lymphoid cell responses and constrains type 2 inflammation. Immunity. 2019;51 (4):682–95.
34. Roberts LB, Schnoeller C, Berkachy R, Darby M, Pillaye J, Oudhoff MJ, et al. Acetylcholine production by group 2 innate lymphoid cells promotes mucosal immunity to helminths. Sci Immunol. 2021;6(57):0359.
35. Chu C, Parkhurst CN, Zhang W, Zhou L, Yano H, Arifuzzaman M, et al. The ChAT-acetylcholine pathway promotes group 2 innate lymphoid cell responses and anti-helminth immunity. Sci Immunol. 2021;6(57):3218.
36. Laffont S, Blanquart E, Savignac M, Cénac C, Laverny G, Metzger D, et al. Androgen signaling negatively controls group 2 innate lymphoid cells. J Exp Med. 2017;214(6):1581–92.
37. Cephus J-Y, Stier MT, Fuseini H, Yung JA, Toki S, Bloodworth MH, et al. Testosterone attenuates group 2 innate lymphoid cell-mediated airway inflammation. Cell Rep. 2017;21(9):2487–99.
38. Kuperman DA, Huang X, Koth LL, Chang GH, Dolganov GM, Zhu Z, et al. Direct effects of interleukin-13 on epithelial cells cause airway hyperreactivity and mucus overproduction in asthma. Nat Med. 2002;8(8):885–9.
39. Kuperman DA, Huang X, Nguyenvu L, Hölscher C, Brombacher F, Erle DJ. IL-4 receptor signaling in Clara cells is required for allergen-induced mucus production. J Immunol. 2005;175(6):3746–52.
40. Ordovas-Montanes J, Dwyer DF, Nyquist SK, Buchheit KM, Vukovic M, Deb C, et al. Allergic inflammatory memory in human respiratory epithelial progenitor cells. Nature. 2018;560(7720):649–54.
41. McKnight CG, Potter C, Finkelman FD. IL-4Rα expression by airway epithelium and smooth muscle accounts for nearly all airway hyperresponsiveness in murine allergic airway disease. Mucosal Immunol. 2020;13(2):283–92.
42. Chang Y-J, Kim HY, Albacker LA, Baumgarth N, McKenzie ANJ, Smith DE, et al. Innate lymphoid cells mediate influenza-induced airway hyperreactivity independently of adaptive immunity. Nat Immunol. 2011;12(7):631–8.

43. Duerr CU, McCarthy CDA, Mindt BC, Rubio M, Meli AP, Pothlichet J, et al. Type I interferon restricts type 2 immunopathology through the regulation of group 2 innate lymphoid cells. Nat Immunol. 2016;17(1):65–75.
44. Keeler SP, Agapov EV, Hinojosa ME, Letvin AN, Wu K, Holtzman MJ. Influenza A virus infection causes chronic lung disease linked to sites of active viral RNA remnants. J Immunol. 2018;201(8): 2354–68.
45. Califano D, Furuya Y, Roberts S, Avram D, McKenzie ANJ, Metzger DW. IFN-γ increases susceptibility to influenza A infection through suppression of group II innate lymphoid cells. Mucosal Immunol. 2018;11(1):209–19.
46. Rane CK, Jackson SR, Pastore CF, Zhao G, Weiner AI, Patel NN, et al. Development of solitary chemosensory cells in the distal lung after severe influenza injury. Am J Physiol Lung Cell Mol Physiol. 2019;316(6):L1141–9.
47. Krasteva G, Canning BJ, Hartmann P, Veres TZ, Papadakis T, Mühlfeld C, et al. Cholinergic chemosensory cells in the trachea regulate breathing. Proc Natl Acad Sci USA. 2011;108(23):9478–83.
48. Tizzano M, Cristofoletti M, Sbarbati A, Finger TE. Expression of taste receptors in solitary chemosensory cells of rodent airways. BMC Pulm Med. 2011;13(11):3.
49. Bankova LG, Dwyer DF, Yoshimoto E, Ualiyeva S, McGinty JW, Raff H, et al. The cysteinyl leukotriene 3 receptor regulates expansion of IL-25-producing airway brush cells leading to type 2 inflammation. Sci Immunol. 2018;3(28):eaat9453.
50. Perniss A, Liu S, Boonen B, Keshavarz M, Ruppert A-L, Timm T, et al. Chemosensory cell-derived acetylcholine drives tracheal mucociliary clearance in response to virulence-associated formyl peptides. Immunity. 2020;52(4):683–99.
51. Molofsky AB, Van Gool F, Liang H-E, Van Dyken SJ, Nussbaum JC, Lee J, et al. Interleukin-33 and interferon-γ counter-regulate group 2 innate lymphoid cell activation during immune perturbation. Immunity. 2015;43(1):161–74.
52. Moro K, Kabata H, Tanabe M, Koga S, Takeno N, Mochizuki M, et al. Interferon and IL-27 antagonize the function of group 2 innate lymphoid cells and type 2 innate immune responses. Nat Immunol. 2016;17(1):76–86.
53. Mchedlidze T, Kindermann M, Neves AT, Voehringer D, Neurath MF, Wirtz S. IL-27 suppresses type 2 immune responses in vivo via direct effects on group 2 innate lymphoid cells. Mucosal Immunol. 2016;9(6):1384–94.
54. Huang Y, Guo L, Qiu J, Chen X, Hu-Li J, Siebenlist U, et al. IL-25-responsive, lineage-negative KLRG1(hi) cells are multipotential "inflammatory" type 2 innate lymphoid cells. Nat Immunol. 2015;16(2):161–9.
55. Silver JS, Kearley J, Copenhaver AM, Sanden C, Mori M, Yu L, et al. Inflammatory triggers associated with exacerbations of COPD orchestrate plasticity of group 2 innate lymphoid cells in the lungs. Nat Immunol. 2016;17(6):626–35.
56. Corral D, Charton A, Krauss MZ, Blanquart E, Levillain F, Lefrançais E, et al. Metabolic control of type 2 innate lymphoid cells plasticity toward protective type 1-like cells during *Mycobacterium tuberculosis* infection. bioRxiv. 2021;2:427257.
57. Simoni Y, Fehlings M, Kløverpris HN, McGovern N, Koo S-L, Loh CY, et al. Human innate lymphoid cell subsets possess tissue-type based heterogeneity in phenotype and frequency. Immunity. 2017;46(1): 148–61.
58. Ghaedi M, Shen ZY, Orangi M, Martinez-Gonzalez I, Wei L, Lu X, et al. Single-cell analysis of RORα tracer mouse lung reveals ILC progenitors and effector ILC2 subsets. J Exp Med. 2020;217(3):20182293.
59. Zeis P, Lian M, Fan X, Herman JS, Hernandez DC, Gentek R, et al. In situ maturation and tissue adaptation of type 2 innate lymphoid cell progenitors. Immunity. 2020;53(4):775–92.
60. Lim AI, Li Y, Lopez-Lastra S, Stadhouders R, Paul F, Casrouge A, et al. Systemic human ILC precursors provide a substrate for tissue ILC differentiation. Cell. 2017;168(6):1086–100.
61. Bando JK, Liang H-E, Locksley RM. Identification and distribution of developing innate lymphoid cells in the fetal mouse intestine. Nat Immunol. 2015;16 (2):153–60.
62. Kiss EA, Vonarbourg C, Kopfmann S, Hobeika E, Finke D, Esser C, et al. Natural aryl hydrocarbon receptor ligands control organogenesis of intestinal lymphoid follicles. Science. 2011;334(6062): 1561–5.
63. Spencer SP, Wilhelm C, Yang Q, Hall JA, Bouladoux N, Boyd A, et al. Adaptation of innate lymphoid cells to a micronutrient deficiency promotes type 2 barrier immunity. Science. 2014;343(6169):432–7.
64. Gury-BenAri M, Thaiss CA, Serafini N, Winter DR, Giladi A, Lara-Astiaso D, et al. The spectrum and regulatory landscape of intestinal innate lymphoid cells are shaped by the microbiome. Cell. 2016;166 (5):1231–46.
65. Schneider C, O'Leary CE, von Moltke J, Liang H-E, Ang QY, Turnbaugh PJ, et al. A metabolite-triggered tuft cell-ILC2 circuit drives small intestinal remodeling. Cell. 2018;174(2):271–84.
66. Lei W, Ren W, Ohmoto M, Urban JF Jr, Matsumoto I, Margolskee RF, et al. Activation of intestinal tuft cell-expressed Sucnr1 triggers type 2 immunity in the mouse small intestine. Proc Natl Acad Sci USA. 2018;115(21):5552–7.
67. Nadjsombati MS, McGinty JW, Lyons-Cohen MR, Jaffe JB, DiPeso L, Schneider C, et al. Detection of succinate by intestinal tuft cells triggers a type

2 innate immune circuit. Immunity. 2018;49(1):33–41.
68. Li S, Bostick JW, Ye J, Qiu J, Zhang B, Urban JF, et al. Aryl hydrocarbon receptor signaling cell intrinsically inhibits intestinal group 2 innate lymphoid cell function. Immunity. 2018;49(5):915–28.
69. Barker N, van Es JH, Kuipers J, Kujala P, van den Born M, Cozijnsen M, et al. Identification of stem cells in small intestine and colon by marker gene Lgr5. Nature. 2007;449(7165):1003–7.
70. Fort MM, Cheung J, Yen D, Li J, Zurawski SM, Lo S, et al. IL-25 induces IL-4, IL-5, and IL-13 and Th2-associated pathologies in vivo. Immunity. 2001;15(6):985–95.
71. Hurst SD, Muchamuel T, Gorman DM, Gilbert JM, Clifford T, Kwan S, et al. New IL-17 family members promote Th1 or Th2 responses in the lung: in vivo function of the novel cytokine IL-25. J Immunol. 2002;169(1):443–53.
72. Fallon PG, Jolin HE, Smith P, Emson CL, Townsend MJ, Fallon R, et al. IL-4 induces characteristic Th2 responses even in the combined absence of IL-5, IL-9, and IL-13. Immunity. 2002;17(1):7–17.
73. Moro K, Yamada T, Tanabe M, Takeuchi T, Ikawa T, Kawamoto H, et al. Innate production of T(H) 2 cytokines by adipose tissue-associated c-Kit(+) Sca-1(+) lymphoid cells. Nature. 2010;463(7280):540–4.
74. Neill DR, Wong SH, Bellosi A, Flynn RJ, Daly M, Langford TKA, et al. Nuocytes represent a new innate effector leukocyte that mediates type-2-immunity. Nature. 2010;464(7293):1367–70.
75. Price AE, Liang H-E, Sullivan BM, Reinhardt RL, Eisley CJ, Erle DJ, et al. Systemically dispersed innate IL-13-expressing cells in type 2 immunity. Proc Natl Acad Sci USA. 2010;107(25):11489–94.
76. Howitt MR, Lavoie S, Michaud M, Blum AM, Tran SV, Weinstock JV, et al. Tuft cells, taste-chemosensory cells, orchestrate parasite type 2 immunity in the gut. Science. 2016;351(6279):1329–33.
77. von Moltke J, Ji M, Liang H-E, Locksley RM. Tuft-cell-derived IL-25 regulates an intestinal ILC2-epithelial response circuit. Nature. 2016;529(7585):221–5.
78. Gerbe F, Sidot E, Smyth DJ, Ohmoto M, Matsumoto I, Dardalhon V, et al. Intestinal epithelial tuft cells initiate type 2 mucosal immunity to helminth parasites. Nature. 2016;529(7585):226–30.
79. Haber AL, Biton M, Rogel N, Herbst RH, Shekhar K, Smillie C, et al. A single-cell survey of the small intestinal epithelium. Nature. 2017;551(7680):333–9.
80. Wilen CB, Lee S, Hsieh LL, Orchard RC, Desai C, Hykes BL Jr, et al. Tropism for tuft cells determines immune promotion of norovirus pathogenesis. Science. 2018;360(6385):204–8.
81. Satoh-Takayama N, Kato T, Motomura Y, Kageyama T, Taguchi-Atarashi N, Kinoshita-Daitoku R, et al. Bacteria-induced group 2 innate lymphoid cells in the stomach provide immune protection through induction of IgA. Immunity. 2020;52(4):635–49.
82. Meyer AR, Engevik AC, Madorsky T, Belmont E, Stier MT, Norlander AE, et al. Group 2 innate lymphoid cells coordinate damage response in the stomach. Gastroenterology. 2020;159(6):2077–91.
83. Hayakawa Y, Sakitani K, Konishi M, Asfaha S, Niikura R, Tomita H, et al. Nerve growth factor promotes gastric tumorigenesis through aberrant cholinergic signaling. Cancer Cell. 2017;31(1):21–34.
84. Goto Y, Obata T, Kunisawa J, Sato S, Ivanov II, Lamichhane A, et al. Innate lymphoid cells regulate intestinal epithelial cell glycosylation. Science. 2014;345(6202):1254009.
85. Pickard JM, Maurice CF, Kinnebrew MA, Abt MC, Schenten D, Golovkina TV, et al. Rapid fucosylation of intestinal epithelium sustains host-commensal symbiosis in sickness. Nature. 2014;514(7524):638–41.
86. Lindemans CA, Calafiore M, Mertelsmann AM, O'Connor MH, Dudakov JA, Jenq RR, et al. Interleukin-22 promotes intestinal-stem-cell-mediated epithelial regeneration. Nature. 2015;528(7583):560–4.
87. Sullivan ZA, Khoury-Hanold W, Lim J, Smillie C, Biton M, Reis BS, et al. γδ T cells regulate the intestinal response to nutrient sensing. Science. 2021;371(6535):8310.
88. Seillet C, Luong K, Tellier J, Jacquelot N, Shen RD, Hickey P, et al. The neuropeptide VIP confers anticipatory mucosal immunity by regulating ILC3 activity. Nat Immunol. 2020;21(2):168–77.
89. Talbot J, Hahn P, Kroehling L, Nguyen H, Li D, Littman DR. Feeding-dependent VIP neuron-ILC3 circuit regulates the intestinal barrier. Nature. 2020;579(7800):575–80.
90. Cardoso V, Chesné J, Ribeiro H, García-Cassani B, Carvalho T, Bouchery T, et al. Neuronal regulation of type 2 innate lymphoid cells via neuromedin U. Nature. 2017;549(7671):277–81.
91. Klose CSN, Mahlakõiv T, Moeller JB, Rankin LC, Flamar A-L, Kabata H, et al. The neuropeptide neuromedin U stimulates innate lymphoid cells and type 2 inflammation. Nature. 2017;549(7671):282–6.
92. Bando JK, Gilfillan S, Di Luccia B, Fachi JL, Sécca C, Cella M, et al. ILC2s are the predominant source of intestinal ILC-derived IL-10. J Exp Med. 2020;217(2):e20191520.
93. Xu H, Ding J, Porter CBM, Wallrapp A, Tabaka M, Ma S, et al. Transcriptional atlas of intestinal immune cells reveals that neuropeptide α-CGRP modulates group 2 innate lymphoid cell responses. Immunity. 2019;51(4):696–708.
94. Jarret A, Jackson R, Duizer C, Healy ME, Zhao J, Rone JM, et al. Enteric nervous system-derived IL-18 orchestrates mucosal barrier immunity. Cell. 2020;180(1):50–63.

95. Nagao K, Kobayashi T, Moro K, Ohyama M, Adachi T, Kitashima DY, et al. Stress-induced production of chemokines by hair follicles regulates the trafficking of dendritic cells in skin. Nat Immunol. 2012;13(8):744–52.
96. Yang J, Hu S, Zhao L, Kaplan DH, Perdew GH, Xiong N. Selective programming of CCR10(+) innate lymphoid cells in skin-draining lymph nodes for cutaneous homeostatic regulation. Nat Immunol. 2016;17(1):48–56.
97. Kobayashi T, Voisin B, Kim DY, Kennedy EA, Jo J-H, Shih H-Y, et al. Homeostatic control of sebaceous glands by innate lymphoid cells regulates commensal bacteria equilibrium. Cell. 2019;176(5): 982–97.
98. Adachi T, Kobayashi T, Sugihara E, Yamada T, Ikuta K, Pittaluga S, et al. Hair follicle-derived IL-7 and IL-15 mediate skin-resident memory T cell homeostasis and lymphoma. Nat Med. 2015;21(11): 1272–9.
99. Roediger B, Kyle R, Yip KH, Sumaria N, Guy TV, Kim BS, et al. Cutaneous immunosurveillance and regulation of inflammation by group 2 innate lymphoid cells. Nat Immunol. 2013;14(6):564–73.
100. Salimi M, Barlow JL, Saunders SP, Xue L, Gutowska-Owsiak D, Wang X, et al. A role for IL-25 and IL-33-driven type-2 innate lymphoid cells in atopic dermatitis. J Exp Med. 2013;210(13): 2939–50.
101. Kim BS, Siracusa MC, Saenz SA, Noti M, Monticelli LA, Sonnenberg GF, et al. TSLP elicits IL-33-independent innate lymphoid cell responses to promote skin inflammation. Sci Transl Med. 2013;5(170): 170ra16.
102. Bernink JH, Ohne Y, Teunissen MBM, Wang J, Wu J, Krabbendam L, et al. c-Kit-positive ILC2s exhibit an ILC3-like signature that may contribute to IL-17-mediated pathologies. Nat Immunol. 2019;20(8):992–1003.
103. Bielecki P, Riesenfeld SJ, Hütter J-C, Torlai Triglia E, Kowalczyk MS, Ricardo-Gonzalez RR, et al. Skin-resident innate lymphoid cells converge on a pathogenic effector state. Nature. 2021;592 (7852):128–32.
104. Rak GD, Osborne LC, Siracusa MC, Kim BS, Wang K, Bayat A, et al. IL-33-dependent group 2 innate lymphoid cells promote cutaneous wound healing. J Investig Dermatol. 2016;136(2):487–96.
105. Leyva-Castillo J-M, Galand C, Kam C, Burton O, Gurish M, Musser MA, et al. Mechanical skin injury promotes food anaphylaxis by driving intestinal mast cell expansion. Immunity. 2019;50(5):1262–75.
106. Wilson SR, Thé L, Batia LM, Beattie K, Katibah GE, McClain SP, et al. The epithelial cell-derived atopic dermatitis cytokine TSLP activates neurons to induce itch. Cell. 2013;155(2):285–95.
107. Derecki NC, Cardani AN, Yang CH, Quinnies KM, Crihfield A, Lynch KR, et al. Regulation of learning and memory by meningeal immunity: a key role for IL-4. J Exp Med. 2010;207(5):1067–80.
108. Fung ITH, Sankar P, Zhang Y, Robison LS, Zhao X, D'Souza SS, et al. Activation of group 2 innate lymphoid cells alleviates aging-associated cognitive decline. J Exp Med. 2020;217(4):e20190915.
109. Russi AE, Ebel ME, Yang Y, Brown MA. Male-specific IL-33 expression regulates sex-dimorphic EAE susceptibility. Proc Natl Acad Sci USA. 2018;115(7):E1520–9.
110. Gadani SP, Smirnov I, Smith AT, Overall CC, Kipnis J. Characterization of meningeal type 2 innate lymphocytes and their response to CNS injury. J Exp Med. 2017;214(2):285–96.
111. Baban B, Braun M, Khodadadi H, Ward A, Alverson K, Malik A, et al. AMPK induces regulatory innate lymphoid cells after traumatic brain injury. JCI Insight. 2021;6(1):e126766.
112. Wu D, Molofsky AB, Liang H-E, Ricardo-Gonzalez RR, Jouihan HA, Bando JK, et al. Eosinophils sustain adipose alternatively activated macrophages associated with glucose homeostasis. Science. 2011;332(6026):243–7.
113. Molofsky AB, Nussbaum JC, Liang H-E, Van Dyken SJ, Cheng LE, Mohapatra A, et al. Innate lymphoid type 2 cells sustain visceral adipose tissue eosinophils and alternatively activated macrophages. J Exp Med. 2013;210(3):535–49.
114. Brestoff JR, Kim BS, Saenz SA, Stine RR, Monticelli LA, Sonnenberg GF, et al. Group 2 innate lymphoid cells promote beiging of white adipose tissue and limit obesity. Nature. 2015;519(7542):242–6.
115. Mahlakõiv T, Flamar A-L, Johnston LK, Moriyama S, Putzel GG, Bryce PJ, et al. Stromal cells maintain immune cell homeostasis in adipose tissue via production of interleukin-33. Sci Immunol. 2019;4(35):eaax0416.
116. Rana BMJ, Jou E, Barlow JL, Rodriguez-Rodriguez-N, Walker JA, Knox C, et al. A stromal cell niche sustains ILC2-mediated type-2 conditioning in adipose tissue. J Exp Med. 2019;216(9):1999–2009.
117. Spallanzani RG, Zemmour D, Xiao T, Jayewickreme T, Li C, Bryce PJ, et al. Distinct immunocyte-promoting and adipocyte-generating stromal components coordinate adipose tissue immune and metabolic tenors. Sci Immunol. 2019;4 (35):eaaw3658.
118. Zhao X-Y, Zhou L, Chen Z, Ji Y, Peng X, Qi L, et al. The obesity-induced adipokine sST2 exacerbates adipose Treg and ILC2 depletion and promotes insulin resistance. Sci Adv. 2020;6(20):eaay6191.
119. Kastenschmidt JM, Coulis G, Farahat PK, Pham P, Rios R, Cristal TT, et al. A stromal progenitor and ILC2 niche promotes muscle eosinophilia and fibrosis-associated gene expression. Cell Rep. 2021;35(2):108997.
120. Knudsen NH, Stanya KJ, Hyde AL, Chalom MM, Alexander RK, Liou Y-H, et al. Interleukin-13 drives

metabolic conditioning of muscle to endurance exercise. Science. 2020;368(6490):eaat3987.
121. Gieseck RL 3rd, Ramalingam TR, Hart KM, Vannella KM, Cantu DA, Lu W-Y, et al. Interleukin-13 activates distinct cellular pathways leading to ductular reaction, steatosis, and fibrosis. Immunity. 2016;45(1):145–58.
122. Koga S, Hozumi K, Hirano K-I, Yazawa M, Terooatea T, Minoda A, et al. Peripheral PDGFRα+ gp38+ mesenchymal cells support the differentiation of fetal liver-derived ILC2. J Exp Med. 2018;215(6): 1609–26.
123. Guo L, Huang Y, Chen X, Hu-Li J, Urban JF Jr, Paul WE. Innate immunological function of TH2 cells in vivo. Nat Immunol. 2015;16(10):1051–9.
124. Van Dyken SJ, Nussbaum JC, Lee J, Molofsky AB, Liang H-E, Pollack JL, et al. A tissue checkpoint regulates type 2 immunity. Nat Immunol. 2016;17 (12):1381–7.
125. Shih H-Y, Sciumè G, Mikami Y, Guo L, Sun H-W, Brooks SR, et al. Developmental acquisition of regulomes underlies innate lymphoid cell functionality. Cell. 2016;165(5):1120–33.
126. Wijeyesinghe S, Beura LK, Pierson MJ, Stolley JM, Adam OA, Ruscher R, et al. Expansible residence decentralizes immune homeostasis. Nature. 2021;592 (7854):457–62.

ILCs and Allergy

6

Hiroki Kabata, Yasutaka Motomura, Tsuyoshi Kiniwa, Tetsuro Kobayashi, and Kazuyo Moro

Abstract

The recent discovery of new innate lymphoid cells (ILCs) has revolutionized the field of allergies. Since most allergic diseases induce a type 2 immune response, Th2 cells, which produce IL-4, IL-5, and IL-13 in an antigen-dependent manner, in addition to basophils and mast cells which are activated by antigen-specific IgE, are thought to play a major role in the pathogenesis. However, since group 2 innate lymphoid cells (ILC2s) produce type 2 cytokines (i.e., IL-2, IL-4, IL-5, IL-6, IL-9, IL-13, GM-CSF, and amphiregulin) in response to various cytokines, including IL-33 in the surrounding environment, the possibility has emerged that there are two types of allergies: allergies induced in an antigen-dependent manner by Th2 cells and allergies induced in an antigen-independent manner by ILC2s. In order to make an impact on the increasing incidence of allergic diseases in the world, it is essential to research and develop new treatments that focus not only on Th2 cells but also on ILC2s. In this chapter, the role of ILCs in allergic diseases, which has rapidly changed with the discovery of ILCs, is discussed, focusing mainly on ILC2s.

Keywords

Asthma · Atopic dermatitis · Contact hypersensitivity · Allergic rhinitis · Chronic rhinosinusitis · Food allergy · Allergic conjunctivitis

H. Kabata
Division of Pulmonary Medicine, Department of Medicine, Keio University School of Medicine, Tokyo, Japan

Laboratory for Innate Immune Systems, RIKEN Center for Integrative Medical Sciences (IMS), Kanagawa, Japan

Y. Motomura
Laboratory for Innate Immune Systems, RIKEN Center for Integrative Medical Sciences (IMS), Kanagawa, Japan

Laboratory for Innate Immune Systems, Department of Microbiology and Immunology, Graduate School of Medicine, Osaka University, Osaka, Japan

Laboratory for Innate Immune Systems, Immunology Frontier Research Center (IFReC), Osaka University, Osaka, Japan

T. Kiniwa · T. Kobayashi
Laboratory for Innate Immune Systems, Department of Microbiology and Immunology, Graduate School of Medicine, Osaka University, Osaka, Japan

K. Moro (✉)
Laboratory for Innate Immune Systems, RIKEN Center for Integrative Medical Sciences (IMS), Kanagawa, Japan

Laboratory for Innate Immune Systems, Department of Microbiology and Immunology, Graduate School of Medicine, Osaka University, Osaka, Japan

Laboratory for Innate Immune Systems, Immunology Frontier Research Center (IFReC), Osaka University, Osaka, Japan

Laboratory for Innate Immune Systems, Graduate School of Frontier Biosciences, Osaka University, Osaka, Japan
e-mail: moro@ilc.med.osaka-u.ac.jp

© The Author(s), under exclusive license to Springer Nature Singapore Pte Ltd. 2022

X.-H. Sun (ed.), *Innate Lymphoid Cells*, Advances in Experimental Medicine and Biology 1365,
https://doi.org/10.1007/978-981-16-8387-9_6

6.1 Introduction

The word "allergy" is coined from the Greek words "allos" (changed) and "ergon" (reaction), meaning that the immune response which evolved to prevent disease turns into a harmful reaction. It also means that since the distant past, people have been aware that an allergy is a strange reaction different from the common immune response, but they had no way of knowing how it had changed. One of the most beneficial roles of the type 2 immune response in organisms is in eliminating helminths. Most helminths are much larger than bacteria and viruses, and are beyond the range of phagocytosis. Therefore, our body has evolved its own defense mechanisms against helminths. IL-5 induces eosinophils to weaken the parasite, and IL-13 induces mucus production from goblet cells to flush helminths out of the body. In this case, since it is difficult for dendritic cells (DCs) to phagocytose a target that is too large, Th2 cells, which are activated based on antigen presentation by DCs, are unable to exert their full power. On the other hand, the larger the target, the more the tissue that will be physically destroyed, and the necrotic epithelial cells will release large amounts of IL-33. In addition to destroying tissue, proteases, which are a component of helminth shells, contribute significantly to IL-33 production by directly inducing cell death in epithelial cells. ILC2s are cells that highly express ST2/IL-1RL1 (the IL-33 receptor) and are located near IL-33-producing cells. ILC2s can quickly respond to IL-33 and produce type 2 cytokines such as IL-5 and IL-13 to flush helminths out of the mouth by sputum or out of the anus by diarrhea in cases of pulmonary or intestinal infection, respectively. This mechanism is primitive compared to the antigen-specific response of acquired immunity, but it is indispensable for humans who coexist with helminths in an uneasy balance. However, our living environment has changed drastically in the last half century, and parasitic infections have decreased dramatically in developed countries.

In countries that are no longer infected with parasites, strange reactions are beginning to occur, just like "allos" plus "ergon." According to the World Allergy Organization (WAO), 30–40% of the world's population suffers from one or more allergy symptoms, making it a serious health problem. The reason for the increase in allergic diseases is not only the decrease in parasitic infections, but also that improved sanitation decreases bacterial and viral infections, which leads to a decrease in the type 1 immune response used to suppress the type 2 immune response. However, if we focus on the fact that parasite infections have decreased, then type 2 immune responses, which are the main cause of allergies, should have also decreased along with the decrease in infections. What needs to be considered here is the mechanism common to both parasitic infections and allergies, whereby ILC2s recognize IL-33 and produces type 2 cytokines. Unlike Th2 cells, ILC2s cannot specifically recognize antigens or distinguish whether IL-33 is produced by a parasite or an allergen that is harmless to the body. As well as parasites, various allergenic antigens such as dust mites, pollen, fungi, and fruits contain proteases which directly induce necrosis of the epithelium, and stimulate it to release IL-33. The days when ILC2s maintain an exquisite coexistence with parasites are long gone, but ILC2s deep inside the body may still be unaware of the changing times and are needlessly initiating type 2 immune responses, relying on IL-33 as an alarm signal (Fig. 6.1).

6.2 Innate Lymphoid Cells in the Lungs

Since the lungs are vital organs for the gas exchange of oxygen and carbon dioxide, they are constantly exposed to various allergens and viruses through breathing. The airway epithelial cells are covered with a layer of mucus to capture allergens and viruses, and ciliated cells eliminate them by ciliary movement. However, when the allergens and viruses are not eliminated and reach airway epithelial cells, cell-derived cytokines are rapidly released and directly activate ILCs and trigger innate immune responses. Among ILCs in the lungs, ILC2s in particular have been

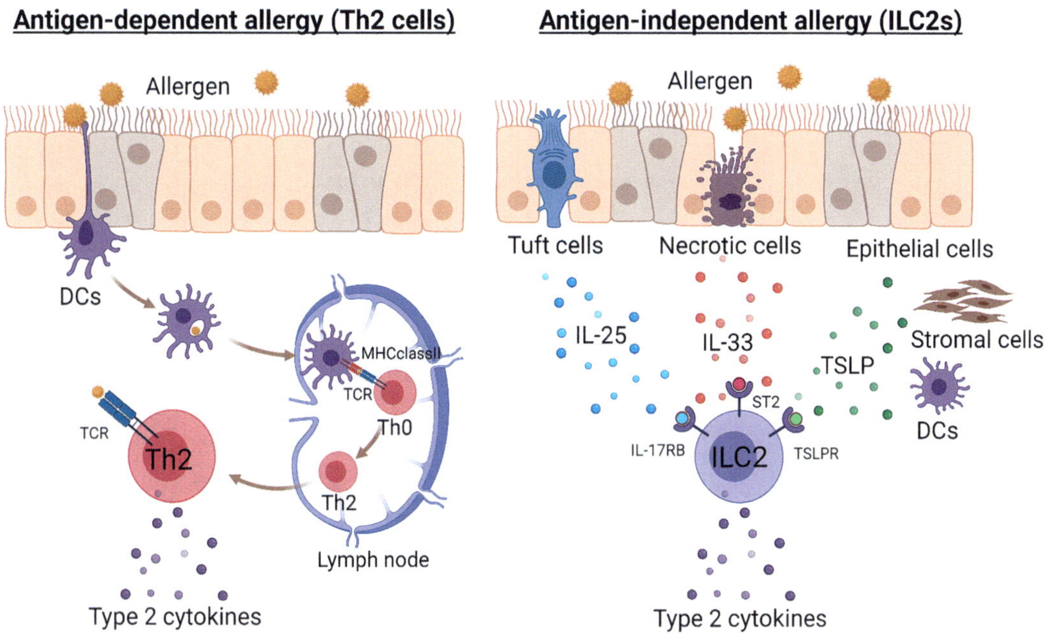

Fig. 6.1 Mechanism of allergy induced by Th2 cells and ILC2s. In antigen-dependent allergy, DCs that have phagocytosed the antigen migrate to lymph nodes and present the antigen to Th0 cells via MHC class II and TCR. Th0 cells differentiate into Th2 cells over a period of about 5 days, and mature Th2 cells migrate to peripheral tissues and produce type 2 cytokines. In contrast, in antigen-independent allergy, the protease activity of the allergen directly causes necrosis in epithelial cells. Since ILC2s are tissue-resident cells, they immediately respond to IL-33 and produce type 2 cytokines. IL-25 secreted by tuft cells and TSLP produced by epithelial cells, stromal cells, and DCs are also involved in the activation of ILC2s

reported to be involved in allergic diseases by inducing type 2 inflammation through the production of IL-5 and IL-13.

The proportion of lung ILCs is small, accounting for only about 0.4–1% of lung cells in naïve mice [1]. Among ILCs, ILC2s are the most abundant (>60%), followed by ILC1s (<20%) and ILC3s (<20%) [2]. On the other hand, the frequency of ILCs in the human lung is only about 0–0.1% of $CD45^+$ cells, and conventional NK (cNK) cells are the major ILCs, followed by ILC1s and ILC3s, while ILC2s are rare [2, 3]. However, the proportion of ILCs has been reported to change dynamically due to plasticity, and the presence of inflammatory cytokines such as IL-1β in the lung due to smoking and obesity increases the frequency of ILC1s and ILC3s [4, 5]. Thus, although human lung ILC2s are relatively rare cell populations, they express high levels of *IL13* as well as epithelial cell-derived cytokine receptors *IL1RL1* and *IL17RB*, compared to blood and tonsil ILC2s [6], and are suggested to have a significant impact on the induction of type 2 inflammation.

6.2.1 Mechanism of IL-33-Induced Asthma

Various allergens and pathogens, such as helminths, house dust mites [7], *Alternaria alternata* [8], papain [9], chitin [10], α-GalCer [11], rhinovirus [12], and influenza virus [13], induce epithelial cell-derived cytokines such as IL-33 via cytotoxicity and ATP release [14] (Fig. 6.2). The mechanism by which IL-33 is released is ingenious considering that allergen-derived proteases not only damage epithelial cells to release IL-33, but also play a role in cleaving full-length IL-33 into an activated form

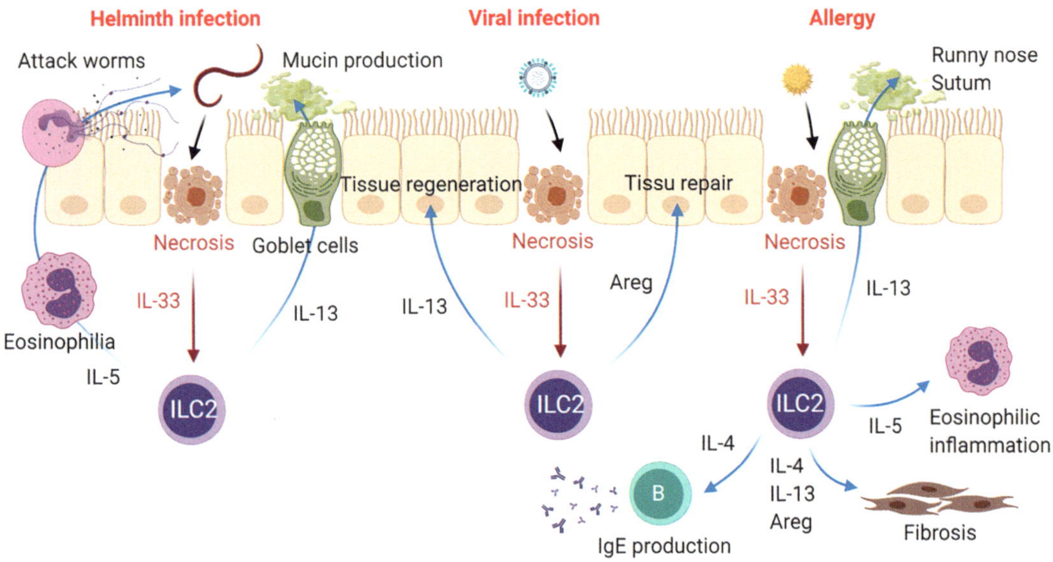

Fig. 6.2 IL-33 induces activation of ILC2s in diverse pathological conditions. In helminth infection, IL-33 induces IL-5 and IL-13 production from ILC2s, which work to expel helminths from the body via worm attack by eosinophils and mucus production by goblet cells, respectively. In viral infections, ILC2s activated by IL-33 play a protective role by producing IL-13 and amphiregulin. IL-13 acts to promote epithelial cell regeneration and amphiregulin promotes tissue repair. In allergy, IL-33 activates ILC2s to produce a variety of cytokines, including IL-4, IL-5, IL-13, and amphiregulin. IL-5 induces eosinophilic inflammation, and IL-13 promotes mucus production by goblet cells, resulting in runny nose and sputum. It has been suggested that IL-4 induces IgE production from B cells, and IL-4, IL-13, and amphiregulin may be involved in fibrosis

[15]. IL-33 stimulates NF-κB and MAPK signaling pathways in ILC2s, and activates the transcription factor GATA3 via p38 MAPK, which strongly induces cell proliferation and production of type 2 cytokines such as IL-5 and IL-13 [16, 17]. IL-5 promotes eosinophil activation, migration, and survival, and IL-13 promotes goblet cell hyperplasia and mucus hypersecretion in the lungs [18]. This IL-33/ILC2-mediated immune response is referred to as "nonallergic eosinophilic inflammation" since it occurs in an antigen-independent manner without acquired immunity [19].

Asthma is a chronic respiratory disease characterized by chronic airway inflammation and reversible airway obstruction, and is considered to be a heterogeneous syndrome consisting of a variety of pathologies. Most people with asthma, except for those with neutrophilic asthma, have type 2 inflammation, and Th2 cells have been considered to play a vital role in inflammation. However, accumulating evidence suggests that IL-33 and ILC2s are also involved in the pathophysiology of asthma. Genome-wide association studies (GWAS) have reported that single-nucleotide polymorphisms (SNPs) in *IL33* and its receptor, *IL1RL1*, are associated with asthma [20], and IL-33 expression is increased in the airways and bronchoalveolar lavage (BAL) fluid in patients with asthma, and is correlated with disease severity and lung function [21–23]. Similarly, the number of ILC2s is increased in the sputum and BAL of patients with asthma, especially in those with severe asthma [22, 24]. However, there are no definitive results on the number of ILC2s in the peripheral blood of patients with asthma [24, 25], which suggests that it is difficult to use the blood to monitor the status of ILC2s in the lungs. Furthermore, house dust mites and rhinoviruses induced IL-33 and increased the number of ILC2s in sputum and BAL, but decreased the number of ILC2s in the blood [26–28]. These findings suggest that even though ILC2s are recognized as tissue-resident cells in mice, a portion of ILC2s are transferred from the peripheral blood to the airways in humans [29, 30]. Until the discovery of ILC2, asthma was thought to be an antigen-specific disease, but it is now clear that IL-33/ILC2-mediated immune responses play an important role, especially in the pathogenesis of virus-induced asthma exacerbation, uncontrolled asthma, and severe asthma [24, 27, 31].

6.2.2 Cytokines and Lipid Mediators that Regulate the Function of ILC2s

IL-33 rapidly and potently activates ILC2s and induces type 2 inflammation, but other epithelial cell-derived cytokines also activate ILC2s in the lungs. For example, IL-25 is an epithelial cell-derived cytokine that activates murine and human ILC2s in combination with IL-2 in vitro [32, 33], and tuft cells in the intestinal tract and bronchi have been reported as IL-25-producing cells [34]. However, IL-25 is less potent than IL-33 in inducing type 2 inflammation and airway contraction in mice [35]. Thymic stromal lymphopoietin (TSLP) is also an epithelial cell-derived cytokine that has modest effects on the activation of ILC2s, but it is rather important in altering the properties of ILC2s; TSLP induces corticosteroid resistance via phosphorylation of STAT5 and expression of Bcl-xL in ILC2s, leading to corticosteroid-resistant type 2 inflammation in mice [36] (Fig. 6.3). Indeed, TSLP expression is increased in the airways of patients with severe asthma [37], and TSLP concentration in BAL correlated with corticosteroid resistance in ILC2s in humans [31]. Although IL-33, IL-25, and TSLP are often described in parallel in reviews as epithelial cell-derived cytokines, more research on the cytokine production mechanism for IL-25 and TSLP is required to understand the regulation of ILC2 by these three cytokines.

ILC2s do not express antigen-specific receptors such as Th2 cells, but ILC2s express

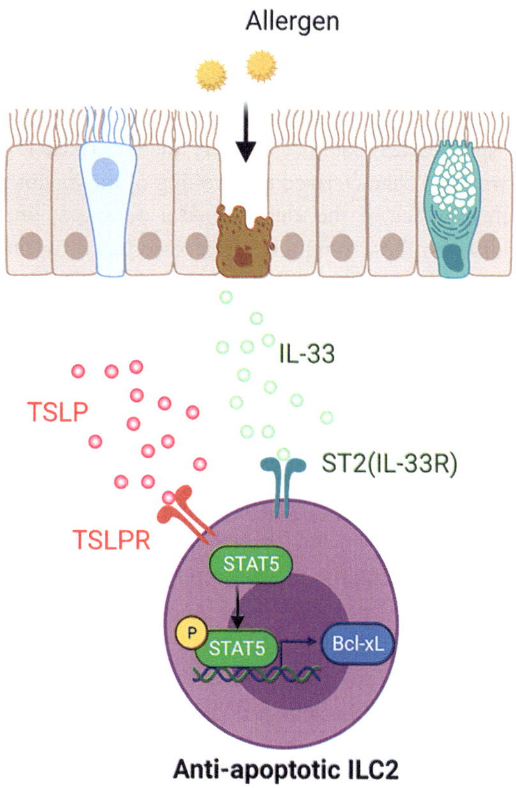

Fig. 6.3 IL-33 and TSLP cooperatively induce steroid resistance in ILC2. Under steady-state and IL-33 stimulation, ILC2s are sensitive to steroids, but under simultaneous stimulation of IL-33 and TSLP, they become resistant to steroid. TSLP induces Bcl-xL expression by phosphorylation of STAT5, resulting in steroid resistance in ILC2s

various receptors for cytokines and lipid mediators, which are influenced by the surrounding environment. For example, proteases in house dust mites and papain induce basophils to produce IL-4, which increases the responsiveness of ILC2s to IL-33 in mice [38]. TNF-like ligand 1A (TL1A, TNFSF15), glucocorticoid-induced TNF receptor ligand (GITRL, TNFSF18), leukotriene, and prostaglandin D2 (PGD2) have been reported to activate ILC2s synergistically with IL-33 in mice and humans [25, 39–47]. Cysteinyl leukotrienes and PGD2 bind to G protein-coupled receptors, which increase the intracellular Ca^{2+} concentrations of ILC2s and promote nuclear transfer of NFAT [48]. Since this signaling pathway is different from that induced by IL-33 or IL-25, lipid mediators can activate ILC2s synergistically with IL-33 and induce IL-4 production from ILC2s [43]. Some patients with asthma have aspirin-exacerbated respiratory disease (AERD), which is characterized by swelling of the mucous membranes of the sinuses, nasal passages, and airways; these symptoms are exacerbated by the ingestion of cyclooxygenase-1 (COX-1) inhibitors, including aspirin and other nonsteroidal anti-inflammatory drugs (NSAIDs) [49]. In patients with AERD, the administration of COX-1 inhibitors increases the production of lipid mediators, such as leukotriene and PGD2, and induces the accumulation of ILC2s in the nasal mucosa, whereas the number of ILC2s is decreased in the blood of patients with AERD [50]. These findings suggest that ILC2s contribute to the pathogenesis of AERD through the production of lipid mediators.

In contrast, some cytokines and lipid mediators inhibit the activation of murine and human ILC2s and suppress type 2 inflammation. Interferons (IFNs) and IL-27 produced by NK cells, T cells, plasmacytoid dendritic cells, and interstitial macrophages suppress ILC2 activation in a STAT1-dependent manner [30, 51–54]. These cytokines are involved in the resolution of ILC2-mediated type 2 inflammation, and without these cytokines, type 2 inflammation would persist for a long time. The bronchial epithelium in patients with asthma has been reported to produce less IFN-β when infected with rhinoviruses [55]. Therefore, replenishment of IFN-β and other suppressive cytokines may be a treatment option for virus-induced exacerbation of asthma [56, 57]. In addition, other lipid mediators such as PGE2, PGI2, and lipoxin A4 have been reported to inhibit the production of IL-5 and IL-13 from ILC2s via cAMP activation [58–61].

Finally, since ILC2s have plasticity, the type of cytokines they produce changes dynamically depending on the surrounding cytokine environment. IL-1β, IL-18, and IL-12 induce IFN-γ production, and retinoic acid induces IL-10 production in murine and human ILC2s [4, 62–64]. Viral infection and smoking provoke inflammatory cytokines, including IL-12, and the frequency of IFN-γ-producing ILC2s is increased in patients with chronic obstructive pulmonary disease [4]. Thus, various regulatory mechanisms exist for ILC2-mediated type 2 inflammation, and a disruption of these regulatory mechanisms results in the exacerbation and progression of allergen-independent type 2 inflammation in vivo.

6.2.3 Neuroimmune Interaction in ILC2s

Neuronal and immune systems have close bidirectional interactions. Neuron-derived neuropeptides and neurotransmitters regulate immune cell functions, whereas inflammatory mediators produced by immune cells enhance neuronal activation [65]. In recent years, several neuropeptides have been shown to directly affect ILC2s. Pulmonary neuroendocrine cells (PNECs) are rare airway epithelial cells that sense various stimulations, including oxygen, stretch, and chemicals, and release neuropeptides, such as calcitonin gene-related peptide (CGRP). ILC2s have been reported to express CGRP receptors and co-localize with PNECs in the lungs of mice [66]. However, the effect of CGRP on ILC2s is complicated; CGRP has induced IL-5 production in murine ILC2s, but constrained IL-13 production and proliferation [67]. ILC2s have also expressed neuromedin U receptor 1 (NmUR1), while neuromedin U (NMU) had a potent effect on inducing the proliferation and cytokine production of IL-5, IL-9, IL-13, and amphiregulin from ILC2s in mice [68–70]. In addition, pulmonary sensory nerve-derived vasoactive intestinal peptide (VIP) activated ILC2s and CD4 T cells to enhance type 2 inflammation in ovalbumin (OVA)- and house dust mite (HDM)-induced asthma mouse models [71]. Therefore, genetic ablation of Nav1.8-positive sensory nerves reduces immune cell infiltration and airway hyperreactivity in OVA-induced asthma mouse models [71].

In contrast, cholinergic neurons produce acetylcholine and α7nAChR, an acetylcholine receptor, which suppresses type 2 cytokine production from murine ILC2s [72]. Furthermore, sympathetic nerves release noradrenalin while β-adrenoceptor stimulations suppress ILC2 proliferation and ILC2-mediated inflammation in mice [73]. These data suggest that the neural system, in addition to the interactions with epithelial cells and other immune cells, may regulate ILC2-induced type 2 inflammation. At this stage, it is not clear how these neural factors are involved in the pathogenesis of asthma, but they may be important factors in understanding the full picture of ILC2-dependent asthma.

6.2.4 The Role of ILC2s in Trained Immunity and Acquired Immunity

ILC2s have been implicated not only in innate immune responses, but also in trained and acquired immunity. The increased number of ILC2s in the lungs and mediastinal lymph nodes of mice treated with IL-33 or papain was maintained even after 4 weeks, indicating that ILC2s are long-lived cells. Furthermore, ILC2s responded more strongly to restimulation and induced more severe type 2 inflammation, suggesting that ILC2s have a memory mechanism [74]. Experiments with $Rag1^{-/-}$ mice treated with *Alternaria* allergen extract have shown that $ICOS^{+}ST2^{+}$ ILC2 generates memory in asthma through epigenetic changes [75].

ILC2s also enhance acquired immune responses. This effect is limited to localized reactions in the lungs and is dispensable in systemic reactions. ILC2s enhance Th2 cell responses but has no effect on Th1 or Th17 cells [76]. Murine ILC2s express MHC class 2, OX40L, CD80, and CD86 and can directly activate $CD4^{+}$ T cells, which may promote the induction of acquired immunity [77–80]. ILC2-derived IL-13 promotes the migration of dendritic cells in the lung to the lymph node, where they promote naïve T cells to differentiate into Th2 cells during allergic lung inflammation [81]. ILC2-derived IL-13 promotes the expression of IL-33 in airway epithelial cells, creating a positive feedback loop and disrupting tight junctions between the cells [82], thereby increasing the penetration of allergens into the epithelium, which may allow for higher penetration of allergens across the epithelium. In summary, ILC2s, which induce innate immune responses, may enhance acquired immune responses and exacerbate asthma.

6.2.5 Treatment of ILC2-Mediated Type 2 Inflammation in the Lungs

Although new asthma therapies targeting ILC2s are expected, few drugs have proven inhibitory effects on ILC2s in patients with asthma. Corticosteroids and leukotriene receptor antagonists, generally used to treat asthma, appear to be somewhat effective against ILC2-mediated type 2 inflammation. Indeed, inhaled corticosteroid treatment (i.e., budesonide) decreased the number of ILC2s in peripheral blood and suppressed type 2 cytokine production from ILC2s [83]. However, other studies have demonstrated that ILC2s in the BAL are resistant to corticosteroids, and continue to produce type 2 cytokines even in the presence of dexamethasone [31]. Importantly, TSLP induces corticosteroid resistance of ILC2s and promotes innate type 2 inflammation, and also induces acquired immunity via DCs and $CD4^{+}$ T cells in asthma model mice [84]. Suppression of TSLP is important for resolving ILC2-induced steroid resistance, while tezepelumab, an anti-TSLP human monoclonal antibody, has been reported to effectively suppress a wide range of inflammation [85–87]. IL-33 is the most potent cytokine that activates ILC2s, and antibodies against IL-33 or its receptor, ST2/IL-1RL1, are currently under development [88]. In addition, among the biologics currently in use, dupilumab, an anti-IL-4/13R antibody, has been suggested to suppress ILC2s in humans [89]. Since ILC2s have a major impact on type 2 inflammation in the lungs, the development of drugs targeting ILC2-mediated inflammation is expected to progress in the future (Fig. 6.4).

	Omalizumab	Mepolizumab	Benralizumab	Dupilumab
Trade names	Xolair	Nucala	fasenra	Dupixent
Targets	Anti-IgE	Anti-IL-5	Anti-IL-5Ra	Anti-IL-4/IL-13R
Indications	• Asthma • Spontaneous urticaria • Nasal polyps • Allergic rhinitis (seasonal / perennial)	• Asthma • EGPA • HES	• Asthma	• Asthma • Atopic dermatitis • CRS with nasal polyps

Fig. 6.4 Biologics targeting type 2 immunity. Approved indications vary by country. *EGPA* eosinophilic granulomatosis with polyangiitis, *HES* hypereosinophilic syndrome

6.3 Innate Lymphoid Cells in the Skin

The skin is the largest organ and provides a barrier to protect the body from bacterial and viral invasion, UV, and external stress damage. Immune cells are strategically placed in the skin to create efficient protective immunity through cellular communication. Similarly to the respiratory and intestinal tracts, the skin also harbors tissue-resident ILCs. Skin ILCs not only have essential functions in maintaining tissue homeostasis, but also play an important role in the pathogenesis of skin diseases, including allergies. The skin consists of three layers: epidermis composed of four different keratinocytes, collagen-rich dermis, and adipocyte-rich subcutaneous tissue. ILCs, which account for 5% of immune cells in murine skin, are present in all three skin layers, but in different proportions. Transcriptome analysis revealed layer-specific heterogeneity of skin ILC subsets (Fig. 6.5) [90]. The subcutaneous tissue is rich in typical $GATA3^+$ ILC2s, which express Sca-1 and IL-33R and produce IL-5 and IL-13, while the epidermal and dermal tissues contain ILCs with mixed phenotypes of ILC3s and ILC2s. Epidermal and dermal ILCs express ICOS and CCR6, but do not express Sca-1 and IL-33R. They express both GATA3 and RORγt and produce IL-13, IL-22, TNF, and lymphotoxins [90]. Skin ILCs exhibit distinct tissue-resident patterns, with ILC1s continuously migrating between the circulation and peripheral lymph nodes in a CD62L- and CCR7-dependent manner, while ILC2s and ILC3s remain in the tissue [91]. With regard to allergic diseases of the skin, ILC2s are thought to play an important role in atopic dermatitis, while ILC1s are thought to cause contact dermatitis together with NK cells.

6.3.1 ILC2s in Atopic Dermatitis

Atopic dermatitis is a chronic inflammatory skin disease in which the patient's quality of life is severely impaired due to endless itching. Since atopic dermatitis is associated with high levels of antigen-specific IgE in the serum and high T cell infiltration in the skin lesions, acquired immunity is thought to be important in the pathogenesis of this disease. In recent years, however, attention has begun to focus on the important role of innate immunity in initiating and promoting atopic inflammation, considering that patients often do not exhibit antigen-specific reactions that are seen in antigen-restricted allergic disorders such as pollen and food allergies. Repeated scratching is known to aggravate symptoms of atopic dermatitis. Tissue damage caused by scratching stimulates the production of cytokines and chemokines in the epidermis, which in turn recruits and activates immune cells, thereby aggravating the condition. From this perspective, alarmins such as IL-33, TSLP, and IL-25 produced by epithelium upon tissue damage that directly activate ILC2s may be deeply involved

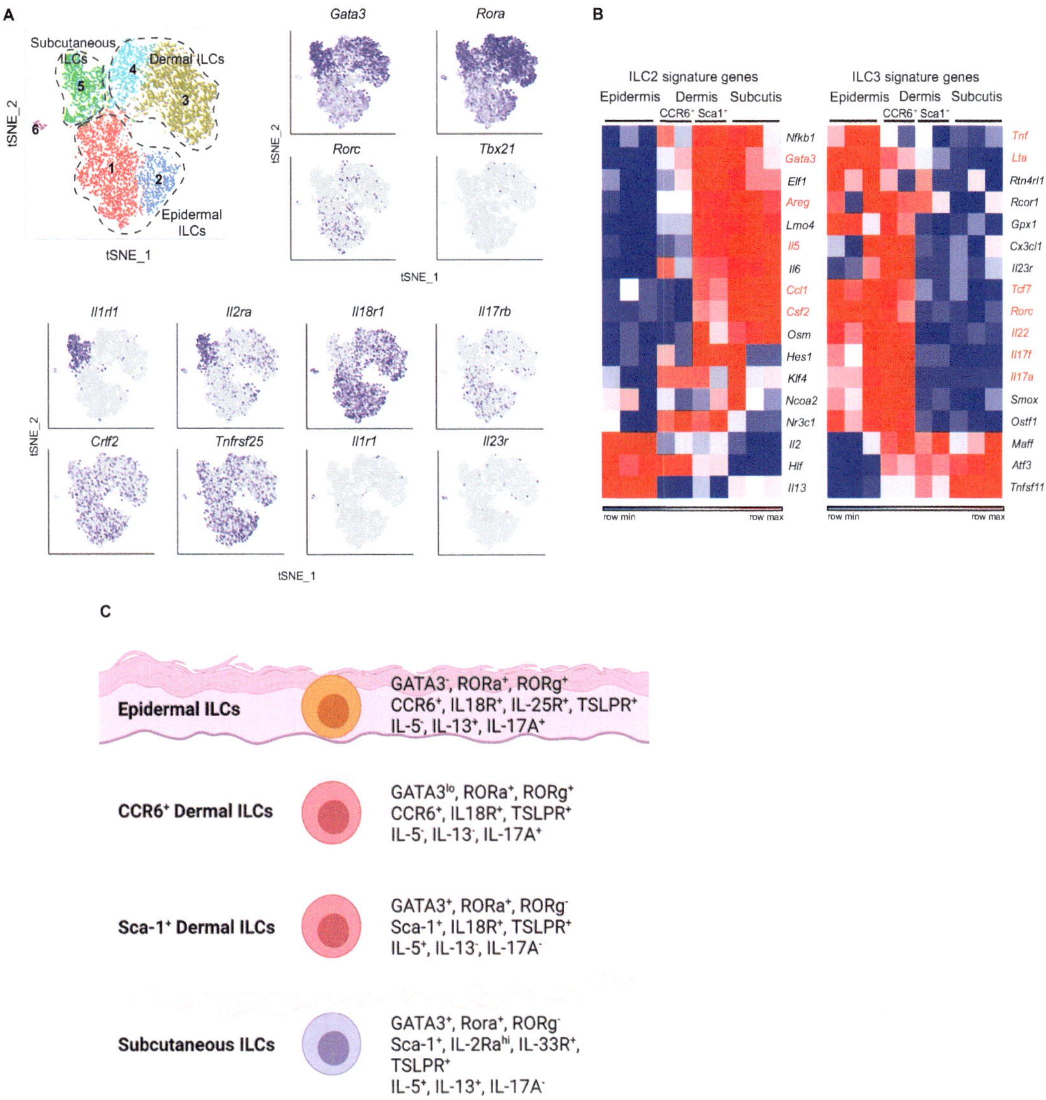

Fig. 6.5 Skin ILCs with distinct transcriptome landscapes. (**a**) Single RNA-seq analysis revealed transcriptome heterogeneity of skin ILC subsets. The subcutaneous ILCs are GATA3$^+$ ILC2s, while the epidermal and dermal tissues contain ILCs with mixed phenotypes of ILC3s and ILC2s. (**b**) Bulk RNA-seq analysis further showed layer-specific identities of skin ILCs. IL-5 is mainly produced by subcutaneous and Sca1$^+$ dermal ILCs, while epidermal ILCs highly express IL-13. (**c**) Distribution of ILCs in the skin. (Fig. 5a and b are adapted from Kobayashi et al., Cell 2019, with permission from Elsevier)

in the pathogenesis of atopic dermatitis. GWAS have identified susceptibility genes for atopic dermatitis, including genes encoding IL-33R, IL-18R, IL-7R, IL-2R, TSLP, IL-4, and IL-13 [92, 93], all of which are associated with ILC2 activation and effector functions. Studies on patients with atopic dermatitis have reported that ILC2s, both in the lesional skin and peripheral blood, are increased compared to ILC2 populations in healthy subjects [94, 95].

To understand the mechanism by which ILC2s are activated and involved in the pathogenesis of atopic dermatitis, an atopic-like dermatitis mouse model with topical application of calcipotriol

(MC903), a vitamin D_3 analog, has been used. Although dermatitis is known to occur in T cell-deficient mice, depletion of T cells and ILCs by intraperitoneal administration of anti-CD25 or anti-CD90.2 antibodies alleviates dermatitis. On the other hand, dermatitis was not alleviated in mice lacking IL-33(*Il33*$^{-/-}$) or IL-25R (*Il17rb*$^{-/-}$), whereas TSLP receptor-deficient mice (*Tslpr*$^{-/-}$) showed impaired responsiveness to ILC2s and improved dermatitis, suggesting that TSLP is important for the activation of skin ILC2s [94]. In addition, IL-13-producing ILC2s interact with mast cells in the dermis of mouse skin and are involved in eosinophil infiltration in response to IL-2 [96].

On the other hand, the importance of IL-33 and IL-25 has also been reported. While the above studies used mice of the C57BL/6 strain background, studies using the BALB/c strain showed that MC903 induced atopic-like dermatitis was reduced in mice lacking IL-25R (*Il17rb*$^{-/-}$) and IL-33R (*Il1rl1*$^{-/-}$) [95]. It has been shown that transgenic mice with forced expression of IL-33, specifically in epidermal keratinocytes, have increased IL-33R$^+$ ILC2s and spontaneous development of atopic-like dermatitis [97]. In these mice, depletion of ILC2s or basophils resulted in an improvement of dermatitis, suggesting that ILC2s and basophils activated by IL-33 are involved in the pathogenesis of atopic inflammation [98]. In the abovementioned study, IL-33R$^+$ ILC2s were enriched in skin lesions of patients with atopic dermatitis, and in vitro stimulation of IL-33 induced the production of type 2 cytokines from human skin-derived ILC2s, further highlighting the role of IL-33 in skin ILC2 activation [95]. The role of IL-25 in ILC2 activation has been shown in an OVA-induced atopic-like dermatitis model. Either IL-25 deficiency, IL-13 deficiency, or ILC2-specific IL-25R deficiency (*Rora*-Cre *Il17rb*$^{flox/flox}$) suppressed dermatitis, suggesting a mechanism by which epidermal derived IL-25 promotes ILC2s to produce IL-13, which leads to chemokine production such as CCL17 and CCL22 from the epidermis and recruits T cells and further promotes dermatitis [99].

It is difficult to conclude which cytokines are most important for the activation of skin ILC2s, because different mouse models of atopic dermatitis and different strains of mice can affect the results. Nonetheless, there is no doubt that type 2 cytokines are central to the pathogenesis of atopic dermatitis, as monoclonal antibodies that block IL-4 and IL-13 signaling markedly improve clinical symptoms of atopic dermatitis [100, 101]. If atopic dermatitis is a more antigen-dependent disease, Th2 cells are likely to be the source of these type 2 cytokines; however, patients often demonstrate increased pruritus by sweating, stress, and diurnal rhythms, but not when exposed to specific antigens, suggesting that ILC2s are a possible source of type 2 cytokines. New therapeutic targets may be identified by studying the pathogenesis of atopic dermatitis from the perspective of ILC2s, which are activated by cytokines produced in response to tissue damage such as scratching.

6.3.2 NK and ILC1 in Contact Hypersensitivity

Allergic contact dermatitis is an inflammatory skin disease caused by the penetration of low-molecular-weight chemicals and metals. In hapten-induced contact hypersensitivity (CHS) in mice [102], antigen-specific or nonspecific innate immune memory may contribute to the induced response. Tissue-resident NK cells in the liver, currently referred to as ILC1s, induce hapten-specific memory responses independent of T and B lymphocytes in the CHS model, suggesting memory-like properties of NK cells or ILC1 [103, 104]. In contrast to the tissue residency of ILC2s and ILC3s, IL-7R$^+$ ILC1s acquire hapten-specific memory in skin-draining lymph nodes, and are recruited to the liver via CXCR6 and maintained by IL-7R signaling [105]. Although hapten-specific memory responses have not been demonstrated in human ILC1s or NK cells, NK cells accumulate in the skin of patients with allergic contact dermatitis, and NK cells release IFN-γ in the presence of T lymphocytes that produce IL-2 in vitro

[106]. These studies suggest the existence of an innate memory response (or trained immunity); however, the underlying mechanism for the long-term persistence of the immune response by ILCs requires further investigation.

6.4 Innate Lymphoid Cells in the Nasal Mucosa

The olfactory system is not only a chemosensory organ, but also a sophisticated immune system that acts as the first line of defense against infections, due to its constant exposure to the open air containing abundant immunogens. ILCs are known to be present in the nasal mucosa, but in healthy individuals, ILC1s and ILC2s are found in less than 0.1% of CD45-positive cells, and considerably less than 0.01% that of ILC3s. However, in inflammatory conditions such as chronic sinusitis, ILC2s are increased by more than 100-fold. Since the nasal mucosa is constantly invaded by antigens from the outside world, it is prone to various allergic symptoms; hence, allergic rhinitis and chronic sinusitis are well-known allergic inflammations of the nose. The discovery of ILC2s has accelerated our understanding of the pathogenesis of these diseases, which is thought to be mainly caused by Th2 cells. Since the nasal polyps, which form in the paranasal sinuses, were one of the first tissues in which human ILC2s were discovered, research on the pathogenesis of nasal allergies, with a focus on ILC2s, has been conducted mainly in humans.

6.4.1 Allergic Rhinitis and ILC2s

Allergic rhinitis is classified as an IgE-mediated type 1 allergy, in which allergen exposure causes a runny nose, sneezing, and nasal congestion. Within 30 min of exposure to an allergen, the cross-linking of IgE and allergen induces histamine release from mast cells, triggering the initial allergic reaction. Subsequently, 6–24 h after allergen exposure, eosinophilic infiltration occurs and triggers tissue destruction and remodeling. Patients with allergic rhinitis have elevated levels of IL-25, IL-33, and TSLP in the serum or nasal lavage fluid [107–109]. GWAS revealed that SNPs in the *IL33* gene are associated with Japanese cedar pollinosis, suggesting that IL-33 is strongly involved in the pathogenesis of allergic rhinitis [109]. IL-33-deficient mice display reduced symptoms of allergic rhinitis [110, 111]. In addition to IL-33, a number of GWAS analyses have demonstrated that SNPs in the *TSLP* gene are also correlated with allergic rhinitis [112–116], and the expression of *TSLP* in the nasal cavity is associated with the severity of the symptoms [117, 118]. In a mouse model of allergic rhinitis, the roles of TSLP and IL-33 were different between acute and chronic models. In the acute allergic rhinitis model, both TSLP and IL-33 contributed to the initial sneeze, and the subsequent eosinophilic infiltration depended on IL-33, whereas in the chronic model, the sneeze response or eosinophilic infiltration depended on TSLP or TSLP and IL-33, respectively. Furthermore, IL-25 was also detected in the nasal lavage fluid of allergic rhinitis patients, and the expression of *Il25* was enhanced in the nasal mucosa of OVA-induced allergic rhinitis model mice [119]. However, in a mouse model of HDM-induced allergic rhinitis, IL-25 deficiency had no effect on the symptoms [111]. Since IL-25 enhances the expression of TSLP in nasal epithelial cells, further analysis of the possible involvement of IL-25 in the pathogenesis of allergic rhinitis is needed [107].

Recently, ILC2s were analyzed in patients with allergic rhinitis. The frequency of ILC2s in peripheral blood was significantly higher in patients with allergic rhinitis than that in healthy subjects [120, 121]. Interestingly, there seems to be a difference in the reactivity of ILC2s depending on the type of allergen: patients with allergic rhinitis to HDM had the highest frequency of ILC2s in the peripheral blood, while patients with allergic rhinitis to other allergens, such as wormwood, had higher ILC2s than those of healthy subjects, but predominantly lower than those of patients with allergic rhinitis to HDM [121, 122]. However, in all patients, there was a correlation between the severity of symptoms and the percentage of ILC2s in the peripheral blood [121], suggesting that ILC2s play an important role in the pathogenesis of allergic rhinitis. Since allergic rhinitis is one of the most antigen-dependent allergic diseases, it is necessary to clarify how ILC2s contribute to its pathogenesis via IL-25, IL-33, and TSLP.

6.4.2 Chronic Rhinosinusitis and ILC2s

Chronic rhinosinusitis (CRS) is an inflammatory disease of the nasal mucosa triggered by upper respiratory tract inflammation, such as from the common cold that spreads to the mucosa and persists for more than 3 months. It has been demonstrated that in CRS with nasal polyps (CRSwNPs), type 2 inflammation, including IgE, IL-5, and IL-13 production, and eosinophil infiltration occur, suggesting that Th2 cells are strongly involved in the pathogenesis. Human ILC2s were first isolated from the nasal polyps of CRSwNP patients in Netherlands, along with the lungs and gut, and ILC2s were found to accumulate in polyps [33]. Subsequently, the accumulation of ILC2s in nasal polyps was confirmed in the United States [123]. It has been proposed that CRSwNPs can be divided into eosinophilic (ECRS) and non-eosinophilic (NECRS) subgroups [124], and ILC2s have been shown to accumulate in nasal polyps in ECRS but not in NECRS in various countries [125–127], suggesting that the mechanism of ILC2-induced ECRS is common worldwide. Despite the accumulation of ILC2s in nasal polyps, there was no difference in ILC2s in peripheral blood in all studies, consistent with the view that ECRS is a local inflammatory disease. Other ILC subsets, including ILC1s and ILC3s, were also analyzed in polyps of CRS; however, while ILC2s were selectively accumulated in polyps and increased 100-fold compared to the ILC2s in the sinus mucosa of healthy subjects, other ILC subsets were unchanged in the polyps [128], suggesting that ILC2s predominantly contribute to polyp development in CRS.

Polyp-derived ILC2s respond to IL-25, IL-33, and TSLP and produce IL-4, IL-5, IL-9, IL-13, and GM-CSF. Many studies have shown that the expression of IL-33 is comparable between the sinus mucosa of CRSwNPs, CRS without nasal polyps (CRSsNPs), and healthy subjects [33, 123, 129]. However, the expression of IL-33 was increased in epithelial cells derived from recurrent CRSwNPs compared with that from first-onset CRSwNP, suggesting that alteration of IL-33 expression in nasal epithelial cells may contribute to the pathogenesis [130]. In contrast, TSLP was shown to be upregulated in the nasal mucosa of CRSwNPs and ECRS [131–134]. Since the expression of TSLP in the nasal mucosa correlated with polyp scores, and GWAS analysis showed that SNPs in the TSLP gene were associated with CRSwNPs, TSLP may play an important role in the pathogenesis of CRSwNPs [135]. The increased expression of IL-25 in CRSwNPs is controversial and therefore inconclusive. However, it has been reported that polyp-derived ILC2s respond to IL-25 [33] and that IL-25 induces differentiation of fibroblasts into myofibroblasts and contributes to polyp remodeling [136], suggesting that IL-25 expression may change with disease progression, or that ECRS pathogenesis is a heterogeneous disease that can be further divided by endotypes.

Recently, the dramatic effect of dupilumab on ECRS has revealed that IL-4 and IL-13 are key factors in the pathogenesis of the disease [137]. It is well known that IgE expression in nasal polyps is high in patients with CRS, regardless of systemic IgE levels [138], suggesting that, unlike allergic rhinitis, inflammation in CRS is not allergen specific. Therefore, in addition to Th2 cells, ILC2s may contribute to the pathogenesis of ECRS through type 2 cytokines in response to IL-33, TSLP, and IL-25. The discovery of ILC2s is expected to accelerate our understanding of CRS pathogenesis and lead to the development of new therapies.

6.4.3 Nasal Allergy and ILC1s, ILC3s

ILC1s, including NK cells, suppress ILC2 function through the production of IFNγ. ILC1s in allergic rhinitis patients produce less IFNγ upon in vitro stimulation than that produced in nonallergic rhinitis patients [139], suggesting that ILC2s may be activated in these patients. In fact, in a study of patients who responded to allergen immunotherapy (AIT), a treatment for allergic rhinitis showed a reduced response of ILC2s to allergen stimulation compared to ILC2s in healthy subjects [140]. In addition, the ratio of ILC2 to ILC1 in patients who responded to AIT was similar to that in healthy controls, suggesting that ILC1s suppress allergic rhinitis by inhibiting the function of ILC2s. The frequency of ILC3s, along with ILC2s, significantly increased only during the grass pollen season in grass-allergic patients, while the frequency of ILC1s did not change. Therefore, ILC3s may have a different function from ILC1s in this condition, but the role of ILC3s in allergic rhinitis requires further analysis. Although ILC2s selectively accumulate in polyps in CRSwNPs, the increased production of IFNγ and IL-17A in the nasal mucosa in CRSsNP supports the possibility that ILC3s may have a different function than ILC1s under these conditions. Although there are reports that increased ILC1s and ILC3s are observed in the nasal mucosa in CRSsNP pathologies, there is no significant difference in the frequency of ILC1s and ILC3s in the nasal mucosa compared to differences in CRSwNPs or healthy individuals, although there is an increasing trend [128]. Therefore, the involvement of ILC1s and ILC3s in the pathogenesis of the disease is still unclear and requires further analysis.

6.5 Food Allergy, Including Anaphylaxis, and ILC2s

Food allergy (FA) is an allergic disease that is increasing worldwide, affecting one in ten people in developed countries. FA is defined as an antigen-specific biological response resulting from exposure to orally ingested food or food-derived components. Most of the symptoms of FA depend on the production of antigen-specific IgE, while anaphylaxis, caused by IgE-mediated mast cell degranulation, is a serious reaction that can lead to death. Various forms of food immunotherapy including oral, sublingual, and epicutaneous delivery routes have been used for the treatment of FA; clinical trials of anti-IgE therapy have also been initiated. Although anti-IgE therapy has not yet been approved for the treatment of FA, the combination of anti-IgE therapy and oral immunotherapy (OIT) is expected to be useful in peanut, milk, and multiple FAs, because anti-IgE treatment decreases the adverse events during OIT and shortens the treatment duration [141, 142].

Various types of ILCs reside in the intestine and contribute to homeostasis. In particular, the intestinal lamina propria (LP) has a high frequency of ILCs compared to that in the intestinal epithelium (IE). In mice, ILCs account for approximately 2.5% of all lymphocytes in the small intestinal LP fraction and 1.8% in the large intestinal LP fraction, which is more than 20 times higher than that in the IE fraction [143]. In the LP fraction of the small intestine, ILC3s are the most abundant, accounting for approximately 60% of all ILCs, while ILC2s and ILC1/NK cells account for approximately 20% each. On the other hand, in the LP fraction of the large intestine, ILC2s are the most abundant, accounting for approximately 50% of all ILCs, while ILC1/NK cells and ILC3s account for approximately 30% and 20%, respectively, which is opposite to the composition of the small intestine. Since FA is a type 2 immune response similar to other allergies, ILC2s among ILCs are thought to be involved in the pathogenesis of FA.

In studies of FA, models in which OVA/alum or peanuts/cholera toxin (PN/CT) are administered orally or peritoneally to mice to induce antigen-specific IgE production are often used. In these FA models, subsequent oral administration of OVA or peanuts to mice can induce type 2 inflammation, gastrointestinal symptoms such as diarrhea, and anaphylactic symptoms such as body temperature decrease through mast cell degranulation. The cytokines involved in the activation of ILC2s, such as IL-25, IL-33, and TSLP, are elevated, and the number of intestinal ILC2s is increased in an IL-33-dependent manner in the FA model mice. Consistent with these data, eosinophil infiltration, IgE production, and anaphylactic symptoms are reduced in IL-33 receptor-deficient mice, suggesting that ILC2s are involved in the pathogenesis of FA [144]. It has been reported that in the OVA/alum-induced FA model, IL-25 production is enhanced and elicits IL-13 production from ILC2s by direct stimulation or indirect activation via IL-25 receptor-positive Th2 cells, resulting in mastocytosis and diarrhea symptoms [145]. Furthermore, in the PN/CT-induced FA model using IL4RαF709 mice, in which the unrestrained form of the IL-4Rα chain lacking the immunoreceptor tyrosine-based inhibitory motif (ITIM) motif is knocked out, ILC2s contribute to the disruption of immune tolerance by suppressing antigen-specific Tregs via IL-4 production and further enhancing IgE reactivity of mast cells via IL-4/IL-13 production, leading to the exacerbation of anaphylaxis [146, 147]. Interestingly, disruption of the skin barrier induces the activation of intestinal ILC2s via IL-33, which enhances mastocytosis and anaphylaxis in OVA-induced FA models [148]. This study indicates that improper activation of ILC2s is involved in the development of allergic marches. On the other hand, Chu et al. reported that ILC2s contribute to type 2 inflammation in the abdominal cavity induced by FA, but not to IgE production, gastrointestinal symptoms, and anaphylaxis, based on the result of ILC2 depletion using Thy1-neutralizing antibody in a PN/CT-induced FA model [149]. These differences may be due to the varying FA models and mouse strains used, or due to the effects of the gut microbiota. Although ILC2s are abundant in the intestinal tracts and can be involved in the pathogenesis of FA as a source of type 2 cytokines, FA studies focusing on ILC2s have not been widely reported, especially in humans, compared to other allergic diseases such as asthma and CRS. Further research in both humans and mice will help to understand the role of ILC2s in the development of FA.

6.6 Allergic Conjunctivitis and ILC2s

Allergic conjunctivitis is an allergic inflammatory disease caused by foreign substances in the eyes, such as pollen, house dust, and contact lenses. The main symptoms of allergic conjunctivitis are itching, redness, and increased tear production, which are thought to be caused by the acquired immune system triggered by foreign substances. However, the involvement of the innate immune system in allergic conjunctivitis has recently attracted attention because of the elevated levels of epithelial cell-derived type 2 initiating cytokines such as IL-33 and TSLP [150, 151]. The mouse line hK14mIL33Tg is a keratin 14-driven transgenic mouse, which overexpresses IL-33 in an epithelial cell-specific manner and spontaneously develops allergic dermatitis and conjunctivitis. In these mice, ILC2s in the conjunctiva were increased by more than 20 times compared to ILC2s in naïve mice, and ILC2-derived IL-5 and IL-13 production was enhanced [152]. It was also reported that ILC2s increased with eosinophils in conjunctivitis induced by papain contact lenses [153]. Interestingly, a similar level of inflammation was induced in *Rag2*$^{-/-}$ mice lacking the acquired immune system, suggesting that ILC2-mediated allergic inflammation plays an essential role in the pathogenesis of allergic conjunctivitis. It is expected that understanding the mechanism by which ILC2s contribute to the maintenance of homeostasis and the induction of allergic inflammation in the conjunctiva, a barrier tissue in contact with the outside world similar to the skin, nasal mucosa, and bronchi would aid in the development of novel therapeutic approaches.

6.7 Conclusion

The word “allergy” is a concept, and the actual diseases, such as asthma, atopic dermatitis, and food allergy, are differentiated according to the organ in which the allergy occurs and are treated by the respective clinical departments. As a result, patients are referred to the Department of Respiratory Medicine for asthma, the Department of Dermatology for atopic dermatitis, and the Department of Gastroenterology for FA, with limited coordination between doctors in each department. On the other hand, in the field of basic research, the existence of ILCs has been recently revealed, and the role of ILC2s in allergic diseases has been rapidly elucidated. As a result, allergies, regardless of the organ, can now be roughly divided into antigen-specific mediated by Th2 cells and antigen-nonspecific mediated by ILC2s, mainly IL-33. Therefore, future research, drug development, and medical treatment for allergies must emphasize antigen and organ specificity to comprehensively understand the diseases. In the face of the explosive increase in allergies worldwide, it is necessary to establish methods that accurately determine whether each patient's allergy is dependent on Th2 cells or ILC2s, and to identify targets that can completely suppress each of these cell types.

References

1. Monticelli LA, Sonnenberg GF, Abt MC, Alenghat T, Ziegler CG, Doering TA, et al. Innate lymphoid cells promote lung-tissue homeostasis after infection with influenza virus. Nat Immunol. 2011;12 (11):1045–54.
2. Yudanin NA, Schmitz F, Flamar AL, Thome JJC, Tait Wojno E, Moeller JB, et al. Spatial and temporal mapping of human innate lymphoid cells reveals elements of tissue specificity. Immunity. 2019;50 (2):505–19.
3. Marquardt N, Kekalainen E, Chen P, Kvedaraite E, Wilson JN, Ivarsson MA, et al. Human lung natural killer cells are predominantly comprised of highly differentiated hypofunctional CD69(−)CD56(dim) cells. J Allergy Clin Immunol. 2017;139(4):1321–30.
4. Silver JS, Kearley J, Copenhaver AM, Sanden C, Mori M, Yu L, et al. Inflammatory triggers associated with exacerbations of COPD orchestrate plasticity of group 2 innate lymphoid cells in the lungs. Nat Immunol. 2016;17(6):626–35.
5. Kim HY, Lee HJ, Chang YJ, Pichavant M, Shore SA, Fitzgerald KA, et al. Interleukin-17-producing innate lymphoid cells and the NLRP3 inflammasome facilitate obesity-associated airway hyperreactivity. Nat Med. 2014;20(1):54–61.
6. Mazzurana L, Czarnewski P, Jonsson V, Wigge L, Ringnér M, Williams TC, et al. Tissue-specific transcriptional imprinting and heterogeneity in human innate lymphoid cells revealed by full-length single-cell RNA-sequencing. Cell Res. 2021;31(5):554–68.
7. Klein Wolterink RG, Kleinjan A, van Nimwegen M, Bergen I, de Bruijn M, Levani Y, et al. Pulmonary innate lymphoid cells are major producers of IL-5 and IL-13 in murine models of allergic asthma. Eur J Immunol. 2012;42(5):1106–16.
8. Bartemes KR, Iijima K, Kobayashi T, Kephart GM, McKenzie AN, Kita H. IL-33-responsive lineage-CD25+ CD44(hi) lymphoid cells mediate innate type 2 immunity and allergic inflammation in the lungs. J Immunol. 2012;188(3):1503–13.
9. Halim TY, Krauss RH, Sun AC, Takei F. Lung natural helper cells are a critical source of Th2 cell-type cytokines in protease allergen-induced airway inflammation. Immunity. 2012;36(3):451–63.
10. Arae K, Ikutani M, Horiguchi K, Yamaguchi S, Okada Y, Sugiyama H, et al. Interleukin-33 and thymic stromal lymphopoietin, but not interleukin-25, are crucial for development of airway eosinophilia induced by chitin. Sci Rep. 2021;11(1):5913.
11. Kim HY, Chang YJ, Subramanian S, Lee HH, Albacker LA, Matangkasombut P, et al. Innate lymphoid cells responding to IL-33 mediate airway hyperreactivity independently of adaptive immunity. J Allergy Clin Immunol. 2012;129(1):216–27.
12. Han M, Rajput C, Hong JY, Lei J, Hinde JL, Wu Q, et al. The innate cytokines IL-25, IL-33, and TSLP cooperate in the induction of type 2 innate lymphoid cell expansion and mucous metaplasia in rhinovirus-infected immature mice. J Immunol. 2017;199(4): 1308–18.
13. Chang YJ, Kim HY, Albacker LA, Baumgarth N, McKenzie AN, Smith DE, et al. Innate lymphoid cells mediate influenza-induced airway hyperreactivity independently of adaptive immunity. Nat Immunol. 2011;12(7):631–8.
14. Kouzaki H, Iijima K, Kobayashi T, O'Grady SM, Kita H. The danger signal, extracellular ATP, is a sensor for an airborne allergen and triggers IL-33 release and innate Th2-type responses. J Immunol. 2011;186(7):4375–87.
15. Cayrol C, Duval A, Schmitt P, Roga S, Camus M, Stella A, et al. Environmental allergens induce allergic inflammation through proteolytic maturation of IL-33. Nat Immunol. 2018;19(4):375–85.
16. Furusawa J, Moro K, Motomura Y, Okamoto K, Zhu J, Takayanagi H, et al. Critical role of p38 and GATA3 in natural helper cell function. J Immunol. 2013;191(4):1818–26.
17. Kabata H, Moro K, Koyasu S. The group 2 innate lymphoid cell (ILC2) regulatory network and its underlying mechanisms. Immunol Rev. 2018;286 (1):37–52.
18. Hammad H, Lambrecht BN. The basic immunology of asthma. Cell. 2021;184(6):1469–85.
19. Lambrecht BN, Hammad H. The immunology of asthma. Nat Immunol. 2015;16(1):45–56.
20. Torgerson DG, Ampleford EJ, Chiu GY, Gauderman WJ, Gignoux CR, Graves PE, et al. Meta-analysis of genome-wide association studies of asthma in ethnically diverse North American populations. Nat Genet. 2011;43(9):887–92.
21. Prefontaine D, Lajoie-Kadoch S, Foley S, Audusseau S, Olivenstein R, Halayko AJ, et al. Increased expression of IL-33 in severe asthma: evidence of expression by airway smooth muscle cells. J Immunol. 2009;183(8):5094–103.
22. Christianson CA, Goplen NP, Zafar I, Irvin C, Good JT Jr, Rollins DR, et al. Persistence of asthma requires multiple feedback circuits involving type

2 innate lymphoid cells and IL-33. J Allergy Clin Immunol. 2015;136(1):59–68.
23. Prefontaine D, Nadigel J, Chouiali F, Audusseau S, Semlali A, Chakir J, et al. Increased IL-33 expression by epithelial cells in bronchial asthma. J Allergy Clin Immunol. 2010;125(3):752–4.
24. Smith SG, Chen R, Kjarsgaard M, Huang C, Oliveria JP, O'Byrne PM, et al. Increased numbers of activated group 2 innate lymphoid cells in the airways of patients with severe asthma and persistent airway eosinophilia. J Allergy Clin Immunol. 2016;137(1): 75–86.
25. Bartemes KR, Kephart GM, Fox SJ, Kita H. Enhanced innate type 2 immune response in peripheral blood from patients with asthma. J Allergy Clin Immunol. 2014;134(3):671–8.
26. Al-Sajee D, Sehmi R, Hawke TJ, El-Gammal A, Howie KJ, Watson RM, et al. Expression of IL-33 and TSLP and their receptors in asthmatic airways after inhaled allergen challenge. Am J Respir Crit Care Med. 2018;198(6):805–7.
27. Jackson DJ, Makrinioti H, Rana BM, Shamji BW, Trujillo-Torralbo MB, Footitt J, et al. IL-33-dependent type 2 inflammation during rhinovirus-induced asthma exacerbations in vivo. Am J Respir Crit Care Med. 2014;190(12):1373–82.
28. Winkler C, Hochdorfer T, Israelsson E, Hasselberg A, Cavallin A, Thorn K, et al. Activation of group 2 innate lymphoid cells after allergen challenge in asthmatic patients. J Allergy Clin Immunol. 2019;144(1):61–9.
29. Mazzurana L, Czarnewski P, Jonsson V, Wigge L, Ringner M, Williams TC, et al. Tissue-specific transcriptional imprinting and heterogeneity in human innate lymphoid cells revealed by full-length single-cell RNA-sequencing. Cell Res. 2021;31(5):554–68.
30. Moro K, Kabata H, Tanabe M, Koga S, Takeno N, Mochizuki M, et al. Interferon and IL-27 antagonize the function of group 2 innate lymphoid cells and type 2 innate immune responses. Nat Immunol. 2016;17(1):76–86.
31. Liu S, Verma M, Michalec L, Liu W, Sripada A, Rollins D, et al. Steroid resistance of airway type 2 innate lymphoid cells from patients with severe asthma: the role of thymic stromal lymphopoietin. J Allergy Clin Immunol. 2018;141(1):257–68 e6.
32. Moro K, Yamada T, Tanabe M, Takeuchi T, Ikawa T, Kawamoto H, et al. Innate production of T(H) 2 cytokines by adipose tissue-associated c-Kit(+) Sca-1(+) lymphoid cells. Nature. 2010;463(7280): 540–4.
33. Mjosberg JM, Trifari S, Crellin NK, Peters CP, van Drunen CM, Piet B, et al. Human IL-25- and IL-33-responsive type 2 innate lymphoid cells are defined by expression of CRTH2 and CD161. Nat Immunol. 2011;12(11):1055–62.
34. Howitt MR, Lavoie S, Michaud M, Blum AM, Tran SV, Weinstock JV, et al. Tuft cells, taste-chemosensory cells, orchestrate parasite type 2 immunity in the gut. Science. 2016;351(6279):1329–33.
35. Barlow JL, Peel S, Fox J, Panova V, Hardman CS, Camelo A, et al. IL-33 is more potent than IL-25 in provoking IL-13-producing nuocytes (type 2 innate lymphoid cells) and airway contraction. J Allergy Clin Immunol. 2013;132(4):933–41.
36. Kabata H, Moro K, Fukunaga K, Suzuki Y, Miyata J, Masaki K, et al. Thymic stromal lymphopoietin induces corticosteroid resistance in natural helper cells during airway inflammation. Nat Commun. 2013;4:2675.
37. Shikotra A, Choy DF, Ohri CM, Doran E, Butler C, Hargadon B, et al. Increased expression of immunoreactive thymic stromal lymphopoietin in patients with severe asthma. J Allergy Clin Immunol. 2012;129(1):104–11.
38. Motomura Y, Morita H, Moro K, Nakae S, Artis D, Endo TA, et al. Basophil-derived interleukin-4 controls the function of natural helper cells, a member of ILC2s, in lung inflammation. Immunity. 2014;40 (5):758–71.
39. Yu X, Pappu R, Ramirez-Carrozzi V, Ota N, Caplazi P, Zhang J, et al. TNF superfamily member TL1A elicits type 2 innate lymphoid cells at mucosal barriers. Mucosal Immunol. 2014;7(3):730–40.
40. Meylan F, Hawley ET, Barron L, Barlow JL, Penumetcha P, Pelletier M, et al. The TNF-family cytokine TL1A promotes allergic immunopathology through group 2 innate lymphoid cells. Mucosal Immunol. 2014;7(4):958–68.
41. Machida K, Aw M, Salter BMA, Ju X, Mukherjee M, Gauvreau GM, et al. The role of the TL1A/DR3 axis in the activation of group 2 innate lymphoid cells in subjects with eosinophilic asthma. Am J Respir Crit Care Med. 2020;202(8):1105–14.
42. Nagashima H, Okuyama Y, Fujita T, Takeda T, Motomura Y, Moro K, et al. GITR cosignal in ILC2s controls allergic lung inflammation. J Allergy Clin Immunol. 2018;141(5):1939–43.
43. Doherty TA, Khorram N, Lund S, Mehta AK, Croft M, Broide DH. Lung type 2 innate lymphoid cells express cysteinyl leukotriene receptor 1, which regulates TH2 cytokine production. J Allergy Clin Immunol. 2013;132(1):205–13.
44. Xue L, Salimi M, Panse I, Mjosberg JM, McKenzie AN, Spits H, et al. Prostaglandin D2 activates group 2 innate lymphoid cells through chemoattractant receptor-homologous molecule expressed on TH2 cells. J Allergy Clin Immunol. 2014;133(4):1184–94.
45. Lund SJ, Portillo A, Cavagnero K, Baum RE, Naji LH, Badrani JH, et al. Leukotriene C4 potentiates IL-33-induced group 2 innate lymphoid cell activation and lung inflammation. J Immunol. 2017;199(3): 1096–104.
46. Wojno ED, Monticelli LA, Tran SV, Alenghat T, Osborne LC, Thome JJ, et al. The prostaglandin D (2) receptor CRTH2 regulates accumulation of group

2 innate lymphoid cells in the inflamed lung. Mucosal Immunol. 2015;8(6):1313–23.
47. Salimi M, Stoger L, Liu W, Go S, Pavord I, Klenerman P, et al. Cysteinyl leukotriene E4 activates human group 2 innate lymphoid cells and enhances the effect of prostaglandin D2 and epithelial cytokines. J Allergy Clin Immunol. 2017;140(4): 1090–100.
48. von Moltke J, O'Leary CE, Barrett NA, Kanaoka Y, Austen KF, Locksley RM. Leukotrienes provide an NFAT-dependent signal that synergizes with IL-33 to activate ILC2s. J Exp Med. 2017;214(1):27–37.
49. White AA, Stevenson DD. Aspirin-exacerbated respiratory disease. N Engl J Med. 2018;379(11): 1060–70.
50. Eastman JJ, Cavagnero KJ, Deconde AS, Kim AS, Karta MR, Broide DH, et al. Group 2 innate lymphoid cells are recruited to the nasal mucosa in patients with aspirin-exacerbated respiratory disease. J Allergy Clin Immunol. 2017;140(1):101–8.
51. Duerr CU, McCarthy CD, Mindt BC, Rubio M, Meli AP, Pothlichet J, et al. Type I interferon restricts type 2 immunopathology through the regulation of group 2 innate lymphoid cells. Nat Immunol. 2016;17(1): 65–75.
52. Molofsky AB, Van Gool F, Liang HE, Van Dyken SJ, Nussbaum JC, Lee J, et al. Interleukin-33 and interferon-gamma counter-regulate group 2 innate lymphoid cell activation during immune perturbation. Immunity. 2015;43(1):161–74.
53. McHedlidze T, Kindermann M, Neves AT, Voehringer D, Neurath MF, Wirtz S. IL-27 suppresses type 2 immune responses in vivo via direct effects on group 2 innate lymphoid cells. Mucosal Immunol. 2016;9(6):1384–94.
54. Okuzumi S, Miyata J, Kabata H, Mochimaru T, Kagawa S, Masaki K, et al. TLR7 agonist suppresses ILC2-mediated inflammation via IL-27-producing interstitial macrophages. Am J Respir Cell Mol Biol. 2021;65(3):309–18.
55. Wark PA, Johnston SL, Bucchieri F, Powell R, Puddicombe S, Laza-Stanca V, et al. Asthmatic bronchial epithelial cells have a deficient innate immune response to infection with rhinovirus. J Exp Med. 2005;201(6):937–47.
56. Djukanovic R, Harrison T, Johnston SL, Gabbay F, Wark P, Thomson NC, et al. The effect of inhaled IFN-beta on worsening of asthma symptoms caused by viral infections. A randomized trial. Am J Respir Crit Care Med. 2014;190(2):145–54.
57. Psallidas I, Backer V, Kuna P, Palmer R, Necander S, Aurell M, et al. A phase 2a, double-blind, placebo-controlled randomized trial of inhaled TLR9 agonist AZD1419 in asthma. Am J Respir Crit Care Med. 2021;203(3):296–306.
58. Maric J, Ravindran A, Mazzurana L, Bjorklund AK, Van Acker A, Rao A, et al. Prostaglandin E2 suppresses human group 2 innate lymphoid cell function. J Allergy Clin Immunol. 2018;141(5):1761–73.
59. Zhou Y, Wang W, Zhao C, Wang Y, Wu H, Sun X, et al. Prostaglandin E2 inhibits group 2 innate lymphoid cell activation and allergic airway inflammation through E-prostanoid 4-cyclic adenosine monophosphate signaling. Front Immunol. 2018;9: 501.
60. Zhou W, Toki S, Zhang J, Goleniewksa K, Newcomb DC, Cephus JY, et al. Prostaglandin I2 signaling and inhibition of group 2 innate lymphoid cell responses. Am J Respir Crit Care Med. 2016;193(1):31–42.
61. Barnig C, Cernadas M, Dutile S, Liu X, Perrella MA, Kazani S, et al. Lipoxin A4 regulates natural killer cell and type 2 innate lymphoid cell activation in asthma. Sci Transl Med. 2013;5(174):174ra26.
62. Ohne Y, Silver JS, Thompson-Snipes L, Collet MA, Blanck JP, Cantarel BL, et al. IL-1 is a critical regulator of group 2 innate lymphoid cell function and plasticity. Nat Immunol. 2016;17(6):646–55.
63. Lim AI, Menegatti S, Bustamante J, Le Bourhis L, Allez M, Rogge L, et al. IL-12 drives functional plasticity of human group 2 innate lymphoid cells. J Exp Med. 2016;213(4):569–83.
64. Morita H, Kubo T, Ruckert B, Ravindran A, Soyka MB, Rinaldi AO, et al. Induction of human regulatory innate lymphoid cells from group 2 innate lymphoid cells by retinoic acid. J Allergy Clin Immunol. 2019;143(6):2190–201.
65. Kabata H, Artis D. Neuro-immune crosstalk and allergic inflammation. J Clin Investig. 2019;129(4): 1475–82.
66. Sui P, Wiesner DL, Xu J, Zhang Y, Lee J, Van Dyken S, et al. Pulmonary neuroendocrine cells amplify allergic asthma responses. Science. 2018;360(6393):8546.
67. Nagashima H, Mahlakoiv T, Shih HY, Davis FP, Meylan F, Huang Y, et al. Neuropeptide CGRP limits group 2 innate lymphoid cell responses and constrains type 2 inflammation. Immunity. 2019;51 (4):682–95.
68. Klose CSN, Mahlakoiv T, Moeller JB, Rankin LC, Flamar AL, Kabata H, et al. The neuropeptide neuromedin U stimulates innate lymphoid cells and type 2 inflammation. Nature. 2017;549(7671):282–6.
69. Wallrapp A, Riesenfeld SJ, Burkett PR, Abdulnour RE, Nyman J, Dionne D, et al. The neuropeptide NMU amplifies ILC2-driven allergic lung inflammation. Nature. 2017;549(7672):351–6.
70. Cardoso V, Chesne J, Ribeiro H, Garcia-Cassani B, Carvalho T, Bouchery T, et al. Neuronal regulation of type 2 innate lymphoid cells via neuromedin U. Nature. 2017;549(7671):277–81.
71. Talbot S, Abdulnour RE, Burkett PR, Lee S, Cronin SJ, Pascal MA, et al. Silencing nociceptor neurons reduces allergic airway inflammation. Neuron. 2015;87(2):341–54.
72. Galle-Treger L, Suzuki Y, Patel N, Sankaranarayanan I, Aron JL, Maazi H, et al. Nicotinic acetylcholine receptor agonist attenuates ILC2-

dependent airway hyperreactivity. Nat Commun. 2016;7:13202.
73. Moriyama S, Brestoff JR, Flamar AL, Moeller JB, Klose CSN, Rankin LC, et al. β2-adrenergic receptor-mediated negative regulation of group 2 innate lymphoid cell responses. Science. 2018;359(6379):1056–61.
74. Martinez-Gonzalez I, Matha L, Steer CA, Ghaedi M, Poon GF, Takei F. Allergen-experienced group 2 innate lymphoid cells acquire memory-like properties and enhance allergic lung inflammation. Immunity. 2016;45(1):198–208.
75. Verma M, Michalec L, Sripada A, McKay J, Sirohi K, Verma D, et al. The molecular and epigenetic mechanisms of innate lymphoid cell (ILC) memory and its relevance for asthma. J Exp Med. 2021;218(7):e20201354.
76. Gold MJ, Antignano F, Halim TY, Hirota JA, Blanchet MR, Zaph C, et al. Group 2 innate lymphoid cells facilitate sensitization to local, but not systemic, TH2-inducing allergen exposures. J Allergy Clin Immunol. 2014;133(4):1142–8.
77. Mirchandani AS, Besnard AG, Yip E, Scott C, Bain CC, Cerovic V, et al. Type 2 innate lymphoid cells drive CD4+ Th2 cell responses. J Immunol. 2014;192(5):2442–8.
78. Oliphant CJ, Hwang YY, Walker JA, Salimi M, Wong SH, Brewer JM, et al. MHCII-mediated dialog between group 2 innate lymphoid cells and CD4(+) T cells potentiates type 2 immunity and promotes parasitic helminth expulsion. Immunity. 2014;41(2):283–95.
79. Drake LY, Iijima K, Kita H. Group 2 innate lymphoid cells and CD4+ T cells cooperate to mediate type 2 immune response in mice. Allergy. 2014;69(10):1300–7.
80. Halim TYF, Rana BMJ, Walker JA, Kerscher B, Knolle MD, Jolin HE, et al. Tissue-restricted adaptive type 2 immunity is orchestrated by expression of the costimulatory molecule OX40L on group 2 innate lymphoid cells. Immunity. 2018;48(6):1195–207.
81. Halim TY, Steer CA, Matha L, Gold MJ, Martinez-Gonzalez I, McNagny KM, et al. Group 2 innate lymphoid cells are critical for the initiation of adaptive T helper 2 cell-mediated allergic lung inflammation. Immunity. 2014;40(3):425–35.
82. Sugita K, Steer CA, Martinez-Gonzalez I, Altunbulakli C, Morita H, Castro-Giner F, et al. Type 2 innate lymphoid cells disrupt bronchial epithelial barrier integrity by targeting tight junctions through IL-13 in asthmatic patients. J Allergy Clin Immunol. 2018;141(1):300–10.
83. Yu QN, Guo YB, Li X, Li CL, Tan WP, Fan XL, et al. ILC2 frequency and activity are inhibited by glucocorticoid treatment via STAT pathway in patients with asthma. Allergy. 2018;73(9):1860–70.
84. Kabata H, Flamar AL, Mahlakoiv T, Moriyama S, Rodewald HR, Ziegler SF, et al. Targeted deletion of the TSLP receptor reveals cellular mechanisms that promote type 2 airway inflammation. Mucosal Immunol. 2020;13(4):626–36.
85. Gauvreau GM, O'Byrne PM, Boulet LP, Wang Y, Cockcroft D, Bigler J, et al. Effects of an anti-TSLP antibody on allergen-induced asthmatic responses. N Engl J Med. 2014;370(22):2102–10.
86. Corren J, Parnes JR, Wang L, Mo M, Roseti SL, Griffiths JM, et al. Tezepelumab in adults with uncontrolled asthma. N Engl J Med. 2017;377(10):936–46.
87. Menzies-Gow A, Corren J, Bourdin A, Chupp G, Israel E, Wechsler ME, et al. Tezepelumab in adults and adolescents with severe, uncontrolled asthma. N Engl J Med. 2021;384(19):1800–9.
88. Porsbjerg CM, Sverrild A, Lloyd CM, Menzies-Gow AN, Bel EH. Anti-alarmins in asthma: targeting the airway epithelium with next-generation biologics. Eur Respir J. 2020;56(5):2000260.
89. Patel G, Pan J, Ye L, Shen X, Rosloff D, D'Souza SS, et al. Blockade of IL-4Rα inhibits group 2 innate lymphoid cell responses in asthma patients. Clin Exp Allergy. 2020;50(2):267–70.
90. Kobayashi T, Voisin B, Kim DY, Kennedy EA, Jo JH, Shih HY, et al. Homeostatic control of sebaceous glands by innate lymphoid cells regulates commensal bacteria equilibrium. Cell. 2019;176(5):982–97.
91. Dutton EE, Gajdasik DW, Willis C, Fiancette R, Bishop EL, Camelo A, et al. Peripheral lymph nodes contain migratory and resident innate lymphoid cell populations. Sci Immunol. 2019;4(35):8082.
92. Paternoster L, Standl M, Waage J, Baurecht H, Hotze M, Strachan DP, et al. Multi-ancestry genome-wide association study of 21,000 cases and 95,000 controls identifies new risk loci for atopic dermatitis. Nat Genet. 2015;47(12):1449–56.
93. Tamari M, Hirota T. Genome-wide association studies of atopic dermatitis. J Dermatol. 2014;41(3):213–20.
94. Kim BS, Siracusa MC, Saenz SA, Noti M, Monticelli LA, Sonnenberg GF, et al. TSLP elicits IL-33-independent innate lymphoid cell responses to promote skin inflammation. Sci Transl Med. 2013;5(170):170ra16.
95. Salimi M, Barlow JL, Saunders SP, Xue L, Gutowska-Owsiak D, Wang X, et al. A role for IL-25 and IL-33-driven type-2 innate lymphoid cells in atopic dermatitis. J Exp Med. 2013;210(13):2939–50.
96. Roediger B, Kyle R, Yip KH, Sumaria N, Guy TV, Kim BS, et al. Cutaneous immunosurveillance and regulation of inflammation by group 2 innate lymphoid cells. Nat Immunol. 2013;14(6):564–73.
97. Imai Y, Yasuda K, Sakaguchi Y, Haneda T, Mizutani H, Yoshimoto T, et al. Skin-specific expression of IL-33 activates group 2 innate lymphoid cells and elicits atopic dermatitis-like inflammation in mice. Proc Natl Acad Sci USA. 2013;110(34):13921–6.

98. Imai Y, Yasuda K, Nagai M, Kusakabe M, Kubo M, Nakanishi K, et al. IL-33-induced atopic dermatitis-like inflammation in mice is mediated by group 2 innate lymphoid cells in concert with basophils. J Investig Dermatol. 2019;139(10):2185–94.
99. Leyva-Castillo JM, Galand C, Mashiko S, Bissonnette R, McGurk A, Ziegler SF, et al. ILC2 activation by keratinocyte-derived IL-25 drives IL-13 production at sites of allergic skin inflammation. J Allergy Clin Immunol. 2020;145(6):1606–14.
100. Beck LA, Thaçi D, Hamilton JD, Graham NM, Bieber T, Rocklin R, et al. Dupilumab treatment in adults with moderate-to-severe atopic dermatitis. N Engl J Med. 2014;371(2):130–9.
101. Simpson EL, Bieber T, Guttman-Yassky E, Beck LA, Blauvelt A, Cork MJ, et al. Two phase 3 trials of dupilumab versus placebo in atopic dermatitis. N Engl J Med. 2016;375(24):2335–48.
102. Honda T, Egawa G, Grabbe S, Kabashima K. Update of immune events in the murine contact hypersensitivity model: toward the understanding of allergic contact dermatitis. J Investig Dermatol. 2013;133(2):303–15.
103. O'Leary JG, Goodarzi M, Drayton DL, von Andrian UH. T cell- and B cell-independent adaptive immunity mediated by natural killer cells. Nat Immunol. 2006;7(5):507–16.
104. Peng H, Jiang X, Chen Y, Sojka DK, Wei H, Gao X, et al. Liver-resident NK cells confer adaptive immunity in skin-contact inflammation. J Clin Investig. 2013;123(4):1444–56.
105. Wang X, Peng H, Cong J, Wang X, Lian Z, Wei H, et al. Memory formation and long-term maintenance of IL-7Rα(+) ILC1s via a lymph node-liver axis. Nat Commun. 2018;9(1):4854.
106. Carbone T, Nasorri F, Pennino D, Eyerich K, Foerster S, Cifaldi L, et al. CD56highCD16-CD62L-NK cells accumulate in allergic contact dermatitis and contribute to the expression of allergic responses. J Immunol. 2010;184(2):1102–10.
107. Xu G, Zhang L, Wang DY, Xu R, Liu Z, Han DM, et al. Opposing roles of IL-17A and IL-25 in the regulation of TSLP production in human nasal epithelial cells. Allergy. 2010;65(5):581–9.
108. Asaka D, Yoshikawa M, Nakayama T, Yoshimura T, Moriyama H, Otori N. Elevated levels of interleukin-33 in the nasal secretions of patients with allergic rhinitis. Int Arch Allergy Immunol. 2012;158(Suppl 1):47–50.
109. Sakashita M, Yoshimoto T, Hirota T, Harada M, Okubo K, Osawa Y, et al. Association of serum interleukin-33 level and the interleukin-33 genetic variant with Japanese cedar pollinosis. Clin Exp Allergy. 2008;38(12):1875–81.
110. Haenuki Y, Matsushita K, Futatsugi-Yumikura S, Ishii KJ, Kawagoe T, Imoto Y, et al. A critical role of IL-33 in experimental allergic rhinitis. J Allergy Clin Immunol. 2012;130(1):184–94.
111. Nakanishi W, Yamaguchi S, Matsuda A, Suzukawa M, Shibui A, Nambu A, et al. IL-33, but not IL-25, is crucial for the development of house dust mite antigen-induced allergic rhinitis. PLoS One. 2013;8(10):e78099.
112. Ramasamy A, Curjuric I, Coin LJ, Kumar A, McArdle WL, Imboden M, et al. A genome-wide meta-analysis of genetic variants associated with allergic rhinitis and grass sensitization and their interaction with birth order. J Allergy Clin Immunol. 2011;128(5):996–1005.
113. Andiappan AK, Wang de Y, Anantharaman R, Suri BK, Lee BT, Rotzschke O, et al. Replication of genome-wide association study loci for allergic rhinitis and house dust mite sensitization in an Asian population of ethnic Chinese in Singapore. J Allergy Clin Immunol. 2013;131(5):1431–3.
114. Birben E, Sahiner UM, Karaaslan C, Yavuz TS, Cosgun E, Kalayci O, et al. The genetic variants of thymic stromal lymphopoietin protein in children with asthma and allergic rhinitis. Int Arch Allergy Immunol. 2014;163(3):185–92.
115. Nilsson D, Henmyr V, Hallden C, Sall T, Kull I, Wickman M, et al. Replication of genome-wide associations with allergic sensitization and allergic rhinitis. Allergy. 2014;69(11):1506–14.
116. Sun Q, Liu Y, Zhang S, Liu K, Zhu X, Liu J, et al. Thymic stromal lymphopoietin polymorphisms and allergic rhinitis risk: a systematic review and meta-analysis with 6351 cases and 11,472 controls. Int J Clin Exp Med. 2015;8(9):15752–8.
117. Zhu DD, Zhu XW, Jiang XD, Dong Z. Thymic stromal lymphopoietin expression is increased in nasal epithelial cells of patients with mugwort pollen sensitive-seasonal allergic rhinitis. Chin Med J. 2009;122(19):2303–7.
118. Mou Z, Xia J, Tan Y, Wang X, Zhang Y, Zhou B, et al. Overexpression of thymic stromal lymphopoietin in allergic rhinitis. Acta Otolaryngol. 2009;129(3):297–301.
119. Li Z, Wang H, Liu L. Interleukin-25 enhances allergic inflammation through p38MAPK and NF-kappaB pathways in mouse models of allergic rhinitis. Iran J Allergy Asthma Immunol. 2014;13(6):412–9.
120. Zhong H, Fan XL, Yu QN, Qin ZL, Chen D, Xu R, et al. Increased innate type 2 immune response in house dust mite-allergic patients with allergic rhinitis. Clin Immunol. 2017;183:293–9.
121. Sun R, Yang Y, Huo Q, Gu Z, Wei P, Tang X. Increased expression of type 2 innate lymphoid cells in pediatric patients with allergic rhinitis. Exp Ther Med. 2020;19(1):735–40.
122. Fan D, Wang X, Wang M, Wang Y, Zhang L, Li Y, et al. Allergen-dependent differences in ILC2s frequencies in patients with allergic rhinitis. Allergy Asthma Immunol Res. 2016;8(3):216–22.
123. Shaw JL, Fakhri S, Citardi MJ, Porter PC, Corry DB, Kheradmand F, et al. IL-33-responsive innate lymphoid cells are an important source of IL-13 in

chronic rhinosinusitis with nasal polyps. Am J Respir Crit Care Med. 2013;188(4):432–9.
124. Payne SC, Early SB, Huyett P, Han JK, Borish L, Steinke JW. Evidence for distinct histologic profile of nasal polyps with and without eosinophilia. Laryngoscope. 2011;121(10):2262–7.
125. Miljkovic D, Bassiouni A, Cooksley C, Ou J, Hauben E, Wormald PJ, et al. Association between group 2 innate lymphoid cells enrichment, nasal polyps and allergy in chronic rhinosinusitis. Allergy. 2014;69(9):1154–61.
126. Ho J, Bailey M, Zaunders J, Mrad N, Sacks R, Sewell W, et al. Group 2 innate lymphoid cells (ILC2s) are increased in chronic rhinosinusitis with nasal polyps or eosinophilia. Clin Exp Allergy. 2015;45(2):394–403.
127. Tojima I, Kouzaki H, Shimizu S, Ogawa T, Arikata M, Kita H, et al. Group 2 innate lymphoid cells are increased in nasal polyps in patients with eosinophilic chronic rhinosinusitis. Clin Immunol. 2016;170:1–8.
128. Poposki JA, Klingler AI, Tan BK, Soroosh P, Banie H, Lewis G, et al. Group 2 innate lymphoid cells are elevated and activated in chronic rhinosinusitis with nasal polyps. Immun Inflamm Dis. 2017;5(3):233–43.
129. Baba S, Kondo K, Kanaya K, Suzukawa K, Ushio M, Urata S, et al. Expression of IL-33 and its receptor ST2 in chronic rhinosinusitis with nasal polyps. Laryngoscope. 2014;124(4):E115–22.
130. Reh DD, Wang Y, Ramanathan M Jr, Lane AP. Treatment-recalcitrant chronic rhinosinusitis with polyps is associated with altered epithelial cell expression of interleukin-33. Am J Rhinol Allergy. 2010;24(2):105–9.
131. Kimura S, Pawankar R, Mori S, Nonaka M, Masuno S, Yagi T, et al. Increased expression and role of thymic stromal lymphopoietin in nasal polyposis. Allergy Asthma Immunol Res. 2011;3 (3):186–93.
132. Mjosberg J, Bernink J, Golebski K, Karrich JJ, Peters CP, Blom B, et al. The transcription factor GATA3 is essential for the function of human type 2 innate lymphoid cells. Immunity. 2012;37(4):649–59.
133. Liu T, Li TL, Zhao F, Xie C, Liu AM, Chen X, et al. Role of thymic stromal lymphopoietin in the pathogenesis of nasal polyposis. Am J Med Sci. 2011;341 (1):40–7.
134. Ouyang Y, Fan E, Li Y, Wang X, Zhang L. Clinical characteristics and expression of thymic stromal lymphopoietin in eosinophilic and non-eosinophilic chronic rhinosinusitis. ORL J Otorhinolaryngol Relat Spec. 2013;75(1):37–45.
135. Nakayama T, Hirota T, Asaka D, Sakashita M, Ninomiya T, Morikawa T, et al. A genetic variant near TSLP is associated with chronic rhinosinusitis with nasal polyps and aspirin-exacerbated respiratory disease in Japanese populations. Allergol Int. 2020;69(1):138–40.
136. Park SK, Jin YD, Park YK, Yeon SH, Xu J, Han RN, et al. IL-25-induced activation of nasal fibroblast and its association with the remodeling of chronic rhinosinusitis with nasal polyposis. PLoS One. 2017;12(8):e0181806.
137. Bachert C, Han JK, Desrosiers M, Hellings PW, Amin N, Lee SE, et al. Efficacy and safety of dupilumab in patients with severe chronic rhinosinusitis with nasal polyps (LIBERTY NP SINUS-24 and LIBERTY NP SINUS-52): results from two multicentre, randomised, double-blind, placebo-controlled, parallel-group phase 3 trials. Lancet. 2019;394(10209):1638–50.
138. Gevaert P, Holtappels G, Johansson SG, Cuvelier C, Cauwenberge P, Bachert C. Organization of secondary lymphoid tissue and local IgE formation to *Staphylococcus aureus* enterotoxins in nasal polyp tissue. Allergy. 2005;60(1):71–9.
139. Lombardi V, Beuraud C, Neukirch C, Moussu H, Morizur L, Horiot S, et al. Circulating innate lymphoid cells are differentially regulated in allergic and nonallergic subjects. J Allergy Clin Immunol. 2016;138(1):305–8.
140. Mitthamsiri W, Pradubpongsa P, Sangasapaviliya A, Boonpiyathad T. Decreased CRTH2 expression and response to allergen re-stimulation on innate lymphoid cells in patients with allergen-specific immunotherapy. Allergy Asthma Immunol Res. 2018;10 (6):662–74.
141. Burks AW, Sampson HA, Plaut M, Lack G, Akdis CA. Treatment for food allergy. J Allergy Clin Immunol. 2018;141(1):1–9.
142. Sampath V, Sindher SB, Alvarez Pinzon AM, Nadeau KC. Can food allergy be cured? What are the future prospects? Allergy. 2019;75(6):1316–26.
143. Kim CH, Hashimoto-Hill S, Kim M. Migration and tissue tropism of innate lymphoid cells. Trends Immunol. 2016;37(1):68–79.
144. Chu DK, Llop-Guevara A, Walker TD, Flader K, Goncharova S, Boudreau JE, et al. IL-33, but not thymic stromal lymphopoietin or IL-25, is central to mite and peanut allergic sensitization. J Allergy Clin Immunol. 2013;131(1):187–200.
145. Lee J-B, Chen C-Y, Liu B, Mugge L, Angkasekwinai P, Facchinetti V, et al. IL-25 and CD4+ TH2 cells enhance type 2 innate lymphoid cell-derived IL-13 production, which promotes IgE-mediated experimental food allergy. J Allergy Clin Immunol. 2016;137(4):1216–25.
146. Burton OT, Medina Tamayo J, Stranks AJ, Miller S, Koleoglou KJ, Weinberg EO, et al. IgE promotes type 2 innate lymphoid cells in murine food allergy. Clin Exp Allergy. 2018;48(3):288–96.
147. Noval Rivas M, Burton OT, Oettgen HC, Chatila T. IL-4 production by group 2 innate lymphoid cells promotes food allergy by blocking regulatory T-cell function. J Allergy Clin Immunol. 2016;138(3): 801–11.

148. Leyva-Castillo J-M, Galand C, Kam C, Burton O, Gurish M, Musser MA, et al. Mechanical skin injury promotes food anaphylaxis by driving intestinal mast cell expansion. Immunity. 2019;50(5):1262–75.
149. Chu DK, Mohammed-Ali Z, Jiménez-Saiz R, Walker TD, Goncharova S, Llop-Guevara A, et al. T helper cell IL-4 drives intestinal Th2 priming to oral peanut antigen, under the control of OX40L and independent of innate-like lymphocytes. Mucosal Immunol. 2014;7(6):1395–404.
150. Matsuda A, Ebihara N, Yokoi N, Kawasaki S, Tanioka H, Inatomi T, et al. Functional role of thymic stromal lymphopoietin in chronic allergic keratoconjunctivitis. Investig Opthalmol Visual Sci. 2010;51 (1):151.
151. Matsuda A, Okayama Y, Terai N, Yokoi N, Ebihara N, Tanioka H, et al. The role of interleukin-33 in chronic allergic conjunctivitis. Investig Opthalmol Vis Sci. 2009;50(10):4646.
152. Imai Y, Hosotani Y, Ishikawa H, Yasuda K, Nagai M, Jitsukawa O, et al. Expression of IL-33 in ocular surface epithelium induces atopic keratoconjunctivitis with activation of group 2 innate lymphoid cells in mice. Sci Rep. 2017;7(1):10053.
153. Sugita J, Asada Y, Ishida W, Iwamoto S, Sudo K, Suto H, et al. Contributions of interleukin-33 and TSLP in a papain-soaked contact lens-induced mouse conjunctival inflammation model. Immun Inflamm Dis. 2017;5(4):515–25.

Innate Lymphoid Cells and Inflammatory Bowel Disease

7

Vincent Peng, Natalia Jaeger, and Marco Colonna

Abstract

The signature hallmark of adaptive immunity is the evolution of somatically rearranged antigen receptors, which confer both diversity and specificity to T and B lymphocytes. For decades, immunologists have observed cells which possess lymphoid characteristics yet lack such antigen-specific receptors. Collectively, these populations are referred to as innate lymphoid cells (ILCs) (Vivier et al. in Cell 174(5):1054–1066, 2018). Cytotoxic natural killer (NK) cells and lymphoid tissue-inducing cells (LTi), which contribute to the formation of lymphoid organs during embryogenesis, are the earliest described ILCs. Subsequently, diverse populations of ILCs have been described based on the signature cytokines they produce. Group 1 ILCs (ILC1) produce IFNγ, group 2 ILCs (ILC2) produce IL-5 and IL-13, and group 3 ILCs (ILC3) produce IL-22 and IL-17. In contrast to adaptive lymphocytes which take several days to undergo clonal expansion and acquire effector functions, ILCs secrete cytokines rapidly in response to activating signals in their tissue of residence. ILCs may also directly regulate adaptive lymphocytes and myeloid cells through co-stimulatory molecules and soluble factors. Thus, ILCs play important roles in both the initiation and amplification of the immune response. When properly regulated, ILCs maintain intestinal homeostasis and protect the host from infection by various pathogens. However, dysregulation of mucosal immunity drives intestinal inflammation and contributes to pathology, such as inflammatory bowel disease (IBD). In this review, we outline the roles that ILCs play in amplifying or regulating intestinal inflammation as well as ongoing efforts to target these disease mechanisms for IBD therapy.

Keywords

Innate lymphoid cells · ILC · Inflammatory bowel disease · IBD · Mucosal immunity

7.1 Intestinal ILCs During Homeostasis

In both mice and humans, ILCs exhibit a diverse array of surface markers and transcription factors which facilitate their identification and classification (Table 7.1). With the exception of NK cells, ILCs are largely tissue resident and are especially enriched at mucosal tissues [2]. The gastrointestinal (GI) tract contains the full diversity of ILCs, where they are highly represented within the

Vincent Peng and Natalia Jaeger contributed equally with all other contributors.

V. Peng · N. Jaeger · M. Colonna (✉)
Department of Pathology and Immunology, Washington University School of Medicine, St Louis, MO, USA
e-mail: mcolonna@wustl.edu

© The Author(s), under exclusive license to Springer Nature Singapore Pte Ltd. 2022

X.-H. Sun (ed.), *Innate Lymphoid Cells*, Advances in Experimental Medicine and Biology 1365,
https://doi.org/10.1007/978-981-16-8387-9_7

Table 7.1 Phenotypic identity of ILC in mice and humans. Adapted from Vivier et al. [1]

Mouse	Marker	NK	ILC1	ILC2	LTi-like ILC3	NKp46+ ILC3
Cell-surface molecules	CD45	+	+	+	int	int
	CD127 (Il-7Ra)	+/−	+/−	+	+	int
	CD161 (NK1.1)	+	+	−	−	+/−
	ST2 (IL-33R)	−	nd	+/−	nd	nd
	CD278 (ICOS)	Int	nd	+	int	int
	IL-17RB (IL25R)	−	nd	+	−	−
	KLRG1	+/−	−	+/−	−	−
	CD117 (c-kit)	−	+/−	+/−	int	int
	CD69	−	+	nd	nd	nd
	CD254 (RANKL)	nd	nd	nd	+	+
	CD196 (CCR6)	−	nd	−	+	−
	CD4	−	−	−	+/−	−
	CD335 (NKp46)	+	+	−	−	+
	CD25 (IL-2Ra)		−	+	+/−	+/−
	MHC-II	−	−	+	+	−
	IL23R	−	−	nd	+	+
	IL-1R	−	+	nd	+	+
	CD122	+	+	−	−	−
	CD314 (NKG2D)	+	nd	−	−	+
	Ly49	+/−	+/−	−	−	−
	CD94	+/−	nd	+/−	−	+/−
	CD253 (TRAIL)	−	+	nd	nd	nd
	Sca-1 (Ly-6a)	−	+	+	nd	+
	CD49d (integrin a4b7)	nd	nd	−	+	+
	CD49a (integrin a1B1)	−	+	nd	nd	nd
	CD90 (Thy1)	+/−	+	+/−	+	+
	CD160	+	+	nd	nd	nd
	CD103	−	−	nd	nd	nd
	CD200R	−	+	nd	nd	nd
	CD304 (NRP-1)	nd	nd	nd	+	−
Transcription factors	T-BET	+	+	−	−	+
	EOMES	+	+/−	−	−	−
	RORyT	−	−	−	+	+
	GATA3	int	int	+	int	int
	AhR	−	+	+	+	+
	RORα	nd	nd	+	nd	+

Human	Marker	NK	ILC1	ILC2	NKp44+ ILC3
Cell-surface molecules	CD45	+	+	+	+
	CD127 (Il-7Ra)	+/−	+/−	+	+
	CD161	+	+	+	+
	CD278 (ICOS)	−	nd	+	+
	IL-17RB (IL25R)	−	−	+	−
	CD294 (CRTH2)	−	−	+	−
	KLRG1	+/−	+/−	+	−
	CD117 (c-kit)	+/−	−	+/−	+
	CD69	int	+/−	nd	nd
	CD254 (RANKL)	−	nd	nd	+
	CD196 (CCR6)	−	+/−	−	+
	CD4	+	+/−	−	−

(continued)

Table 7.1 (continued)

Human	Marker	NK	ILC1	ILC2	NKp44+ ILC3
	CD335 (NKp46)	+	+/–	–	+
	CD25 (IL-2Ra)	+/–	int	+	+/–
	IL-23R	+/–	+/–	–	+
	IL-1R	+/–	+/–	int	+
	CD122	+	nd	nd	int
	CD314 (NKG2D)	+	+	nd	–
	CD94	+/–	+/–	–	–
	IL-12RB	+	+	–	–
	CD56	+	+/–	–	+/–
	CD183 (CXCR3)	+/–	+	nd	–
	CD337 (NKp30)	+	+	+/–	+/–
	CD336 (NKp44)	–	+/–	–	+
	CD103	–	+/–	nd	–
	CD16	+/–	–	–	–
	NKp80	+	–	nd	nd
	CD300LF	+/–	+/–	nd	+
	CD160	+/–	+/–	–	–
	CD39	–	+/–	–	+/–
	CXCR6	–	+/–	–	+/–
	TIGIT	+/–	+/–	–	–
Transcription factors	T-BET	+	+	–	–
	EOMES	+	+/–	–	–
	RORγT	–	int/–	int/–	+
	GATA3	int	int	+	int
	AhR	+	+	+	+
	IKAROS	+	+	+	+
	AIOLOS	+/–	+/–	–	–
	HELIOS	–	–	–	+/–

+ positive expression, int intermediate or low expression, +/– heterogeneous expression, – negative expression, nd expression not determined

intestinal lamina propria (LP). Along the murine GI tract, ILC2 and ILC3 are the most prevalent groups of ILC in the small intestine LP with ILC3 in greater abundance, whereas ILC2 are more enriched in the stomach and colon LP [3]. Compared to the LP, NK cells and ILC1 are the most abundant population of ILCs in the intraepithelial lymphocyte (IEL) compartment of the small intestine [4]. In humans, ILC3 are the most abundant type of ILC in the ileum and colon with the lowest frequency in the esophagus, stomach, and duodenum [5]. In contrast, ILC1 are most enriched in these areas, potentially reflecting a difference in tissue-specific signaling which biases ILC differentiation or recruitment. Compared with ILC1 and ILC3, ILC2 are much less abundant across the entire GI tract.

ILC1 were first characterized as TRAIL^{+} NK1.1^{+} NK cells in the liver of adult mice and were thought to be immature precursors to conventional NK cells based on their CD127^{+} CD27^{+} CD11b^{-} phenotype and lack of expression of most Ly49 receptors, CD49b (DX5), and EOMES [6]. However, adoptive transfer studies using ILC progenitors have established that ILC1 and NK emerge from distinct committed precursors [7–9]. These developmental studies also established that ILC1 share a universal requirement for the transcription factor T-BET, but, unlike NK cells, develop independently of EOMES. In humans and mice, ILC1 can be found in both the LP and the intraepithelial lymphocyte (IEL) layer of the small intestine. The surface phenotype of these ILC1 is largely overlapping with NK cells; both cell types express classical NK lineage markers NK1.1 and NKp46 and produce the signature cytokine IFNγ. The identification of ILC1 is performed using integrin α_1 (CD49a) and CD200R1, which are preferentially expressed by ILC1, and EOMES, which is expressed in NK cells but not ILC1 [10, 11]. A recent study using single-cell RNA-sequencing of ILC1 from multiple murine tissues has demonstrated several tissue-imprinting signatures among ILC1 [12]. Specifically, intestinal ILC1 from the IEL or LP expressed higher amounts of the transcription factors *Nfkbiz*, *Nfkbia*, *Rora*, *Ahr*, and *Hic1* when compared to ILC1 from other tissues in the body. *Rora* and *Hic1* are significantly involved in retinoic acid signaling, and *Ahr*, which is important for the generation of ILC3, regulates multiple transcriptional programs in response to metabolites produced from the diet and intestinal microflora [13–15]. Thus, the phenotype and function of ILC1 are sculpted by the intestinal environment, especially by the metabolites that are abundant in this compartment.

Human ILC1 encompass two major phenotypes depending on the intestinal compartment from which they originate. Intraepithelial ILC1s (IEL-ILC1) are the most abundant ILC subset in the IEL compartment of the small intestine. In addition to CD49a, IEL-ILC1 are identified based on the positive expression of CD56, NKp44, and CD103 [16]. CD103, encoded by *ITGAE*, forms a heterodimer with β_7 integrin to bind to E-cadherin, a cellular adhesion molecule expressed by intestinal epithelial cells, thus establishing IEL-ILC1 residence in the IEL compartment. Both human and mouse IEL-ILC1 express the activating receptor CD160, which recognizes HVEM on IECs and mediates defense against acute bacterial infections [17]. LP ILC1 express IL-7Rα (CD127) and the C-type lectin CD161 but are negative for CD56, NKp44, c-Kit, and CD160 [16, 18]. IEL-ILC1 express IL-15 receptor and IL-18 receptor whereas LP ILC1 express receptors for IL-12 and IL-1β [18, 19]. The expression of different chemokine and interleukin receptors reveals that each population may rely on different factors to migrate to each layer of the intestines, survive, and execute their effector functions.

ILC2 are identified as lineage-negative lymphocytes which express high levels of the transcription factor GATA-3 and possess the capacity to produce IL-4, IL-5, and IL-13 [20–22]. Along the murine GI tract, ILC2 are highly abundant with pronounced enrichment in the stomach and small intestine. The surface phenotype of ILC2 is highly heterogenous based on the tissue of residence [23]. In the intestine, ILC2 are KLRG1$^+$ IL-25R$^+$ and express lower levels of ST2 and CD25 than their counterparts in other tissues. ILC2 can also be identified by high constitutive expression of IL-5 and IL-13, and reporter mice for these cytokines have been valuable for studying this cell type [24, 25]. ILC2 homeostasis and activation are significantly regulated by cytokines IL-25, IL-33, and TSLP, which are largely produced by epithelial cells [26]. In the small intestine, tuft cells, a rare population of chemosensory epithelial cells, constitutively produce IL-25 and promote ILC2 expansion during helminth infection [27, 28]. Emerging work has demonstrated an unanticipated role for neuro-immune cross talk in regulating ILC2 function—these interactions have been thoroughly reviewed by others [29]. The primary effector function of ILC2 is the rapid and robust production of type 2 cytokines such as IL-4, IL-5, IL-9, and IL-13. In the intestine, these cytokines are highly relevant in promoting the classical “weep and sweep” response to intestinal parasites [30]. This response involves massive eosinophil recruitment, goblet cell hyperplasia resulting in increased mucus secretion, and rapid smooth muscle contractions leading to heightened intestinal motility. In animal models of helminth infection, ILC2 have been definitively shown to play a significant and nonredundant role in mediating this response and expulsion of parasites [31, 32].

In humans, ILC2 also lack T, B, and NK lineage markers and express CD127. ILC2 can be distinguished from other human ILC subsets by surface expression of the chemoattractant receptor-homologous molecule expressed on Th2 lymphocytes (CRTH2, also known as prostaglandin D_2 receptor 2), encoded by *PTGDR2* [33, 34]. Human ILC2 also express CD7, c-Kit (CD117), ICOS, CD161, and IL-2 high-affinity receptor (IL-2Rα or CD25) [33]. Like their mouse counterparts, ILC2 express the highest levels of GATA-3 compared to other ILCs. Human ILC2, when activated with IL-25 and IL-33, produce robust amounts of IL-5 and IL-13 [35]. In addition to type 2 cytokines, murine and human ILC2 are both capable of producing the epidermal growth factor receptor ligand amphiregulin (AREG), which promotes wound healing and epithelial remodeling [36]. Interestingly, it has been shown that T regulatory cells (Tregs) express EGFR and that AREG enhances their effector regulatory functions [37]. Collectively, these findings suggest an important role for ILC2 in epithelial regeneration and resolution of tissue inflammation.

ILC3 are the most abundant ILC type in the GI tract, which is their central tissue of residence. Murine ILC3 can be identified as lineage-negative $CD45^{low}$ Thy1+ lymphocytes which express the lineage-defining transcription factor RORγT [38, 39]. ILC3 have been classified into three subsets based on the expression of the NK cell receptor NKp46 and the chemokine receptor CCR6: $NKp46^+$ ILC3, $CCR6^+$ ILC3, and double-negative (DN) ILC3. $CCR6^+$ ILC3 are also referred to as adult LTi-like ILC3 based on their phenotypic similarity to fetal LTi, which are found during embryogenesis. Despite being considered as part of the same family, $NKp46^+$ ILC3 and $CCR6^+$ ILC3 develop from distinct precursors and exhibit differences in effector function, spatial distribution, and regulation [9]. $NKp46^+$ ILC3 predominantly produce IL-22 and GM-CSF, whereas adult LTi-like ILC3 are capable of producing IL-22, IL-17A/F, and lymphotoxins [38, 40–42]. Both IL-22 and IL-17 play important roles in intestinal epithelial barrier function. IL-22 promotes intestinal barrier function through multiple mechanisms such as (1) promoting antimicrobial peptide production; (2) inducing tight junction proteins, such as claudins; and (3) promoting fucosylation and glycosylation of the intestinal epithelium, which nurtures the development of symbiotic commensal microbes [43–46]. In addition to preventing the intrusion of intestinal microbiota, IL-22 is also involved in promoting intestinal epithelial proliferation and survival which is beneficial for the host during intestinal injury but deleterious in the context of malignancy [47–49]. $NKp46^+$ ILC3 are dispersed throughout the LP and intestinal villi whereas $CCR6^+$ ILC3 are localized in clusters, or solitary isolated lymphoid tissue (SILT), near the base of intestinal crypts [50]. Production of IL-22 and IL-17 is markedly enhanced by IL-23 and IL-1β. In addition to regulation by cytokines, ILC3 are also under dynamic control by metabolites such as AHR ligands and short-chain fatty acids (SCFA) as well as neuronal signals such as vasoactive intestinal peptide (VIP)—these interactions have been comprehensively reviewed by others [29, 51, 52]. Lastly, the activity of $CCR6^+$ ILC3 is modulated by the location and density of SILT which are controlled by homotypic RANK–RANKL interactions, the chemokine receptor CXCR5, and the oxysterol sensor GPR183 [53–55].

Human ILC3 were first identified in the tonsil and the small intestine as $CD56^+$ $NKp44^+$ IL-22-producing cells [56]. Despite the expression of some NK markers, these ILC3 can be distinguished from NK cells by their restricted expression of the transcription factor RORγt. These ILC3 express IL-23R and IL-1Rα and are capable of producing IL-22, IL-26, leukemia inhibitory factor (LIF), GM-CSF, B cell-activating factor (BAFF), and TNFα. Like their murine counterparts, human ILC3 are significantly activated by IL-23 and IL-1β, but they are also positively regulated by IL-2 and IL-7 [57]. $NKp44^+$ ILC3 in the gut can express CD96, CD103, CD69, and CD39, reminiscent of the tissue-residency phenotype of IEL-ILC1 [56]. $NKp44^+$ ILC3 also express the chemokine CCL20 and its cognate receptor CCR6, indicating that they can regulate the recruitment and spatial distribution of other immune cells, as well as their own. Another subset of $NKp44^-$ ILC3 has also been described in the small intestine LP [58, 59]. Unlike $NKp44^+$ ILC, $NKp44^-$ ILC3 are capable of producing IL-17A/F. While the relationship between these two subsets remains unclear, this NKp44^{-} subset may be analogous to the murine DN ILC3 as it can differentiate into NKp44^{+} ILC3 in vitro [58, 59].

A special attribute of ILC3 with particular relevance to intestinal inflammation is the capacity to undergo plasticity toward ILC1 [60]. NKp46^{+} ILC3 were observed to express the ILC1-defining transcription factor T-BET, and this factor was shown to be required for their development [61, 62]. It has been demonstrated that graded acquisition of T-BET expression is an essential feature for the development of NKp46^{+} ILC3 from DN ILC3 in vitro and in vivo [63]. Fate-mapping studies using the *Rorc* locus have determined that ~30% of intestinal ILC1 are derived from ILC3 [8]. ILC3 → ILC1 plasticity is accompanied by profound functional changes. Commensurate with the acquisition of T-BET, NKp46^{+} ILC3 gain the capacity to produce IFNγ in response to stimulation with IL-12. Studies in mice have found ILC3 → ILC1 plasticity to be positively regulated by microbiota, TLR signaling, and IL-23, but IL-12 signaling was found to be dispensable [63–65]. Subsequent studies have identified the transcription factor c-MAF to be a negative regulator of ILC3 → ILC1 plasticity [65, 66].

The first evidence that human ILCs can interconvert to different subsets was shown in vitro by long-term culture of tonsil NKp44^{+} ILC3 in media supplemented with IL-7, IL-2, and/or IL-1β [57]. Under these conditions, ILC3 show reduced production of IL-22 and IL-17A and increased production of IFNγ and LIF. Subsequent studies found that the plasticity of NKp44^{+} ILC3 is driven by coordinated expression of the transcription factors T-BET and AIOLOS, where the former works to drive transcription of *IFNG* and the latter represses the ILC3 transcriptional program [67, 68]. The same study found that IL-23 and TGF-β promote the expression of both these factors and also reduce the level of RORγT. Another study has described CD117^{+} NKp44^{-} ILC3 to be plastic toward CD127^{+} NKp44^{-} ILC1, which is mediated by IL-12 and IL-18 [69]. Recently, evidence has shown that the plasticity of human NKp44^{+} ILC3 exists in vivo. High-dimensional single-cell analyses have shown that NKp44^{+} ILC3 and CD103^{+} ILC1 exist on opposite sides of a spectrum with multiple distinct intermediates which can be distinguished by different levels of CD103, CCR6, and CD300LF expression [67]. RNA velocity analysis has demonstrated a cellular trajectory which originates from NKp44^{+} ILC3 and passes through these intermediates to terminate in CD103^{+} ILC1. Taken together, ILC3 → ILC1 plasticity is a conserved biological process that occurs during homeostasis, but the rate and equilibrium of this process can be modulated by several environmental signals. The selective pressure behind the evolution of plasticity is unknown; since ILCs are tissue resident, plasticity may allow a given subset of ILCs to adapt the immune response to simultaneously counter diverse pathogens without having to rely on de novo differentiation of naïve cells from the bone marrow. Whatever the reason, in our modern world, this mechanism of plasticity may now be a consequential source of ILC pathogenicity in intestinal inflammation.

7.2 Roles of ILCs in Inflammatory Bowel Disease

Inflammatory bowel disease (IBD) is comprised of two chronic intestinal inflammatory disorders: Crohn's disease (CD) and ulcerative colitis (UC). CD frequently affects the terminal ileum but can involve the entire GI tract, whereas UC is restricted to the colon and rectum [70]. Furthermore, inflammation in CD is typically transmural and occurs in discontinuous "skip lesions" while inflammation in UC is restricted to the mucosa and often spans the colon in an uninterrupted pattern. CD is associated with additional features not typically observed in UC, such as intestinal granulomas, strictures, fistulas, and fibrofatty infiltration known as "creeping fat." Current evidence suggests that the pathogenesis of these diseases involves the interaction of host genetics with gut microbiota and their associated products [71]. Since ILCs are highly enriched in the GI mucosa, they are likely to affect IBD

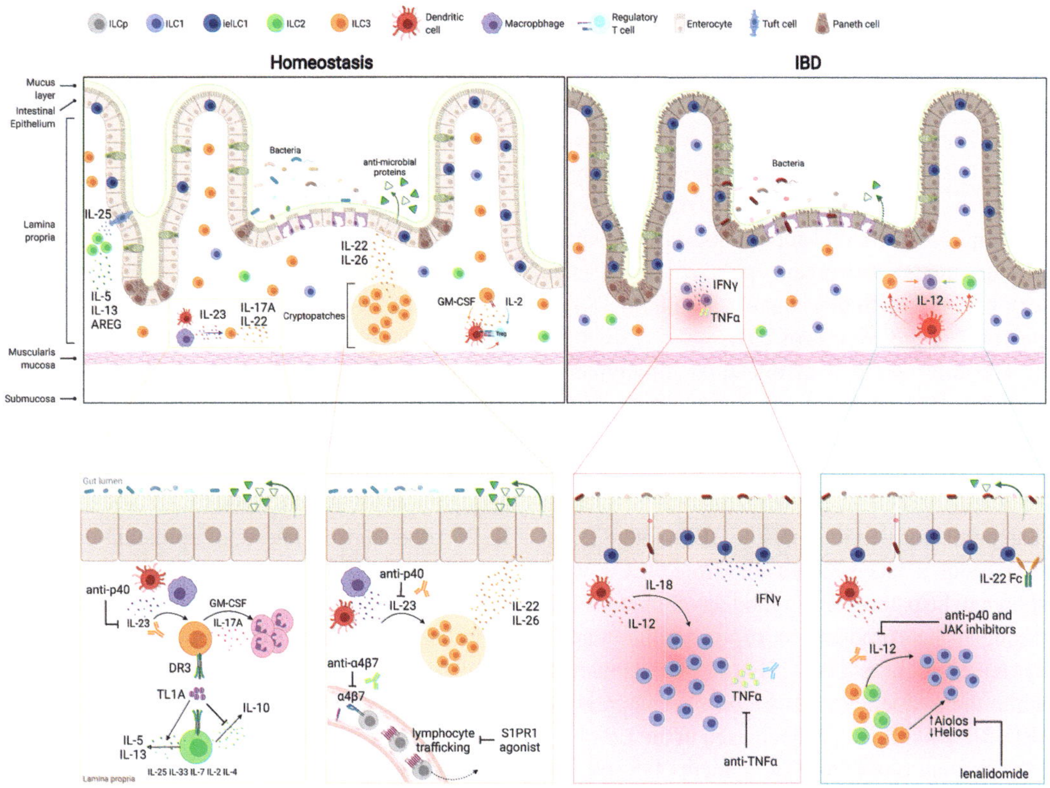

Fig. 7.1 ILCs in intestinal homeostasis and inflammation. All major subsets of ILCs are present in mouse and human intestine and contribute toward intestinal homeostasis. In IBD, ILCs undergo distinct biological changes including an increased ILC1–ILC3 ratio. Several candidate therapies for IBD act on ILC-associated pathways, either to dampen inflammatory signaling or to restore ILC-derived homeostatic functions. Adapted from "Intestinal Immune System (Small Intestine)" and "Immune Response in IBD," by BioRender.com (2021). Retrieved from https://app.biorender.com/biorender-templates

pathogenesis through the modulation of ongoing inflammation or through the provision of critical barrier functions that, when lost, lead to disease flares (Fig. 7.1). As the comprehensive identification of human ILCs can be challenging with traditional flow cytometry, recent studies have utilized high-dimensional mass cytometry or single-cell RNA sequencing to track how the intestinal immune compartment changes in IBD [19, 72–77]. Collectively, these studies have examined the lamina propria and intraepithelial layer of the small intestine and colon as well as how these compartments change during disease and response to therapy.

IEL-ILC1 are strategically located at the interphase between the luminal side and the intestinal barrier, where they play an essential role in maintaining intestinal homeostasis and act as the first line of defense against pathogens. The observation that all different populations of IEL, including IEL-ILC1, share similar effector programs suggests an evolutionary pressure to generate lymphocytes capable of protecting the intestinal epithelial layer through cytolytic function. IEL-ILC1 are key players in the IEL as they are constantly poised to respond in case of danger and can produce a copious amount of IFNγ, and in certain cases TNFα. However, in the context of IBD, these effector functions may exacerbate disease. Innate producers of these pro-inflammatory cytokines, such as ILC1, have been shown by several different studies to accumulate in the LP of the inflamed ileum and colon of patients with inflammatory bowel disease (IBD) [18, 69, 78, 79]. Furthermore, a study in mice has shown that depletion of IEL-ILC1 can ameliorate intestinal histopathology in an experimental model of innate colitis [16]. Taken together, these studies suggest a pathologic role for ILC1 in the context of IBD.

The role of ILC3 in intestinal inflammatory diseases is more ambiguous than ILC1. Genome-wide association studies have identified several Th17/ILC3-associated genes (e.g., *RORC, NFIL3, IL22, IL23R, IL17A, IL26, CCR6, STAT3*) that predispose individuals to IBD [80–82]. In support of a pro-inflammatory role, ILC3 were shown to be pathogenic in models of innate colitis through the production of pro-inflammatory cytokines such as GM-CSF [83–85]. However, while the frequency of Th17 cells is increased in IBD, NKp44^{+} ILC3 are decreased in the LP of the inflamed gut of CD and UC patients [18, 69, 86, 87]. Compared to Th17, ILC3 produce larger amounts of cytokine at baseline, including IL-22 and IL-26, that act on epithelial cells, to maintain intestinal homeostasis [38]. Furthermore, ILC3 can regulate T cell responses directly through MHC-II and IL-2 and indirectly through maintenance of tolerogenic DC via GM-CSF [42, 88–90]. A major theory on the disappearance of NKp44^{+} ILC3 is based on the increased conversion of ILC3–ILC1. This theory is consistent with the increased numbers of ILC1, decreased numbers of ILC3, and high levels of IL-12 present in the LP during inflammation. As there are no methods to fate-map cells in humans, the possibility cannot be ruled out that increased ILC1 differentiation and loss of ILC3 are biologically separate events. An additional question that remains is whether the altered ILC1–ILC3 ratio is a pathogenic cause underlying IBD or a consequence of the hyper-inflammatory milieu. One future approach to address this question is by examining ILC1 and ILC3 in patients in the early stages of disease where there is relatively decreased inflammation. Thus, ILC3 may play opposing contextual roles in IBD; aberrant regulation of ILC3 may promote the development of the disease, but the ensuing loss of ILC3-derived effector functions may prevent the restoration of intestinal homeostasis as inflammation progresses.

Due to their relatively low abundance in the intestinal mucosa, ILC2 have not been investigated in IBD to the same extent as ILC1 and ILC3. However, several studies have elucidated a potential role for ILC2 in the pathogenesis of IBD. A pathologic role for ILC2 has been suggested by studies using an oxazolone-induced model of colitis in mice, which has histologic resemblance to human UC [91]. Colonic histopathology in this model was dependent on IL-25 produced by intestinal epithelial cells and associated with IL-13 produced by colonic ILC2 and NKT cells [92]. A recent study also postulated a role for ILC2 in the pathogenesis of CD [93]. Using a genetically inbred mouse strain SAMP1/YitFc, which develops spontaneous ileitis with similar features to human CD, it was found that the onset of inflammation was associated with a large accumulation of ILC2, whereas ILC1 and ILC3 were minimally affected. Treatment with anti-ST2-blocking antibodies inhibited ILC2 expansion and protected against spontaneous ileitis. While these studies show a pathologic role for ILC2, ILC2 may also be beneficial during the resolution of intestinal inflammation. As discussed earlier, ILC2 also produce the tissue regenerative factor AREG. A study showed that AREG protected against acute dextran sulfate sodium (DSS)-induced colitis and that recombinant IL-33 was protective in this model [94]. Thus, in certain settings, ILC2 may be important for the repair and homeostasis of the intestinal tissue following inflammation. Studies regarding human ILC2 and IBD have been conducted in the context of both CD and UC. The frequency of ILC2 is increased in newly diagnosed UC patients [86]. Whether ILC2 are numerically altered in CD is unclear as some studies show an increase whereas others show no change [69, 79, 86, 93, 95, 96]. Like ILC3, human ILC2 were demonstrated to undergo plasticity toward ILC1 in the context of

IBD. A group recently identified a subset of IL-13^{+} IFNγ^{+} ILC2 present in the lamina propria of CD patients [96]. Moreover, they analyzed peripheral blood ILC2 from patients with a loss-of-function *IL12RB1* mutation and showed that ILC2 → ILC1 plasticity only occurs when *IL12RB1* is expressed. Together, these in vivo results suggest a role for IL-12 in ILC2 → ILC1 plasticity and a general mechanism for ILC1-directed plasticity in IBD pathogenesis.

7.3 Current and Future Therapies Targeting ILCs in IBD

The clinical translation of biologic agents targeting key immunologic molecules has revolutionized the treatment of IBD [97, 98]. Historically, an anti-TNF-α monoclonal antibody (mAb) was the first biologic approved for IBD management, followed by other mAbs targeting the IL-12/IL-23 signaling pathways and α4β7 integrin. Both IL-12 and IL-23 have been associated with the pathogenesis of several autoimmune diseases; thus, concurrent inhibition of these cytokines has become a promising therapeutic target [99]. Recently, ustekinumab, a mAb that targets the p40 subunit of IL-12 and IL-23, has been approved by the FDA to treat CD and UC [100, 101]. Moreover, multiple mAbs specifically targeting the p19 subunit of IL-23 are in advance phase trials for CD [102]. A clinical study has found a restored ILC subset ratio in the mucosa of CD patients treated with anti-TNF or anti-p40. Interestingly, treatment with these biologics resulted in an increase of NKp44^{+} ILC3 in the circulation over time [95].

Conversely, the use of mAbs against IL-17A was ineffective for IBD in humans [103], despite IL-17A being implicated in the pathogenesis of IBD and its blockage having become a successful treatment for other conditions [103, 104]. This failure in IBD therapeutic development remains a mystery, but it has been hypothesized that blockade of IL-17A may lead to impaired barrier function and integrity. Corroborating this, patients with other inflammatory conditions have reported developing CD while on treatment with anti-IL-17A mAbs [105]. These data indicate a context-specific role for IL-17A in mucosal homeostasis and autoimmunity, and agents targeting upstream regulators such as ustekinumab may be more therapeutically effective.

The development and function of ILCs are tightly controlled by cytokine signaling through the JAK-STAT pathway [106]. JAK3 is required for the efficient differentiation of all ILCs and chemical inhibition of JAK3 in human ILC block proliferation in vitro [107]. The receptors for IL-12 and IL-23 signal through the same combination of JAK (JAK2/Tyk2) proteins. JAK inhibitors are currently being investigated for the treatment of other inflammatory diseases, such as atopic dermatitis, and represent a promising approach for IBD [108, 109]. Tofacitinib, an inhibitor of JAK1/3 with minor inhibition of JAK2, is approved for moderate to severe UC but was unsuccessful in showing clinical efficacy against CD [109, 110]. Intriguingly, filgotinib, a selective JAK1 inhibitor, appears to be efficacious for the treatment of both UC and CD [111, 112]. Lastly, selective inhibition of JAK2 or TYK2 may be an attractive target to blunt IL-12 and IL-23 signaling while minimizing side effects [113]. As these cytokines are critical for ILC3 → ILC1 plasticity, this class of JAK inhibitors may be an exciting approach to specifically target this process in IBD.

As discussed earlier, the transcription factor AIOLOS plays a critical role in promoting ILC3 → ILC1 plasticity in humans. Since ILC1 accumulation has been associated with CD, inhibition of AIOLOS may block ILC3 → ILC1 trans-differentiation and reduce the number of pathogenic ILC1. Based on this concept, lenalidomide, a small-molecule inhibitor targeting AIOLOS and IKAROS, is currently a therapeutic candidate for CD treatment. Indeed, treatment with lenalidomide was shown to suppress ILC1 differentiation and promote the transcription factor HELIOS, which is typically associated with ILC3 identity [68].

Treatments targeting the homing, recruitment, and tissue residency of intestinal lymphocytes are also being explored. The anti-$\alpha_4\beta_7$ mAb vedolizumab is approved for use in UC and CD therapy and has shown remarkable clinical benefit for patients with these conditions [114, 115]. Vedolizumab works by blocking interaction between $\alpha_4\beta_7$ integrin and the mucosal vascular addressin cell adhesion molecule (MAdCAM-1) which is highly expressed by intestinal venules, thereby blocking the trafficking of circulating immune cells into intestinal tissues. In addition to gut-homing adaptive lymphocytes, circulating ILC progenitors have been shown to highly express $\alpha_4\beta_7$ integrin. Furthermore, treatment with vedolizumab has been associated with a decrease in intestinal ILC1 and increase in NKp44$^+$ ILC3, thus restoring the ILC1–ILC3 ratio observed in healthy controls [86]. Whether blocking of $\alpha_4\beta_7$ achieves this effect through direct inhibition of ILC differentiation or whether these observations are due to reduced intestinal inflammation remains unknown.

Given the reproducible and robust reduction of ILC3 in IBD, it has been proposed that crucial homeostatic functions provided by ILC3 are lost, leading to the exacerbation of intestinal inflammation. Several studies are focusing on restoring this balance, for instance, by developing an IL-22-Fc fusion protein [116]. Additionally, the role of IL-26, an IL-10 family member cytokine, has been understudied in IBD due to its lack of expression in mice. However, IL-26 is highly expressed by human Th17 and ILC3 [117]. IL-26 binds to IL10R2/IL20R1 expressed by epithelial cells and signals via the canonical STAT3/JAK1/TYK2 pathway that induces the secretion of pro-inflammatory cytokines, such as TNFα and IL-6, and antimicrobial proteins. Peculiarly, IL-26 has been shown to have DNA carrier capacity that induces the release of type I interferon and IL-1β [118], and UC patients have detectable cell-free DNA that is correlated with disease severity [119, 120]. GWAS studies identified a single-nucleotide polymorphism in the IL-26 gene associated with increased severity in IBD [82, 121], and IL-26 is elevated in the serum of CD patients. However, there is an indication that this polymorphism leads to reduced function of the protein, so perhaps the increase seen in patients is a compensatory response.

GWAS IBD studies identified another gene, *TNFSF15*, that has recently been associated with ILC3 biology [71, 122]. *TNFSF15*, which encodes the protein TL1A, is a TNF superfamily member which is expressed by antigen-presenting cells upon stimulation and intestinal CX_3CR1^+ mononuclear phagocytes [123]. TL1A binds to its receptor DR3, which is expressed by ILC2 and ILC3 as well as Tregs. The exact role of TL1A in intestinal inflammation is unclear. Mice lacking DR3 or TL1A are more susceptible to DSS-induced colitis through either decreased Tregs or loss of IL-22-producing ILC3 [124, 125]. Furthermore, mice overexpressing TL1A display spontaneous small intestinal inflammation attributed to IL-5 and IL-13 from over-activated ILC2 independent of T cells, implicating a central role for ILC2 regulation in these models [126, 127]. TL1A was also shown to inhibit ILC2 production of the immunosuppressive cytokine IL-10 in vitro, which may also explain the spontaneous inflammatory phenotype [128]. Lastly, a study found that agonistic antibodies directed against DR3 exacerbated experimental innate colitis in an ILC3-dependent manner [129]. This effect was attributed to increased production of GM-CSF by ILC3 which promotes the recruitment of neutrophils and eosinophils. In humans, treatment with TL1A or agonistic anti-DR3 antibodies elicits increased GM-CSF production from tonsil ILC3 in vitro. Taken together, it is likely that TL1A plays a context-specific role, where it positively regulates ILC3-derived IL-22 to promote barrier function during homeostasis but also amplifies ongoing inflammation through ILC3-derived GM-CSF. Because of its pro-inflammatory effects, DR3 is an attractive therapeutic target to dampen the inflammatory response in CD [130].

While the advent of IBD biologic agents has produced substantial and meaningful benefits in patient outcomes, the majority of IBD patients are refractory to complete remission by any single therapy. This likely reflects the heterogeneous nature of IBD and advocates for the development of meaningful biomarkers which can accurately predict response to therapy and expansion of the current arsenal of therapeutics to comprehensively target the diverse array of disease pathways. ILCs occupy a critical juncture at these two goals and better understanding of their biology will instruct new paradigms in our comprehension and treatment of IBD (Fig. 7.1).

References

1. Vivier E, Artis D, Colonna M, Diefenbach A, Di Santo JP, Eberl G, et al. Innate lymphoid cells: 10 years on. Cell. 2018;174(5):1054–66.
2. Gasteiger G, Fan X, Dikiy S, Lee SY, Rudensky AY. Tissue residency of innate lymphoid cells in lymphoid and nonlymphoid organs. Science. 2015;350(6263):981–5.
3. Kim CH, Hashimoto-Hill S, Kim M. Migration and tissue tropism of innate lymphoid cells. Trends Immunol. 2016;37(1):68–79.
4. Cortez VS, Colonna M. Diversity and function of group 1 innate lymphoid cells. Immunol Lett. 2016;179:19–24.
5. Krämer B, Goeser F, Lutz P, Glässner A, Boesecke C, Schwarze-Zander C, et al. Compartment-specific distribution of human intestinal innate lymphoid cells is altered in HIV patients under effective therapy. PLoS Pathog. 2017;13(5): e1006373.
6. Takeda K, Cretney E, Hayakawa Y, Ota T, Akiba H, Ogasawara K, et al. TRAIL identifies immature natural killer cells in newborn mice and adult mouse liver. Blood. 2005;105(5):2082–9.
7. Daussy C, Faure F, Mayol K, Viel S, Gasteiger G, Charrier E, et al. T-bet and Eomes instruct the development of two distinct natural killer cell lineages in the liver and in the bone marrow. J Exp Med. 2014;211(3):563–77.
8. Klose CSN, Flach M, Möhle L, Rogell L, Hoyler T, Ebert K, et al. Differentiation of type 1 ILCs from a common progenitor to all helper-like innate lymphoid cell lineages. Cell. 2014;157(2):340–56.
9. Constantinides MG, McDonald BD, Verhoef PA, Bendelac A. A committed precursor to innate lymphoid cells. Nature. 2014;508(7496):397–401.
10. Weizman O-E, Adams NM, Schuster IS, Krishna C, Pritykin Y, Lau C, et al. ILC1 confer early host protection at initial sites of viral infection. Cell. 2017;171(4):795–808.
11. Peng H, Jiang X, Chen Y, Sojka DK, Wei H, Gao X, et al. Liver-resident NK cells confer adaptive immunity in skin-contact inflammation. J Clin Investig. 2013;123(4):1444–56.
12. McFarland AP, Yalin A, Wang S-Y, Cortez VS, Landsberger T, Sudan R, et al. Multi-tissue single-cell analysis deconstructs the complex programs of mouse natural killer and type 1 innate lymphoid cells in tissues and circulation. Immunity. 2021;54(6): 1320–37.
13. Burrows K, Antignano F, Chenery A, Bramhall M, Korinek V, Underhill TM, et al. HIC1 links retinoic acid signalling to group 3 innate lymphoid cell-dependent regulation of intestinal immunity and homeostasis. PLoS Pathog. 2018;14(2):e1006869.
14. Larange A, Cheroutre H. Retinoic acid and retinoic acid receptors as pleiotropic modulators of the immune system. Annu Rev Immunol. 2016;34(1): 369–94.
15. Lee JS, Cella M, McDonald KG, Garlanda C, Kennedy GD, Nukaya M, et al. AHR drives the development of gut ILC22 cells and postnatal lymphoid tissues via pathways dependent on and independent of Notch. Nat Immunol. 2012;13(2):144–51.
16. Fuchs A, Vermi W, Lee JS, Lonardi S, Gilfillan S, Newberry RD, et al. Intraepithelial type 1 innate lymphoid cells are a unique subset of cytokine responsive interferon-γ-producing cells. Immunity. 2013;38(4): 769–81.
17. Shui J-W, Larange A, Kim G, Vela JL, Zahner S, Cheroutre H, et al. HVEM signalling at mucosal barriers provides host defence against pathogenic bacteria. Nature. 2012;488(7410):222–5.
18. Bernink JH, Peters CP, Munneke M, te Velde AA, Meijer SL, Weijer K, et al. Human type 1 innate lymphoid cells accumulate in inflamed mucosal tissues. Nat Immunol. 2013;14(3):221–9.
19. Simoni Y, Fehlings M, Kløverpris HN, McGovern N, Koo S-L, Loh CY, et al. Human innate lymphoid cell subsets possess tissue-type based heterogeneity in phenotype and frequency. Immunity. 2017;46(1): 148–61.
20. Moro K, Yamada T, Tanabe M, Takeuchi T, Ikawa T, Kawamoto H, et al. Innate production of T(H) 2 cytokines by adipose tissue-associated c-Kit(+) Sca-1(+) lymphoid cells. Nature. 2010;463(7280): 540–4.
21. Neill DR, Wong SH, Bellosi A, Flynn RJ, Daly M, Langford TKA, et al. Nuocytes represent a new innate effector leukocyte that mediates type-2-immunity. Nature. 2010;464(7293):1367–70.
22. Price AE, Liang H-E, Sullivan BM, Reinhardt RL, Eisley CJ, Erle DJ, et al. Systemically dispersed innate IL-13-expressing cells in type 2 immunity. PNAS. 2010;107(25):11489–94.
23. Ricardo-Gonzalez RR, Van Dyken SJ, Schneider C, Lee J, Nussbaum JC, Liang H-E, et al. Tissue signals imprint ILC2 identity with anticipatory function. Nat Immunol. 2018;19(10):1093–9.
24. Liang H-E, Reinhardt RL, Bando JK, Sullivan BM, Ho I-C, Locksley RM. Divergent expression patterns of IL-4 and IL-13 define unique functions in allergic immunity. Nat Immunol. 2012;13(1):58–66.
25. Nussbaum JC, Van Dyken SJ, von Moltke J, Cheng LE, Mohapatra A, Molofsky AB, et al. Type 2 innate

lymphoid cells control eosinophil homeostasis. Nature. 2013;502(7470):245–8.
26. Roan F, Obata-Ninomiya K, Ziegler SF. Epithelial cell-derived cytokines: more than just signaling the alarm. J Clin Investig. 2019;129(4):1441–51.
27. Gerbe F, Sidot E, Smyth DJ, Ohmoto M, Matsumoto I, Dardalhon V, et al. Intestinal epithelial tuft cells initiate type 2 mucosal immunity to helminth parasites. Nature. 2016;529(7585):226–30.
28. Howitt MR, Lavoie S, Michaud M, Blum AM, Tran SV, Weinstock JV, et al. Tuft cells, taste-chemosensory cells, orchestrate parasite type 2 immunity in the gut. Science. 2016;351(6279):1329–33.
29. Godinho-Silva C, Cardoso F, Veiga-Fernandes H. Neuro-immune cell units: a new paradigm in physiology. Annu Rev Immunol. 2019;37(1):19–46.
30. Douglas B, Oyesola O, Cooper MM, Posey A, Tait Wojno E, Giacomin PR, et al. Immune system investigation using parasitic helminths. Annu Rev Immunol. 2021;39(1):639–65.
31. Fallon PG, Ballantyne SJ, Mangan NE, Barlow JL, Dasvarma A, Hewett DR, et al. Identification of an interleukin (IL)-25-dependent cell population that provides IL-4, IL-5, and IL-13 at the onset of helminth expulsion. J Exp Med. 2006;203(4):1105–16.
32. Oeser K, Schwartz C, Voehringer D. Conditional IL-4/IL-13-deficient mice reveal a critical role of innate immune cells for protective immunity against gastrointestinal helminths. Mucosal Immunol. 2015;8(3):672–82.
33. Mjösberg JM, Trifari S, Crellin NK, Peters CP, van Drunen CM, Piet B, et al. Human IL-25- and IL-33-responsive type 2 innate lymphoid cells are defined by expression of CRTH2 and CD161. Nat Immunol. 2011;12(11):1055–62.
34. Xue L, Salimi M, Panse I, Mjösberg JM, McKenzie ANJ, Spits H, et al. Prostaglandin D2 activates group 2 innate lymphoid cells through chemoattractant receptor-homologous molecule expressed on TH2 cells. J Allergy Clin Immunol. 2014;133(4):1184–94.
35. Mjösberg J, Bernink J, Golebski K, Karrich JJ, Peters CP, Blom B, et al. The transcription factor GATA3 is essential for the function of human type 2 innate lymphoid cells. Immunity. 2012;37(4):649–59.
36. Monticelli LA, Sonnenberg GF, Abt MC, Alenghat T, Ziegler CGK, Doering TA, et al. Innate lymphoid cells promote lung-tissue homeostasis after infection with influenza virus. Nat Immunol. 2011;12(11):1045–54.
37. Zaiss DMW, van Loosdregt J, Gorlani A, Bekker CPJ, Gröne A, Sibilia M, et al. Amphiregulin enhances regulatory T cell-suppressive function via the epidermal growth factor receptor. Immunity. 2013;38(2):275–84.
38. Satoh-Takayama N, Vosshenrich CAJ, Lesjean-Pottier S, Sawa S, Lochner M, Rattis F, et al. Microbial flora drives interleukin 22 production in intestinal NKp46+ cells that provide innate mucosal immune defense. Immunity. 2008;29(6):958–70.
39. Satoh-Takayama N, Lesjean-Pottier S, Vieira P, Sawa S, Eberl G, Vosshenrich CAJ, et al. IL-7 and IL-15 independently program the differentiation of intestinal CD3-NKp46+ cell subsets from Id2-dependent precursors. J Exp Med. 2010;207(2):273–80.
40. Takatori H, Kanno Y, Watford WT, Tato CM, Weiss G, Ivanov II, et al. Lymphoid tissue inducer-like cells are an innate source of IL-17 and IL-22. J Exp Med. 2009;206(1):35–41.
41. Tumanov AV, Koroleva EP, Guo X, Wang Y, Kruglov A, Nedospasov S, et al. Lymphotoxin controls the IL-22 protection pathway in gut innate lymphoid cells during mucosal pathogen challenge. Cell Host Microbe. 2011;10(1):44–53.
42. Mortha A, Chudnovskiy A, Hashimoto D, Bogunovic M, Spencer SP, Belkaid Y, et al. Microbiota-dependent crosstalk between macrophages and ILC3 promotes intestinal homeostasis. Science. 2014;343(6178):1249288.
43. Goto Y, Obata T, Kunisawa J, Sato S, Ivanov II, Lamichhane A, et al. Innate lymphoid cells regulate intestinal epithelial cell glycosylation. Science. 2014;345(6202):1254009.
44. Zheng Y, Valdez PA, Danilenko DM, Hu Y, Sa SM, Gong Q, et al. Interleukin-22 mediates early host defense against attaching and effacing bacterial pathogens. Nat Med. 2008;14(3):282–9.
45. Tsai P-Y, Zhang B, He W-Q, Zha J-M, Odenwald MA, Singh G, et al. IL-22 upregulates epithelial claudin-2 to drive diarrhea and enteric pathogen clearance. Cell Host Microbe. 2017;21(6):671–81.
46. Pickard JM, Maurice CF, Kinnebrew MA, Abt MC, Schenten D, Golovkina T, et al. Rapid fucosylation of intestinal epithelium sustains host-commensal symbiosis in sickness. Nature. 2014;514(7524):638–41.
47. Grivennikov S, Karin E, Terzic J, Mucida D, Yu G-Y, Vallabhapurapu S, et al. IL-6 and Stat3 are required for survival of intestinal epithelial cells and development of colitis-associated cancer. Cancer Cell. 2009;15(2):103–13.
48. Kirchberger S, Royston DJ, Boulard O, Thornton E, Franchini F, Szabady RL, et al. Innate lymphoid cells sustain colon cancer through production of interleukin-22 in a mouse model. J Exp Med. 2013;210(5):917–31.
49. Lindemans CA, Calafiore M, Mertelsmann AM, O'Connor MH, Dudakov JA, Jenq RR, et al. Interleukin-22 promotes intestinal-stem-cell-mediated epithelial regeneration. Nature. 2015;528(7583):560–4.
50. Satoh-Takayama N, Serafini N, Verrier T, Rekiki A, Renauld J-C, Frankel G, et al. The chemokine receptor CXCR6 controls the functional topography of interleukin-22 producing intestinal innate lymphoid cells. Immunity. 2014;41(5):776–88.
51. Withers DR, Hepworth MR. Group 3 innate lymphoid cells: communications hubs of the intestinal immune system. Front Immunol. 2017;8:1298.

52. Willinger T. Metabolic control of innate lymphoid cell migration. Front Immunol. 2019;10:2010.
53. Emgård J, Kammoun H, García-Cassani B, Chesné J, Parigi SM, Jacob J-M, et al. Oxysterol sensing through the receptor GPR183 promotes the lymphoid-tissue-inducing function of innate lymphoid cells and colonic inflammation. Immunity. 2018;48(1):120–32.
54. Bando JK, Gilfillan S, Song C, McDonald KG, Huang SC-C, Newberry RD, et al. The tumor necrosis factor superfamily member RANKL suppresses effector cytokine production in group 3 innate lymphoid cells. Immunity. 2018;48(6):1208–19.
55. Sécca C, Bando JK, Fachi JL, Gilfillan S, Peng V, Luccia BD, et al. Spatial distribution of LTi-like cells in intestinal mucosa regulates type 3 innate immunity. PNAS. 2021;118(23):e2101668118.
56. Cella M, Fuchs A, Vermi W, Facchetti F, Otero K, Lennerz JKM, et al. A human NK cell subset provides an innate source of IL-22 for mucosal immunity. Nature. 2009;457(7230):722–5.
57. Cella M, Otero K, Colonna M. Expansion of human NK-22 cells with IL-7, IL-2, and IL-1β reveals intrinsic functional plasticity. PNAS. 2010;107(24): 10961–6.
58. Cupedo T, Crellin NK, Papazian N, Rombouts EJ, Weijer K, Grogan JL, et al. Human fetal lymphoid tissue–inducer cells are interleukin 17-producing precursors to RORC + CD127 + natural killer-like cells. Nat Immunol. 2009;10(1):66–74.
59. Crellin NK, Trifari S, Kaplan CD, Satoh-Takayama-N, Di Santo JP, Spits H. Regulation of cytokine secretion in human CD127+ LTi-like innate lymphoid cells by Toll-like receptor 2. Immunity. 2010;33(5):752–64.
60. Bal SM, Golebski K, Spits H. Plasticity of innate lymphoid cell subsets. Nat Rev Immunol. 2020;27: 1–14.
61. Sciumé G, Hirahara K, Takahashi H, Laurence A, Villarino AV, Singleton KL, et al. Distinct requirements for T-bet in gut innate lymphoid cells. J Exp Med. 2012;209(13):2331–8.
62. Rankin LC, Groom JR, Chopin M, Herold MJ, Walker JA, Mielke LA, et al. The transcription factor T-bet is essential for the development of NKp46 + innate lymphocytes via the Notch pathway. Nat Immunol. 2013;14(4):389–95.
63. Klose CSN, Kiss EA, Schwierzeck V, Ebert K, Hoyler T, d'Hargues Y, et al. A T-bet gradient controls the fate and function of CCR6-RORγt+ innate lymphoid cells. Nature. 2013;494(7436): 261–5.
64. Vonarbourg C, Mortha A, Bui VL, Hernandez PP, Kiss EA, Hoyler T, et al. Regulated expression of nuclear receptor RORγt confers distinct functional fates to NK cell receptor-expressing RORγt(+) innate lymphocytes. Immunity. 2010;33(5):736–51.
65. Tizian C, Lahmann A, Hölsken O, Cosovanu C, Kofoed-Branzk M, Heinrich F, et al. c-Maf restrains T-bet-driven programming of CCR6-negative group 3 innate lymphoid cells. eLife. 2020;9:e52549.
66. Parker ME, Barrera A, Wheaton JD, Zuberbuehler MK, Allan DSJ, Carlyle JR, et al. c-Maf regulates the plasticity of group 3 innate lymphoid cells by restraining the type 1 program. J Exp Med. 2019;217(1):e20191030. https://doi.org/10.1084/jem.20191030.
67. Cella M, Gamini R, Sécca C, Collins PL, Zhao S, Peng V, et al. Subsets of ILC3–ILC1-like cells generate a diversity spectrum of innate lymphoid cells in human mucosal tissues. Nat Immunol. 2019;20(8): 980–91.
68. Mazzurana L, Forkel M, Rao A, Acker AV, Kokkinou E, Ichiya T, et al. Suppression of Aiolos and Ikaros expression by lenalidomide reduces human ILC3−ILC1/NK cell transdifferentiation. Eur J Immunol. 2019;49(9):1344–55.
69. Bernink JH, Krabbendam L, Germar K, de Jong E, Gronke K, Kofoed-Nielsen M, et al. Interleukin-12 and -23 control plasticity of CD127(+) group 1 and group 3 innate lymphoid cells in the intestinal lamina propria. Immunity. 2015;43(1):146–60.
70. Abraham C, Cho JH. Inflammatory bowel disease. N Engl J Med. 2009;361(21):2066–78.
71. Jostins L, Ripke S, Weersma RK, Duerr RH, McGovern DP, Hui KY, et al. Host–microbe interactions have shaped the genetic architecture of inflammatory bowel disease. Nature. 2012;491(7422): 119–24.
72. Smillie CS, Biton M, Ordovas-Montanes J, Sullivan KM, Burgin G, Graham DB, et al. Intra- and intercellular rewiring of the human colon during ulcerative colitis. Cell. 2019;178(3):714–30.
73. Martin JC, Chang C, Boschetti G, Ungaro R, Giri M, Grout JA, et al. Single-cell analysis of Crohn's disease lesions identifies a pathogenic cellular module associated with resistance to anti-TNF therapy. Cell. 2019;178(6):1493–508.
74. Huang B, Chen Z, Geng L, Wang J, Liang H, Cao Y, et al. Mucosal profiling of pediatric-onset colitis and IBD reveals common pathogenics and therapeutic pathways. Cell. 2019;179(5):1160–76.
75. Corridoni D, Antanaviciute A, Gupta T, Fawkner-Corbett D, Aulicino A, Jagielowicz M, et al. Single-cell atlas of colonic CD8 + T cells in ulcerative colitis. Nat Med. 2020;26:1480–90.
76. West NR, Hegazy AN, Owens BMJ, Bullers SJ, Linggi B, Buonocore S, et al. Oncostatin M drives intestinal inflammation and predicts response to tumor necrosis factor-neutralizing therapy in patients with inflammatory bowel disease. Nat Med. 2017;23 (5):579–89.
77. Jaeger N, Gamini R, Cella M, Schettini JL, Bugatti M, Zhao S, et al. Single-cell analyses of Crohn's disease tissues reveal intestinal intraepithelial T cells heterogeneity and altered subset distributions. Nat Commun. 2021;12(1):1921.

78. Geremia A, Arancibia-Cárcamo CV, Fleming MPP, Rust N, Singh B, Mortensen NJ, et al. IL-23-responsive innate lymphoid cells are increased in inflammatory bowel disease. J Exp Med. 2011;208(6):1127–33.
79. Gwela A, Siddhanathi P, Chapman RW, Travis S, Powrie F, Arancibia-Cárcamo CV, et al. Th1 and innate lymphoid cells accumulate in primary sclerosing cholangitis-associated inflammatory bowel disease. J Crohns Colitis. 2017;11(9):1124–34.
80. Barrett JC, Hansoul S, Nicolae DL, Cho JH, Duerr RH, Rioux JD, et al. Genome-wide association defines more than 30 distinct susceptibility loci for Crohn's disease. Nat Genet. 2008;40(8):955–62.
81. Khor B, Gardet A, Xavier RJ. Genetics and pathogenesis of inflammatory bowel disease. Nature. 2011;474(7351):307–17.
82. Wang K, Baldassano R, Zhang H, Qu H-Q, Imielinski M, Kugathasan S, et al. Comparative genetic analysis of inflammatory bowel disease and type 1 diabetes implicates multiple loci with opposite effects. Hum Mol Genet. 2010;19(10):2059–67.
83. Buonocore S, Ahern PP, Uhlig HH, Ivanov II, Littman DR, Maloy KJ, et al. Innate lymphoid cells drive interleukin-23-dependent innate intestinal pathology. Nature. 2010;464(7293):1371–5.
84. Song C, Lee JS, Gilfillan S, Robinette ML, Newberry RD, Stappenbeck TS, et al. Unique and redundant functions of NKp46+ ILC3s in models of intestinal inflammation. J Exp Med. 2015;212(11):1869–82.
85. Pearson C, Thornton EE, McKenzie B, Schaupp A-L, Huskens N, Griseri T, et al. ILC3 GM-CSF production and mobilisation orchestrate acute intestinal inflammation. Elife. 2016;5:e10066.
86. Forkel M, van Tol S, Höög C, Michaëlsson J, Almer S, Mjösberg J. Distinct alterations in the composition of mucosal innate lymphoid cells in newly diagnosed and established Crohn's disease and ulcerative colitis. J Crohns Colitis. 2019;13(1):67–78.
87. Takayama T, Kamada N, Chinen H, Okamoto S, Kitazume MT, Chang J, et al. Imbalance of NKp44+-NKp46− and NKp44−NKp46+ natural killer cells in the intestinal mucosa of patients with Crohn's disease. Gastroenterology. 2010;139(3):882–92.
88. Hepworth MR, Monticelli LA, Fung TC, Ziegler CGK, Grunberg S, Sinha R, et al. Innate lymphoid cells regulate CD4+ T cell responses to intestinal commensal bacteria. Nature. 2013;498(7452):113–7.
89. Hepworth MR, Fung TC, Masur SH, Kelsen JR, McConnell FM, Dubrot J, et al. Group 3 innate lymphoid cells mediate intestinal selection of commensal bacteria-specific CD4+ T cells. Science. 2015;348(6238):1031–5.
90. Zhou L, Chu C, Teng F, Bessman NJ, Goc J, Santosa EK, et al. Innate lymphoid cells support regulatory T cells in the intestine through interleukin-2. Nature. 2019;568(7752):405–9.
91. Heller F, Fuss IJ, Nieuwenhuis EE, Blumberg RS, Strober W. Oxazolone colitis, a Th2 colitis model resembling ulcerative colitis, is mediated by IL-13-producing NK-T cells. Immunity. 2002;17(5):629–38.
92. Camelo A, Barlow JL, Drynan LF, Neill DR, Ballantyne SJ, Wong SH, et al. Blocking IL-25 signalling protects against gut inflammation in a type-2 model of colitis by suppressing nuocyte and NKT derived IL-13. J Gastroenterol. 2012;47(11):1198–211.
93. Salvo CD, Buela K-A, Creyns B, Corridoni D, Rana N, Wargo HL, et al. NOD2 drives early IL-33-dependent expansion of group 2 innate lymphoid cells during Crohn's disease-like ileitis. J Clin Investig. 2021;131(5):e140624.
94. Monticelli LA, Osborne LC, Noti M, Tran SV, Zaiss DMW, Artis D. IL-33 promotes an innate immune pathway of intestinal tissue protection dependent on amphiregulin–EGFR interactions. Proc Natl Acad Sci USA. 2015;112(34):10762–7.
95. Creyns B, Jacobs I, Verstockt B, Cremer J, Ballet V, Vandecasteele R, et al. Biological therapy in inflammatory bowel disease patients partly restores intestinal innate lymphoid cell subtype equilibrium. Front Immunol. 2020;11:1847.
96. Lim AI, Menegatti S, Bustamante J, Le Bourhis L, Allez M, Rogge L, et al. IL-12 drives functional plasticity of human group 2 innate lymphoid cells. J Exp Med. 2016;213(4):569–83.
97. Uhlig HH, Powrie F. Translating immunology into therapeutic concepts for inflammatory bowel disease. Annu Rev Immunol. 2018;36(1):755–81.
98. Cobb LM, Verneris MR. Therapeutic manipulation of innate lymphoid cells. JCI Insight. 2021;6(6):e146006.
99. Leppkes M, Neurath MF. Cytokines in inflammatory bowel diseases—update 2020. Pharmacol Res. 2020;158:104835.
100. Feagan BG, Sandborn WJ, Gasink C, Jacobstein D, Lang Y, Friedman JR, et al. Ustekinumab as induction and maintenance therapy for Crohn's disease. N Engl J Med. 2016;375(20):1946–60.
101. Sands BE, Sandborn WJ, Panaccione R, O'Brien CD, Zhang H, Johanns J, et al. Ustekinumab as induction and maintenance therapy for ulcerative colitis. N Engl J Med. 2019;381(13):1201–14.
102. Schmidt C, Grunert PC, Stallmach A. An update for pharmacologists on new treatment options for inflammatory bowel disease: the clinicians' perspective. Front Pharmacol. 2021;12:655054. https://doi.org/10.3389/fphar.2021.655054.
103. Hueber W, Sands BE, Lewitzky S, Vandemeulebroecke M, Reinisch W, Higgins PDR, et al. Secukinumab, a human anti-IL-17A monoclonal antibody, for moderate to severe Crohn's disease: unexpected results of a randomised, double-blind placebo-controlled trial. Gut. 2012;61(12):1693–700.
104. Blanco FJ, Möricke R, Dokoupilova E, Codding C, Neal J, Andersson M, et al. Secukinumab in active rheumatoid arthritis: a phase III randomized, double-

blind, active comparator- and placebo-controlled study. Arthritis Rheumatol. 2017;69(6):1144–53.
105. Fauny M, Moulin D, D'Amico F, Netter P, Petitpain N, Arnone D, et al. Paradoxical gastrointestinal effects of interleukin-17 blockers. Ann Rheum Dis. 2020;79(9):1132–8.
106. Stabile H, Scarno G, Fionda C, Gismondi A, Santoni A, Gadina M, et al. JAK/STAT signaling in regulation of innate lymphoid cells: the gods before the guardians. Immunol Rev. 2018;286(1):148–59.
107. Robinette ML, Cella M, Telliez JB, Ulland TK, Barrow AD, Capuder K, et al. Jak3 deficiency blocks innate lymphoid cell development. Mucosal Immunol. 2018;11(1):50–60.
108. Rogler G. Efficacy of JAK inhibitors in Crohn's disease. J Crohns Colitis. 2020;14(Supplement_2): S746–54.
109. Panés J, Sandborn WJ, Schreiber S, Sands BE, Vermeire S, D'Haens G, et al. Tofacitinib for induction and maintenance therapy of Crohn's disease: results of two phase IIb randomised placebo-controlled trials. Gut. 2017;66(6):1049–59.
110. Sandborn WJ, Su C, Sands BE, D'Haens GR, Vermeire S, Schreiber S, et al. Tofacitinib as induction and maintenance therapy for ulcerative colitis. N Engl J Med. 2017;376(18):1723–36.
111. Feagan BG, Danese S, Loftus EV, Vermeire S, Schreiber S, Ritter T, et al. Filgotinib as induction and maintenance therapy for ulcerative colitis (SELECTION): a phase 2b/3 double-blind, randomised, placebo-controlled trial. Lancet. 2021;397(10292):2372–84.
112. Vermeire S, Schreiber S, Petryka R, Kuehbacher T, Hebuterne X, Roblin X, et al. Clinical remission in patients with moderate-to-severe Crohn's disease treated with filgotinib (the FITZROY study): results from a phase 2, double-blind, randomised, placebo-controlled trial. Lancet. 2017;389(10066):266–75.
113. Danese S, Peyrin-Biroulet L. Selective tyrosine kinase 2 inhibition for treatment of inflammatory bowel disease: new hope on the rise. Inflamm Bowel Dis. 2021;27(12):2023–30. https://doi.org/10.1093/ibd/izab135.
114. Feagan BG, Rutgeerts P, Sands BE, Hanauer S, Colombel J-F, Sandborn WJ, et al. Vedolizumab as induction and maintenance therapy for ulcerative colitis. N Engl J Med. 2013;369(8):699–710.
115. Sandborn WJ, Feagan BG, Rutgeerts P, Hanauer S, Colombel J-F, Sands BE, et al. Vedolizumab as induction and maintenance therapy for Crohn's disease. N Engl J Med. 2013;369(8):711–21.
116. Rothenberg ME, Wang Y, Lekkerkerker A, Danilenko DM, Maciuca R, Erickson R, et al. Randomized phase I healthy volunteer study of UTTR1147A (IL-22Fc): a potential therapy for epithelial injury. Clin Pharmacol Ther. 2019;105(1): 177–89.
117. Stephen-Victor E, Fickenscher H, Bayry J. IL-26: an emerging proinflammatory member of the IL-10 cytokine family with multifaceted actions in antiviral, antimicrobial, and autoimmune responses. PLoS Pathog. 2016;12(6):e1005624.
118. Larochette V, Miot C, Poli C, Beaumont E, Roingeard P, Fickenscher H, et al. IL-26, a cytokine with roles in extracellular DNA-induced inflammation and microbial defense. Front Immunol. 2021;10: 204. https://doi.org/10.3389/fimmu.2019.00204.
119. Koike Y, Uchida K, Tanaka K, Ide S, Otake K, Okita Y, et al. Dynamic pathology for circulating free DNA in a dextran sodium sulfate colitis mouse model. Pediatr Surg Int. 2014;30(12):1199–206.
120. Rauh P, Rickes S, Fleischhacker M. Microsatellite alterations in free-circulating serum DNA in patients with ulcerative colitis. Digest Dis. 2003;21(4):363–6.
121. Silverberg MS, Cho JH, Rioux JD, McGovern DPB, Wu J, Annese V, et al. Ulcerative colitis-risk loci on chromosomes 1p36 and 12q15 found by genome-wide association study. Nat Genet. 2009;41(2): 216–20.
122. Yamazaki K, McGovern D, Ragoussis J, Paolucci M, Butler H, Jewell D, et al. Single nucleotide polymorphisms in TNFSF15 confer susceptibility to Crohn's disease. Hum Mol Genet. 2005;14(22): 3499–506.
123. Meylan F, Richard AC, Siegel RM. TL1A and DR3, a TNF family ligand-receptor pair that promotes lymphocyte costimulation, mucosal hyperplasia, and autoimmune inflammation. Immunol Rev. 2011;244 (1):188–96.
124. Longman RS, Diehl GE, Victorio DA, Huh JR, Galan C, Miraldi ER, et al. CX3CR1+ mononuclear phagocytes support colitis-associated innate lymphoid cell production of IL-22. J Exp Med. 2014;211(8):1571–83.
125. Castellanos JG, Woo V, Viladomiu M, Putzel G, Lima S, Diehl GE, et al. Microbiota-induced TNF-like ligand 1A drives group 3 innate lymphoid cell-mediated barrier protection and intestinal T cell activation during colitis. Immunity. 2018;49(6): 1077–89.
126. Yu X, Pappu R, Ramirez-Carrozzi V, Ota N, Caplazi P, Zhang J, et al. TNF superfamily member TL1A elicits type 2 innate lymphoid cells at mucosal barriers. Mucosal Immunol. 2014;7(3):730–40.
127. Meylan F, Hawley ET, Barron L, Barlow JL, Penumetcha P, Pelletier M, et al. The TNF-family cytokine TL1A promotes allergic immunopathology through group 2 innate lymphoid cells. Mucosal Immunol. 2014;7(4):958–68.
128. Bando JK, Gilfillan S, Di Luccia B, Fachi JL, Sécca C, Cella M, et al. ILC2s are the predominant source of intestinal ILC-derived IL-10. J Exp Med. 2019;217:e20191520. https://doi.org/10.1084/jem.20191520.
129. Li J, Shi W, Sun H, Ji Y, Chen Y, Guo X, et al. Activation of DR3 signaling causes loss of ILC3s and exacerbates intestinal inflammation. Nat Commun. 2019;10(1):3371.

130. Li Z, Buttó LF, Buela K-A, Jia L-G, Lam M, Ward JD, et al. Death receptor 3 signaling controls the balance between regulatory and effector lymphocytes in SAMP1/YitFc mice with Crohn's disease-like ileitis. Front Immunol. 2021;9:362. https://doi.org/10.3389/fimmu.2018.00362.

Coordination of Mucosal Immunity by Innate Lymphoid Cells

8

Jordan Z. Zhou and Gregory F. Sonnenberg

Abstract

Mucosal barrier surfaces of the mammalian body are frequent sites of pathogen colonization or entry and are also densely colonized with trillions of normally beneficial microbes, termed the microbiota. Therefore, it is paramount that the host immune system recognizes these microbes and is capable of differentiating between them. To this end, a multitude of mechanisms have evolved to carefully balance the need for immune activation in the face of infections while maintaining an appropriate level of tolerance to protect both the host and the beneficial microbes from hyperactivation. These mechanisms include the deployment of an emerging class of tissue-resident innate immune cells, innate lymphoid cells (ILCs), that are enriched at mucosal barriers such as the lungs and intestines, and are critical mediators of tissue homeostasis, tolerance, repair, and innate immunity. Recent findings have provided insight into the regulation of these cells and their interactions, not only with microbes, both commensal and foreign, but also with other systems of the body to prevent disease and promote tissue health. Here, we discuss recent findings in the regulation and function of ILCs, including a focus on their interactions with bodily systems, such as the nervous system, and how these interactions affect their functionality in states of health, infection, and disease.

Keywords

Innate lymphoid cells · Neuro-immune interactions · Cytokine biology · Mucosal immunity

8.1 Introduction

Mucosal barrier tissues such as the lungs and intestines are colonized by trillions of symbiotic microbes, termed the microbiota [1], which are essential for maintaining homeostasis, assisting in nutrient absorption, and metabolism [2, 3] as well as coordinating immunomodulation by working cooperatively with the host immune system to protect against invading pathogens [4, 5]. These functions are particularly important due to the continuous exposure of these tissues to the outside environment. As the microbiota is critical for

J. Z. Zhou · G. F. Sonnenberg (✉)
Division of Gastroenterology, Joan and Sanford I. Weill Department of Medicine, Weill Cornell Medicine, Cornell University, New York, NY, USA

Department of Microbiology and Immunology, Jill Roberts Institute for Research in Inflammatory Bowel Disease, Weill Cornell Medicine, Cornell University, New York, NY, USA

Jill Roberts Institute for Research in Inflammatory Bowel Disease, Weill Cornell Medicine, Cornell University, New York, NY, USA
e-mail: gfsonnenberg@med.cornell.edu

© The Author(s), under exclusive license to Springer Nature Singapore Pte Ltd. 2022

X.-H. Sun (ed.), *Innate Lymphoid Cells*, Advances in Experimental Medicine and Biology 1365,
https://doi.org/10.1007/978-981-16-8387-9_8

maintaining homeostasis, tolerogenic mechanisms must be employed to prevent overactive immune responses against these beneficial microbes; however, the immune system must still be poised to act in the case of pathogenic infection. Therefore, the immune system must maintain a delicate balance between tolerance and activation to maintain the beneficial interactions with the microbiota while also responding appropriately to invading pathogens. Critical to maintaining this balance are the interactions between the microbiota, pathogenic microbes, and mucosal immunity, particularly the innate immune system, which rapidly senses changes in the environment to subsequently calibrate an adaptive immune response.

One of the more recently appreciated members of the innate immune system, which have been shown to be essential for both the establishment and maintenance of tolerance to the microbiome as well as robust responses against pathogens, is the family of innate lymphoid cells (ILCs). ILCs are predominantly tissue-resident lymphocytes that are abundant at mucosal surfaces, where, during development, they make up a significant portion of the total immune cells [6]. While ILCs exhibit phenotypic and functional similarities to T cells, they lack the diversified antigen-specific T cell receptor (TCR) and rather respond rapidly to cytokines expressed by other immune cells or other environmental signals [7–10]. Additionally ILCs are dependent on the common gamma chain (γ_c), otherwise known as IL-2Rγ, the transcriptional repressor, inhibitor of DNA binding 2 (ID2) [11, 12], GATA3 [13], and, in some cases, zinc finger and BTB domain-containing protein 16 (PLZF) for their development [14]. Furthermore, ILCs commit to an effector state upon differentiation and are genetically poised to rapidly respond to environmental signals [15, 16], whereas T cells develop into a naïve state and later differentiate after antigen-presenting cell-mediated activation, generally in the secondary lymphoid organs.

The family of ILCs includes not only cytotoxic natural killer (NK) cells and lymphoid tissue inducer (LTi) cells, which were discovered in 1975 in the context of anticancer immunity [17] and 1997 in the context of lymphoid organ development during embryogenesis [18], respectively, but also several subsets of more recently defined noncytotoxic helper-like ILC subsets, including group 1 ILCs (ILC1s), group 2 ILCs (ILC2s), and group 3 ILCs (ILC3s). These helper-like ILC subsets, similar to their $CD4^+$ T cell counterparts, can be characterized by the expression of signature transcription factors and effector cytokines.

ILC1s are defined by their expression of the Th1 cell-associated transcription factor, T-box 21 (T-bet), and the production of interferon-γ. This group includes both the classical NK cells, which express high levels of IFN-γ, eomesodermin, and cytotoxic molecules such as granzyme and perforin after activation, and another subset of noncytotoxic or weakly cytotoxic IFN-γ and tumor necrosis factor (TNF)-producing ILC1s, which develop from an innate lymphoid cell precursor rather than an NK cell precursor in both humans and mice [7, 12, 14, 19, 20]. Functionally, ILC1s respond to IL-12, IL-15, and IL-18 [21, 22] and assist in antiviral responses [23] as well as protection against intracellular pathogens such as *Salmonella enterica* and *Toxoplasma gondii* [12, 24]. ILC2s are the innate counterparts to Th2 cells which are characterized by the high expression of transcription factors, GATA3 and, in mice, the retinoic acid receptor-related orphan receptor α (RORα) [25–27]. These cells respond to cytokine signals such as IL-25, IL-33, and thymic stromal lymphopoietin (TSLP) [28, 29] and produce classical Th2 cell cytokines in both humans and mice including IL-4, IL-5, IL-9, and IL-13 [30–32]. Additionally, mouse studies have shown that ILC2s express amphiregulin (AREG) to promote tissue recovery after viral infection [31] as well as IL-10 after stimulation with cytokines such as IL-2, IL-4, IL-10, IL-27, IL-33, the neuropeptide, neuromodulin (NMU), or the allergen, papain [33, 34]. Due to their production of Th2 cell-associated cytokines and ability to promote eosinophilia at sites of infection, ILC2s are critical for defense against helminth pathogens [35]; however, dysregulation of these cells has also been associated with chronic respiratory diseases such as asthma [36]. ILC3s, similar to Th17 cells, are defined by their expression of the transcription factor, RAR-related orphan receptor gamma (RORγt), and cytokines IL-17 and IL-22. ILC3s are abundant at mucosal surfaces, particularly the intestines, and are involved in responses to extracellular bacteria and regulation of the microbiome [10]. ILC3s can be segregated into a $NKp46^+$ population, which produce granulocyte-monocyte-colony-stimulating factor (GM-CSF) and, in a T-bet dependent manner, IFN-γ, as well as a $NKp46^-$ population, which are largely $CCR6^+$ LTi-like cells, that develop very early on and are necessary for the formation of lymphoid organs [18]. Clinically, due to their abundance in the intestinal tract, dysregulation of multiple ILC3 subsets has been associated with inflammatory bowel disease (IBD) and colorectal cancer (CRC).

Mouse models and studies in human samples indicate an essential role for ILCs in coordinating mucosal immunity. ILCs sense and interact with microbes, both commensal and pathogenic, to direct immunological responses, including direct cytokine-dependent immunity to pathogens, or more complex pathways of tolerance, repair, or inflammation, which involves dynamic interactions with multiple other hematopoietic or non-hematopoietic cell types. In this chapter, we discuss these host–microbe interactions and how they impact the regulation of the major ILC subsets in the context of mucosal immunity. We also review recent literature and discuss emerging avenues of research regarding the interaction between ILCs, microbes, and other physiological systems, such as the nervous system and cellular metabolism, which control the outcome of mucosal infections. Finally, we identify current gaps in knowledge among these areas and propose that additional research on these sophisticated cellular interactions could yield novel opportunities for therapeutic intervention.

8.2 ILCs and Microbes

ILCs exhibit robust and context-dependent interactions with different classes of microbes. For example, some subsets of ILCs require the microbiota for development, others only become activated following infection, and each subset differentially interacts with other cell types or receives direct signals from microbes in order to coordinate their effector functions. Below, we discuss the complex interactions between ILC subsets and microbes during mucosal immunity.

8.2.1 Infectious Microbes and ILC Function

Due to the localization of ILCs at barrier tissues, it is inevitable that they come in contact with infecting microbes from the outside environment and it is critical that these cell types are capable of contributing to host defense against these pathogens to prevent disease. However, this antimicrobial, protective function of ILCs must be tightly regulated to prevent aberrant inflammation.

ILC1s respond to infectious microbes in the intestine in a process typically driven by myeloid cell-derived IL-12, to subsequently induce IFN-γ-dependent mucosal immunity (Fig. 8.1). Additionally, follicular reticular cells (FRCs) in the mesenteric lymph nodes (mLN) and Peyer's patches secrete IL-15 to maintain ILC1 populations, though the production of this cytokine is limited in a MyD88-dependent manner to prevent overactivation of ILCs during infection [37]. Indeed, deletion of MyD88, increased IL-15 expression, and subsequent ILC1 expansion result in more severe disease pathology after viral infection. Importantly, ILC1 function cannot be completely inhibited since they are an essential source of IFN-γ in the tissues during the early stages of antiviral response to mouse cytomegalovirus (MCMV) infection [23]. However, this production of IFN-γ could become pathogenic as a recent study has shown that, during MCMV infection, microglia in the brain increase the expression of chemokines, C–X–C chemokine ligand (CXCL)9 and CXCL10, which engage CXCR3 on ILC1s and NK cells, targeting them to the brain, causing neuroinflammation in an IFN-γ-dependent manner [38]. Further, in the intestines, intraepithelial ILC1s were shown to

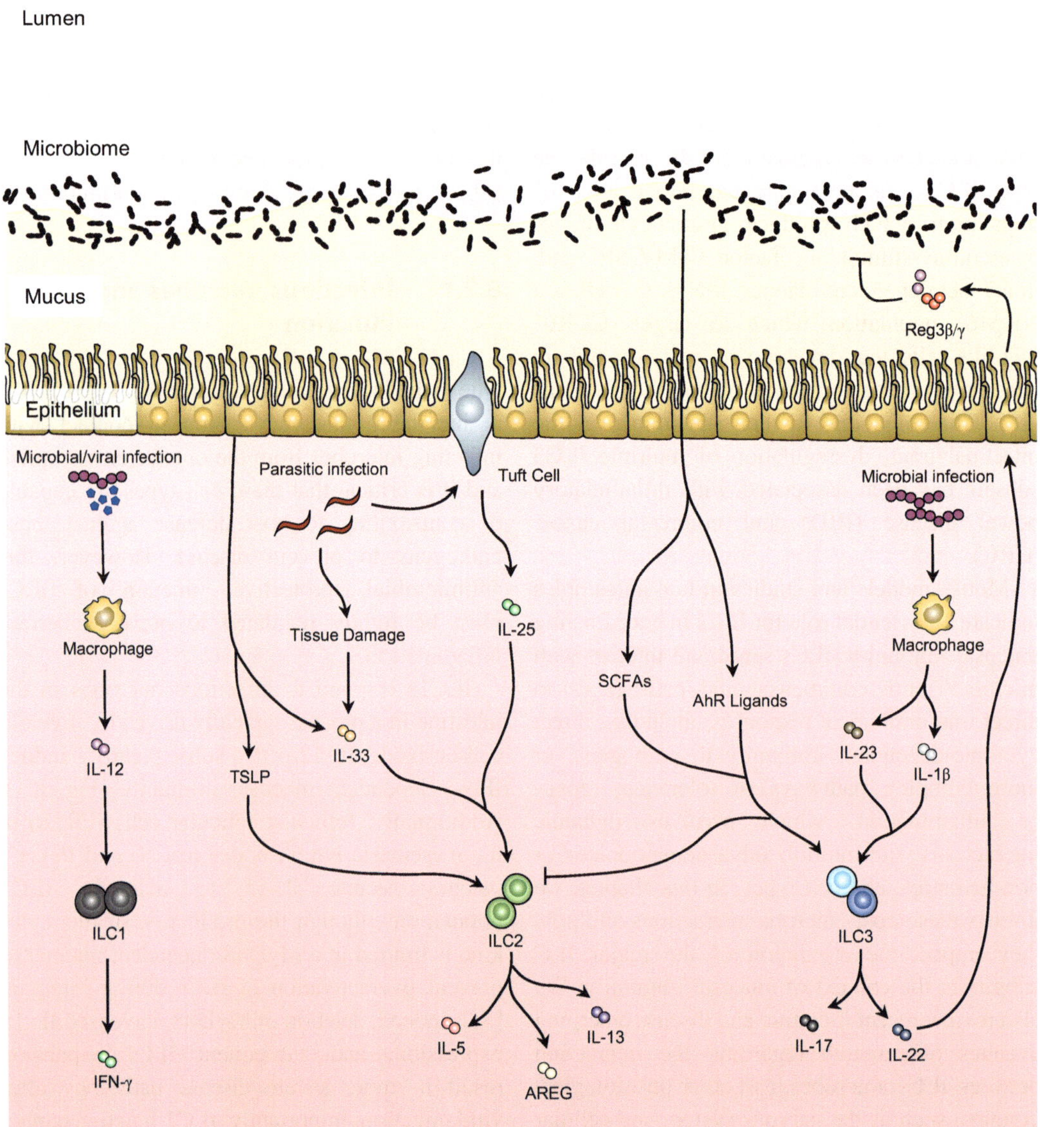

Fig. 8.1 Microbial- and cytokine-dependent circuits that orchestrate ILC-dependent mucosal immunity. ILCs are enriched at mucosal barriers and poised to respond to environmental signals. These signals are derived from microbes, diet, or host-derived factors that form distinct tissue circuits and promote ILC1, ILC2, or ILC3 responses. These ILCs interpret environmental signals to subsequently coordinate numerous other cell types and orchestrate diverse outcomes of tissue immunity, inflammation, repair, tolerance, and homeostasis

be dysregulated in IBD patients and increased IFN-γ production was associated with an anti-CD40 mouse model of colitis [22]. In addition to viral infections, ILC1s have also been implicated in protection against the intracellular parasite, *Toxoplasma gondii* [12]. In this study, *Rag1*$^{-/-}$*Il2rg*$^{-/-}$ mice, which lack T cells , B cells, and ILCs, reconstituted with purified ILC1s were found to initiate a significantly stronger inflammatory response after challenge with

T. gondii and were able to better control the infection when compared with mice that did not receive any donor ILCs. Interestingly, a separate study suggested that infection with *T. gondii* induces the conversion of NK cells to ILC1-like cells in a mechanism involving IL-12, which persists even after pathogen clearance [39], though whether this transition leads to pathologies, similar to what has been observed in the case of MCMV, remains unclear.

ILC2s are critical for host defense against parasitic infections, most commonly being associated with reaction to helminths, or worms (Fig. 8.1). Previous work had shown that IL-25, an IL-17 family member cytokine, is upregulated in mouse models of *Aspergillus fumigatus* and *Nippostrongylus brasiliensis* infection and subsequently induced eosinophilia in an IL-5- and IL-13-dependent manner [40, 41], although, it was not until several years later that IL-25 was found to be an activating signal for ILC2s [42] and that transfer of ILCs into IL-25R-deficient mice was able to partially restore IL-25-dependent innate immunity to *N. brasiliensis* infection through the production of IL-13 [35]. More recently, it was discovered that tuft cells uniquely secrete IL-25 in response to *N. brasiliensis* infection in mice and that tuft cell-derived IL-25 initiates intestinal ILC2 circuit of activation, which subsequently amplifies further tuft cell expansion [43–46]. The mechanisms by which IL-25 expression is induced in these cells are unclear, though the succinate receptor has been implicated as an activator of tuft cells that triggers type 2 immunity [47, 48]. Moreover, the cells that produce IL-25 in other parts of the body, such as the respiratory tract, to activate ILC2s are still uncertain, though tuft cells in the thymus have also been shown to produce IL-25 to promote ILC2 residency and function [49]. In addition to IL-25, infection with *N. brasiliensis* as well as the nematode, *Strongyloides venezuelensis*, also induces ILC2 production of IL-13 by promoting a strong IL-33 response, which is necessary for the accumulation of ILC2s during early phases of infection. Deficiency in IL-33 showed significant reductions in eosinophil infiltration and pathogen clearance [50, 51]. The activation of ILC2s by these alarmin molecules is not exclusive to parasitic infections, however, as a recent report showed that IL-33 is upregulated in response to *C. difficile* and that this induction of IL-33 can drive ILC2 activation and protection against disease in mouse models. Additionally, in patients with *C. difficile* infection, survival seems to be correlated with the expression of the IL-33 receptor, ST2 [52]. The reduction in the survival of mice deficient in ST2 seems somewhat modest in comparison to mice deficient in IFN-γ, a classical ILC1 cytokine that can also be produced by ILC3s [53], indicating a potentially greater role for these cell types [54]. Interestingly, ILC2s have also been associated with induction of adaptive immunity, specifically Th2 cells, through the expression of major histocompatibility complex class II (MHC-II) in mouse models of parasitic infections [55, 56], as well as through the expansion of DCs in response to allergens [57], making them critical mediators of both adaptive and innate immunity.

In the context of ILCs, much of the research regarding response to bacterial infections has been primarily focused on the functions of RORγt$^+$ ILC3s and their ability to produce cytokines such as IL-17, which promotes neutrophil infiltration, as well as IL-22, which can induce the expression of antimicrobial peptides from epithelial cells or reconstruction of the epithelium after infection (Fig. 8.1). This permits ILC3s to execute incredibly diverse responses and serve as potent mediators of barrier immunity. The expression of these cytokines, particularly IL-17 has been found to be crucial for the clearance of respiratory pathogens such as *Mycobacterium tuberculosis*, *Klebsiella pneumoniae*, and *Pseudomonas aeruginosa* [58–60] and, although IL-17 is also expressed by T cell subsets, it has been observed that production of IL-17A by ILC3s is essential for clearance of *K. pneumoniae* in mice by promoting antimicrobial functions of Ly6C^{hi} inflammatory monocytes, enhancing their phagocytic and bactericidal capabilities [61]. Importantly, Rag2-deficient mice, which lack B and T cells, did not show any significant difference in bacterial clearance relative to wild-type mice when infected with *K. pneumoniae* whereas mice with both Rag2 and γ_c deficiency had a significantly reduced antimicrobial response and survival. This phenomenon is consistent in mouse models of *M. tuberculosis* infection wherein reconstitution of Rag2$^{-/-}$ *Il2rg*$^{-/-}$ mice with lung RORγt^{+} ILC3s could rescue the wild-type phenotype [60], highlighting the importance of ILC3s in these models. In addition to promoting antimicrobial immunity in the respiratory tract, ILC3s have also been shown to promote immunity to pathogens in the intestine. *Citrobacter rodentium* is a well-studied mouse model of enterohemorrhagic infections and it has been shown that ILC3s are critical mediators of IL-22-dependent innate immunity [62–64]. Many original studies focused on the role of the NKp46^{+} ILC3 subset in promoting immunity to *C. rodentium* and their dependence on the microbiota [63, 64]. However, it was subsequently demonstrated that the NKp46^{+} subset is dispensable for innate immunity to this pathogen, and that the CCR6^{+} LTi-like subset is critical for IL-22-dependent innate immunity [65, 66]. Further studies showed that this process is dependent on IL-23 [62, 65, 67], likely generated by intestinal macrophages [68–70], and promotes antimicrobial response in epithelial cells. However, several fundamental questions remain regarding the role of ILC subsets in the context of mucosal infection, including a differential role for IL-23 in mucosal immunity that has been shown to be dependent upon pathogen dose [71], as well as potential redundancy with T cells in humans and mice [72–74].

Finally, ILC3s may impact many other aspects of mucosal immunity to pathogens that extend beyond IL-22 and IL-17. For example, LTi-like ILC3s are critical for the development and restoration of lymphoid tissues, which impact primary and secondary immune responses to viral infections [75, 76]. These cells can potentially also influence diverse functions like T cell selection in the thymus through the expression of receptor activator of nuclear factor kappa-B ligand (RANKL), which promotes the expression of autoimmune regulator (AIRE) in thymic epithelial cells [77]. NKp46^{+} ILC3s are also a potent source of IL-2 to support Tregs in the small intestine [78], which could also be a mechanism for contracting the immune response after clearance of the infectious microbes. Therefore, there is a complex array of pathways by which ILCs respond to and orchestrate an effective immune response against invading pathogens at mucosal barrier sites.

8.2.2 Direct Microbiota-Dependent Regulation of ILC Development and Function

Although several functions of ILCs are similar to those of T cells, they undergo a unique developmental path. In adults, these cells arise from common lymphoid progenitors (CLPs) in the bone marrow that differentiate into common helper ILC precursors [12], whereas, during fetal development, ILCs can differentiate from progenitors in the fetal liver [14, 79]. Due to their shared occupation of mucosal barrier tissues with the microbiota, many groups sought to describe the impact of these microbes on the development of ILCs. Studies utilizing germ-free (GF) mice revealed that both NK cells and ILC2s are able to develop in the absence of a microbiome [31, 80]. The importance of microbial colonization for ILC3 development has been subject to some debate as some reports have shown no significant changes in ILC3 numbers in the small intestines of either GF or antibiotic-treated mice [81], while others have shown a selective reduction in IL-22-producing NKp46^{+} ILC3s in the intestine of GF mice [64, 81, 82]. This is consistent with the dependence of this ILC3 subset on the aryl hydrocarbon receptor (Ahr), a transcription factor that is activated by a number of ligands produced in part by microbiota-derived metabolites [83–85]. However, this may be more nuanced and not necessarily dependent only on postnatal exposure to microbiota. A study using mice transiently colonized during pregnancy has implicated microbial colonization of the mother in the development of NKp46^{+} ILC3s [86]. In contrast to NKp46^{+} ILC3s, the presence of secondary lymphoid structures in mouse fetuses was thought to suggest normal development of LTi cells in the absence of a microbiome since the fetal environment was considered to be sterile [18]. However, maternal colonization also increased the number of LTi cell progenitors but had little impact on ILC1s and ILC2s. Furthermore, recent studies have provided evidence to suggest that the fetal environment may not be sterile, and that microbial colonization can occur as early as the second trimester in human fetuses, leading to priming of the fetal immune system [87]. This provides some clarification as Ahr has been shown to promote the development of isolated lymphoid follicles (ILFs) and cryptopatches (CPs) [84, 85], and microbiota-augmented retinoic acid [88] has been shown to be critical for maintaining LTi cells during fetal development [75]. Finally, following birth, Nod1-dependent signals from commensal bacteria can also promote the functional maturation of CPs and ILFs, indicating a critical role in supporting the functions of LTi-like ILC3s [89]. Based on these more recent studies, the importance of the microbiome in promoting the development of ILC3s is becoming less controversial.

Unlike their development, the functions of virtually all ILC subsets depend heavily on the status of the microbiome. Single-cell transcriptional profiling of mouse small intestines revealed a number of distinct ILC clusters featuring unique transcriptional signatures that are significantly altered upon ablation of the microbiome, particularly in ILC1s and ILC3s [90]. With these data having been generated, future research will work to describe the mechanisms by which the microbiome regulates ILC function and stability. One such mechanism may be the direct engagement of microbial products with pattern recognition receptors (PRRs) such as toll-like receptors (TLRs). Previous work has described the expression of TLRs on human peripheral blood NK cells [91], and activation of TLR3 and TLR9 promotes cytotoxic activity [92] while activation of TLR2 promotes antiviral immunity [93]. Furthermore, human circulating ILC2s have been shown to express several TLRs that, when activated, lead to the expression of IL-5 and IL-13 as well as CD40L, which promotes IgE production from B cells [94]. Conversely, TLR9 activation of murine ILC2s by CpG suppresses IL-33-mediated airway inflammation through the stimulation of NK cells and subsequent production of IFN-γ [95]. Finally, TLR transcripts have also been observed in ILC3s from human tonsils with TLR2 being the dominantly expressed receptor and little to no expression of TLR3 or TLR4 [96]; however, another study showed increased apoptosis of ILC3s after in vitro stimulation of total small intestine lymphocytes from rhesus macaques with LPS [97], although, as these experiments did not use purified ILC3s, it is uncertain whether this is a direct or indirect effect. Additionally, the sensing of microbial products via TLRs may be somewhat specific for human ILC3s since stimulation of purified mouse LTi-like cells and NKp46$^+$ ILC3s, which did not express TLRs, could not reproduce the increase in cytokine production, mainly IL-22, as was observed in human ILCs [96].

Other than TLRs, both NK cells and ILC3s express natural cytotoxicity receptors (NCRs), such as NKp30, NKp44, and NKp46, that are capable of recognizing microbial components and activating pro-inflammatory responses upon engagement [98, 99]. Moreover, microbiome-derived metabolites can signal directly to ILCs to regulate their function. For example, as previously mentioned, Ahr produced by the microbiome is essential for postnatal development of ILFs and CPs by ILC3s in mice [84, 85]. Additionally, Ahr signaling is critical for the production of cytokines by ILC3s, as it binds directly to the mouse *Il22* locus together with RORγt to promote gene expression, while also inhibiting the function of intestinal ILC2s by modifying the chromatin landscape to suppress the IL-33-ST2 signaling pathway [83, 100]. Importantly, these metabolites can have an endocrine effect and do not necessarily act only on cells in the local environment since alterations in Ahr ligands produced by the gut microbiota have been shown to alter the expression of IL-22 in pancreatic ILCs, which has been associated with protection against autoimmune diabetes in mouse models [101]. Furthermore, the microbiome is a key producer of short-chain fatty acids (SCFA), bile acids, and tryptophan metabolites. By feeding mice a high-fiber diet, it was observed that the microbiome fermented the dietary fibers into SCFAs, such as butyrate, which inhibited ILC2 function and suppressed ILC2-mediated airway hypersensitivity [102]. In the intestines, microbiome-derived SCFAs were found to promote ILC expansion for ILC1s, ILC2s, and ILC3s [103], which express the receptor, Ffar2, and produce increased IL-22 in the presence of butyrate via AKT and STAT3 activation [104]. Consistently, Ffar2 deficiency in ILC3s led to decreased expression of effector cytokines and reduced protection against *C. rodentium* infection as well as increased sensitivity to dextran sodium sulfate (DSS)-induced colitis, highlighting another key mechanism by which the microbiome can influence ILC3 function. Over the last decade, many pathways for direct interaction between the microbiome and ILCs have been described; however, our understanding of these interactions is still incomplete as novel, high-throughput technologies reveal additional, unexplored mechanisms.

8.2.3 Indirect Interactions Between the Microbiota and ILCs

In contrast to the direct activation of ILCs, the microbiome can also indirectly regulate ILC function by modifying cytokine production from other cells in the environment such as macrophages [105, 106]. Conversely, ILC functions can alter antimicrobial responses by other cells to promote either tolerance or activation against commensal and environmental microbes [107]. In the case of ILC1s, previous reports have shown that, in mice, NK cell function, but not development, is dependent on the microbiome even in non-mucosal tissues [80], likely through the production of metabolites that are released into circulation. Alterations in the microbiome have also been associated with colitis in mouse models, which was driven by the MyD88-dependent production of IL-12 [18], a key activating cytokine for ILC1s. Consistently, accumulation of ILC1s was observed in inflamed tissue from patients suffering from Crohn's disease [21]; however, in mouse models of bacteria-induced colitis, T-bet expression in ILCs seems to be protective against dysbiosis-associated intestinal inflammation by preventing the expression of IL-17A in response to TNF produced by colonic DCs, which drives the production of IL-23, similar to what other groups observed in patients with Crohn's disease [108–111].

GATA3$^+$ ILC2s have been associated with protection of intestinal tissues and containment of the microbiome as signaling via the alarmin, IL-33, which signals through its receptor, IL1RL1 (ST2), and is released upon tissue damage and can be regulated by the microbiota, ameliorates DSS-induced colitis in mice through the production of the growth factor, AREG [52, 110]. In addition to IL-33, ILC2s are also activated by TSLP expressed by epithelial cells downstream of NF-κB and induce the expression of Th2 cell-associated cytokines such as IL-4 and IL-13 [112–114]. Due to the regulation of TSLP by NF-κB, and the expression of TLRs on epithelial cells in both the intestine and airway [115, 116], induction of TSLP expression by microbial components that signal through PRRs can potentially act as an activating signal for ILC2s while limiting ILC3-dependent cytokine production [117]. However, these signals must be carefully balanced as dysregulation of ILC2s is associated with exacerbated allergic inflammation and asthma [36] and, similar to ILC1s, expansion of ILC2s has been associated with intestinal inflammation in both humans and mice [18].

RORγt$^+$ ILC3s have been the most extensively studied in the context of intestinal homeostasis and inflammation. In the intestines, microbiota indirectly influence ILC3 function through the activation of local myeloid cells that produce IL-1β and IL-23 [118, 119]. Additionally, in mice, the microbiome enhances the expression of IL-7, which is critical for the stabilization of RORγt expression in adoptively transferred ILC3s as well as secretion of effector cytokines such as IL-22 [82], which bind to IL-22R expressed by intestinal epithelial cells (IECs) and promote gut homeostasis through wound healing via STAT3 activation after DSS-induced intestinal damage and inflammation [120]. This can also occur in other barrier tissues at which ILCs are likely to be present such as the skin [121]. Signaling of IL-22 in epithelial cells has the added effect of promoting fucosylation of IECs through the fucosyltransferase, *Fut2* [122–124]. These fucosylated proteins can then be shed and taken up by the microbiome to help prevent dysbiosis and opportunistic infections in mouse models. Alternatively, IL-22 signaling in epithelial cells induces the expression of antimicrobial peptides such as Reg3γ and Reg3β [62], expression of which is necessary to limit microbial association with the intestinal surface and prevent dysregulation of IFN-γ production in CD4 T cells [125]. ILC3s also express other factors that directly protect the intestinal barrier from damage, including the HB-EGF, which is important in limiting TNF-mediated cell death in the epithelial (Zhou et al. PMID: 35102343).

Interestingly, this indirect regulation of T cells is not the only mechanism by which ILC3s exert their influence over the adaptive immune system. Previous studies have shown that $CCR6^{+}$ ILC3s are a major subset in the mLN and express MHCII in both mice and humans, leading to the deletion of microbiome-specific $CD4^{+}$ T cells, and that this expression of MHCII is altered in Crohn's disease patients, which correlates with aberrant activation of Th17 cells [126–128]. Within the mLN and colon, expression of MHCII in ILC3s is constitutive, regulated in a manner that is comparable to thymic epithelial cells, and is independent of microbiota or inflammatory stimulus [128]. However, in the small intestine or spleen, ILC3s appear to upregulate MHCII and co-stimulatory molecules in response to inflammatory stimuli and could promote T cell responses, but the biological significance of these findings and whether these represent the same or distinct ILC3 subsets remain unclear. In addition, T cell selection does not seem to be the only mechanism by which MHCII expression on ILCs can regulate homeostasis with the microbiome. A recent study showed that disruption of MHCII expression on ILC3s leads to increased T follicular helper cell activity and promotes B cell class switch recombination to IgA, which is associated with alterations in microbial populations and metabolites [129]. This ILC3 control over microbiota-specific T cells critically supports homeostasis with a diverse microbiota and this can have important consequences in other contexts. For example, it was recently demonstrated that these interactions support microbiota colonization that is poised to elicit type 1 immunity, which subsequently supports antitumor immunity and responsiveness to immunotherapies in CRC [130]. ILC3s can also sense the microbiota to become a dominant source of IL-2 in the small intestine of mice [78]. This production of IL-2 is essential for the maintenance of intestinal Tregs and tolerance to dietary antigens. Furthermore, the dysregulation of IL-2 expression in ILC3s was shown to be associated with Crohn's disease, further highlighting the importance of ILC3 functions in the maintenance of intestinal homeostasis. Finally, ILC3s also express MHCII in the airway-draining lymph nodes and limit inflammatory Th17 and Th2 cell responses to allergens (Teng et al. PMID: 34818549), suggesting that these regulatory properties extend to other mucosal barrier surfaces.

8.3 Regulation of ILC Metabolism by Microbes

Of late, the field of immunometabolism has shown to hold exciting insight into the inner workings of immunity. Over the last several years, there has been a growing appreciation for the role of cellular metabolic states and the ability of cellular metabolism to regulate immune cell functions. This is particularly important during infection as proper activation of immune cells such as ILCs is critical for the clearance of a number of pathogens; however, the literature surrounding the mechanisms by which ILC metabolism is controlled during infection is incomplete and little is known about how these cells interact with pathogens to alter their metabolism in support of protective immunity. Current understanding of T cells may provide some clues to the role of metabolism in ILCs during infection, though, due to the functional differences between T cells and ILCs, there are likely many differences between the two cell types. The relationship between the metabolism of T cells and ILCs has been nicely reviewed elsewhere [131]. More recently, in ILC2s, programmed cell death protein-1 (PD-1) has been shown to be upregulated after IL-33 stimulation in mice. In PD-1-deficient models, ILC2s have increased glycolysis and catabolism of methionine and glutamine, which enhances ILC2 proliferation and expression of cytokines such as IL-5 and IL-13 [132], suggesting a reduced capacity to clear helminth infections; however, the mechanism by which the expression of PD-1 in ILC2s is controlled during infection is still unclear. Additionally, in ILC3s, hypoxia-induced stabilization of HIF-1α was critical for clearing *C. difficile* infection in mice as HIF-1α ablation in *Rorc*-expressing cells led to impaired IL-22, IL-17, CSF2, and TNF expression in ILC3s [133]. Previous studies have also shown the importance of HIF-1α stabilization, though in an mTORC1-dependent manner, which was associated with an increase in the production of reactive oxygen species (ROS) and metabolic reprograming [134]. This dependence on mTORC1 was critical for the production of IL-22 and IL-17A in mouse models of *C. rodentium* infection, highlighting the importance of a metabolic shift toward glycolytic pathways in ILC3 activation. Although much progress has been made in the last few years, there is still much work to be done in this field to determine the relevance of context-dependent metabolic shifts in ILC populations during infection as different infectious agents may require that the ILC populations responding to the infections have different metabolic profiles to promote unique cellular functions.

8.4 ILC Interaction with the Nervous System

An emerging paradigm in ILC biology is that these cell types are uniquely poised for communication with the nervous system. This can occur in the central nervous system (CNS), or locally, with the peripheral nervous system, and these interactions have substantial implications for immunity, inflammation, and tissue homeostasis. Furthermore, a number of specific targets in neuronal-ILC interactions have been identified to either promote or inhibit ILC responses and, in some contexts, ILCs even employ pathways typically associated with the nervous system to execute their rapid effector functions (Fig. 8.2).

8.4.1 ILCs and the Central Nervous System

The CNS is an immunologically specialized site and is uniquely protected from external solutes by the blood–brain barrier (BBB) and blood–cerebrospinal fluid (CSF) barrier, which act as filters to control the contents entering the CNS via circulation. In recent years, it has become increasingly clear that ILCs participate in complex interactions with the nervous system and the roles of these interactions in health and disease have only begun to be appreciated. In addition, the presence of ILCs in the CNS has been observed, though the mechanisms by which they traffic to the CNS have not yet been fully described. Due to the lack of traditional lymphatics in the CNS parenchyma, immune cells must find alternative routes to travel in and out of the tissues. Currently, there are three major mechanisms by which cells can migrate into the CNS: (1) exiting the endothelium of the blood–CSF barrier at the choroid plexus, a region that produces CSF and harbors a diverse population of immune cells, particularly cDCs, which may participate in the reactivation of local T cells [135]; (2) exiting circulation into the subarachnoid space and Virchow–Robin perivascular spaces, which have also been suggested to be sites of antigen presentation and immune surveillance [136]; and (3) exiting circulation directly into the parenchymal perivascular spaces by crossing through the BBB [137]. In addition, recent evidence suggests that some immune cells derived from skull cap and vertebral bone marrow niches migrate directly into the meninges, the

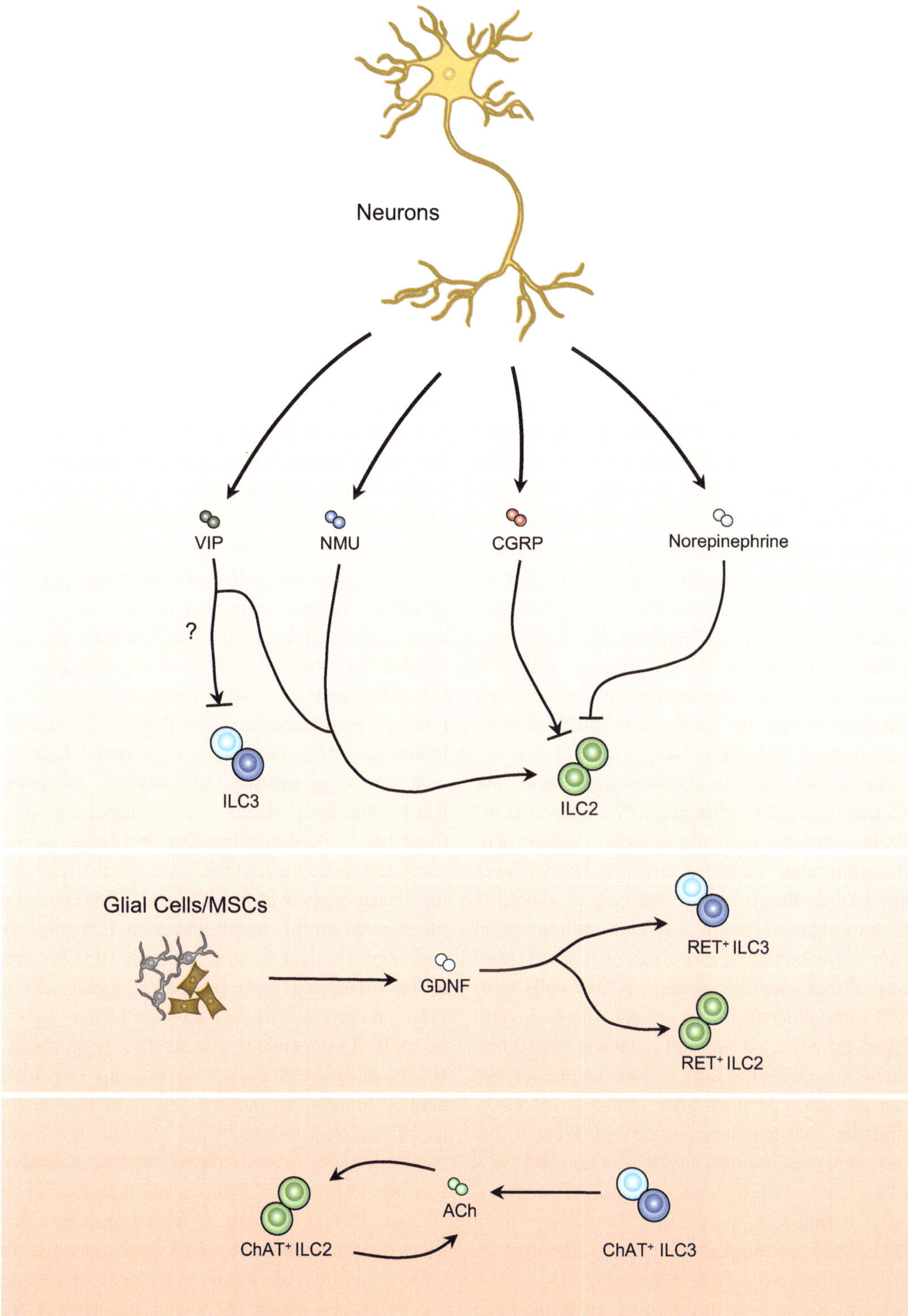

Fig. 8.2 Neuronal pathways control ILC-dependent mucosal immunity. Numerous interactions between ILCs and the nervous system have been identified. These include neuropeptides that can either promote or inhibit ILC effector functions. ILCs can also utilize pathway typically associated with the nervous system to coordinate their effector functions, such as RET or ChAT

membranous layers that envelop and protect the CNS, through specialized vessels and independent of the blood circulation. These cells then have the potential to migrate into the CNS parenchyma in contexts of inflammation or injury [138, 139]. Conversely, cells and other materials can leave the CNS through the CSF, which drains into the deep cervical lymph nodes via meningeal lymphatics or potentially through perineural outflow routes to lymphatics outside of the CNS, though whether ILCs utilize these migratory mechanisms has yet to be explored [138, 140–142]. Moreover, the functional significance of the presence of ILCs in the CNS is only beginning to be examined and their contribution to maintaining tissue homeostasis is currently an active area of investigation. Additionally, the effector functions of ILCs on the CNS may not be limited to immediate signaling events with neighboring cells, but their influence on cells in distant tissues such as the gut can potentially have substantial consequences on the CNS [143]. ILCs are intimately connected to the microbiota, which is known to have an impact on CNS health and neurological diseases, such as multiple sclerosis and Parkinson's disease [144].

As of yet, little is known about how the functions of ILCs influence CNS homeostasis. Previous findings in mouse models of experimental autoimmune encephalomyelitis (EAE) have shown that the transcription factor, T-bet, is an important mediator of neuroinflammation [145, 146]; however, more recently, it was discovered that adoptive transfer of Th17 cells from TCR transgenic mice that are specific to myelin oligodendrocyte glycoprotein (MOG) into T-bet-deficient mice did not fully restore the phenotypes seen in wild-type mice after induction of EAE, indicating that the pathogenicity of T-bet is not solely due to its expression and function in T cells [147]. Consistent with these findings, as previously mentioned, production of IFN-γ from ILC1s has been implicated in neuroinflammation [38]. Furthermore, T-bet-expressing ILCs were found to localize to the meninges during EAE and transfer of myelin-specific T cells into *Tbx21*$^{fl/fl}$ NKp46-Cre mice lacking ILC1s and NKp46^{+} ILC3s showed a significant reduction in EAE pathology by preventing the infiltration and cytokine profile of Th17 cells in the CNS. Conversely, production of IFN-γ by meningeal NK cells may be protective as it can promote an anti-inflammatory subset of astrocytes in the CNS expressing the death receptor ligand, tumor necrosis factor-related apoptosis-inducing ligand or Apo 2 ligand (TRAIL), and subsequently drive cell death of CNS-associated T cells to prevent neuroinflammation [148]. Intriguingly, these effector functions appear to require or be licensed by the intestinal microbiota, demonstrating a profound influence of the microbiota on ILC functions that occur distal to the gut. A unique subset of inflammatory ILC3s also migrate into the CNS during EAE and are essential for re-stimulating myelin-specific CD4 T cells via MHCII and promoting disease pathogenesis (Grigg et al. PMID: 34853467). This is in striking contrast to tissue-resident ILC3s in the gut and peripheral lymph nodes that remain tolerogenic during EAE and can be harnessed to prevent disease progression. Additionally, infection of ILC2-deficient mice with herpes simplex virus type 1 constitutively expressing IL-2, another model of CNS demyelination, showed little to no neurodegenerative phenotype; however, when wild-type ILC2s are reintroduced into these mice, the demyelination phenotype is rescued, suggesting a role for ILC2s in demyelinating disease as well [149] and, although the mechanism for this is still uncertain, it is possibly context dependent as, in some cases, ILC2s seem to be associated with protection against EAE [150] and promote healing after spinal cord injury in an IL-33-dependent manner [151]. Similarly, ILC2s have been shown to prevent cognitive decline in aging as intracerebroventricular transfer of ILC2s, activated by IL-33, IL-2, IL-7, and stem cell factor, was shown to improve cognitive function in mice [152]. Based on these studies, it is clear that ILCs play a pivotal role in CNS homeostasis and future work will continue to build upon these novel observations to determine the extent to which ILCs can influence CNS immunity and inflammation. Moreover, future work may also focus on the mechanisms by which components of the CNS such as neurons or glial cells may influence CNS-localized ILC function.

The CNS may conversely impact ILC3 homeostasis within the intestine. This mechanistically occurs at multiple levels, including fine-tuning from brain cues and interpreting light–dark cycles and feeding cycles [153]. Critically involved in this homeostatic regulation is ILC3-specific expression of circadian regulators, such as BMAL1 and NR1D1 [153–155]. Mice with ILC3-specific deletion in BMAL1 exhibit altered diurnal rhythms, reduced population frequencies, and increased sensitivity to both *C. rodentium* infection and DSS-induced intestinal damage and inflammation. Lineage-specific deletion of BMAL1 or NR1D1 also resulted in microbiota-dependent hyperactivation of ILC3s, which may account for their cellular depletion in the context and could be relevant to the loss of ILC3s from the inflamed intestine of IBD patients [154, 155]. In support of this latter possibility, ILC3s isolated from the inflamed intestine exhibited altered circadian clock genes relative to ILC3s isolated from matched non-inflamed intestine. These data suggest that there are mechanisms by which the CNS acts on ILC3s in the intestines through long-range neural circuits that connect the brain, and this fine-tunes ILC3 homeostasis in the context of dietary intake and colonization with a diverse microbiota.

8.4.2 ILCs and the Peripheral Nervous System

Outside of the brain and spinal cord, the peripheral nervous system (PNS) provides connections between the CNS and peripheral tissues. The PNS can be divided into two major subsystems, the somatic system, which provides sensory information about the environment, and the autonomic system, which maintains bodily functions that are critical for maintaining homeostasis and includes, within its classification, the enteric nervous system (ENS), a complex network of neurons completely contained within the gastrointestinal tract that orchestrates intestinal physiology separately from the CNS. These subsystems can then be further divided into sensory or afferent neurons that carry signals toward the CNS from the periphery as well as motor or efferent neurons, which carry signals from the CNS out into the periphery to promote effector functions of cells within the tissues [156]. Over 100 years ago, it was discovered that, when sympathetic neurons, members of the autonomic nervous system, are resected from the ears of rabbits, experimental inflammation is exacerbated; however, when sensory nerves are resected, inflammation is inhibited, leading to reduced clearance of bacterial pathogens, providing early evidence for the regulation of immune responses by the PNS [156, 157]. It has not been until recently that many of the mechanisms underlying these observations have been elucidated, and not until the past several years have the effects of these neurons on the regulation of ILC functions been examined.

Key mechanisms by which peripheral neurons control the function of ILCs are through activation of surface receptors and production of neuropeptides such as calcitonin gene-related peptide (CGRP), neuromedin U (NMU), and vasoactive intestinal peptide (VIP) (Fig. 8.2). In a recent study, ILC2s were found to express β_2-adrenergic receptor (β_2AR) and these β_2AR-expressing ILC2s in mice co-localize with adrenergic neurons, indicating a neuronal regulatory circuit for ILC2s [158]. Indeed, it was found that activation of β_2AR in ILC2s limits type 2 inflammation. Consistently, β_2AR-deficient mice as well as *Adrb2*$^{fl/fl}$ *Il7r*$^{cre/+}$ mice *were* able to more effectively control *N. brasiliensis* infection compared to wild-type mice. This protection was associated with the enhanced production of type 2 cytokines such as IL-5 and IL-13 by ILC2s, which led to increased eosinophil infiltration. These data provide the first evidence of a negative regulatory mechanism by which neurons can promote tissue homeostasis by restricting ILC function. A series of subsequent studies showed another negative regulatory mechanism by which CGRP suppresses type 2 inflammation in mice through the inhibition of ILC2 activation [159–161]. In the intestines, it was found that CGRP was produced by choline acetyltransferase

(ChAT)-expressing neurons to maintain tissue homeostasis at steady state, though another study showed that expression of CGRP by pulmonary neuroendocrine cells suggested that CGRP acts to stimulate ILC2 cytokine production in the lungs [162]. Finally, enteric VIP-expressing neurons restrain ILC3 function by releasing VIP in response to feeding, which signals through its receptor, VIPR2, expressed on $CCR6^+$ ILC3s, to improve lipid absorption by inhibiting IL-22 production in mice [163]; however, this mechanism may still be somewhat controversial as another study found that deficiency of VIPR2 in ILC3s leads to an impairment in IL-22 production and protection against DSS-induced colitis [164]. In the context of ILC2s, VIP production by sensory neurons has been associated with allergic airway inflammation by inducing the expression of inflammatory signals through the VPAC2 receptor [165, 166]. These studies have shed light on a complex regulatory network in which neurons restrict ILC functions to promote tissue homeostasis. Additionally, due to the negative regulation imposed by these neurons, there may also be alternative mechanisms that regulate neuronal functions to release ILC cytokine production in the case of infection, though this avenue of research still needs to be explored. Future work may also focus on describing additional mechanisms by which neurons can control ILC function, including ILC1s and ILC3s, to control inflammation and pathogen clearance or whether signals from the microbiome can influence the utilization of these mechanisms by neurons at barrier tissues.

In contrast to the inhibitory effects of neurons on ILCs via β_2AR and VIP, the production of NMU has been shown as a stimulatory signal for ILCs. Through the use of *Nmur1*$^{LacZ/+}$ reporter mice, a large fraction of $NMUR1^+$ cells in the small intestine showed phenotypes consistent with ILC2s [167–169]. Further experimentation revealed that NMU is a potent activator of ILC2 production of cytokines such as IL-5, IL-13, CSF2, and AREG in cells from both the gut and the lungs and that NMUR deficiency leads to impaired clearance of helminth infections whereas treatment of mice with exogenous NMU improves eosinophil infiltration and worm expulsion. Additionally, it was found that NMU promotes lung inflammation induced by ILC2s, which was later confirmed and shown to act synergistically with IL-25 to promote activation of ILC2s, leading to exacerbated inflammation in mouse models of allergic airway inflammation [169]. Interestingly, NMU also synergizes with IL-33 to enhance IL-2, IL-6, IL-9, IL-13, and IL-5 production as well as AREG when combined with CGRP in mouse cells [159], though the functional significance of many of these interactions is still unclear, though it is possible to speculate a role in modifying Treg populations as IL-2 production from ILC3s has been shown to promote Treg stability [78] and AREG has been shown to maintain the immunosuppressive functions of Tregs [170]. Conversely, IL-6 production from enteric neurons has been shown to increase total Treg populations while inhibiting the RORγt-expressing subset in the gut in response to microbial signals [171].

Interestingly, ILCs may also utilize pathways typically associated with the nervous system to exert their influence (Fig. 8.2). For example, both ILC2s and ILC3s have been shown to express ChAT, the enzyme responsible for the biosynthesis of acetylcholine, in mouse models of protease-induced airway inflammation and helminth infection, respectively, to induce inflammatory phenotypes [172, 173]. It was also found that ILC2s express acetylcholine receptors and that stimulation with the neurotransmitter promotes cytokine production and parasite expulsion. Additionally, ILC2s in mouse adipose tissue have recently been shown to express the neuroregulatory receptor, RET, which recognizes glial-derived neurotrophic factor (GDNF), produced by mesenchymal stromal cells (MSCs), and deficiency of RET in ILC2s led to increased susceptibility to high-fat-diet-induced obesity and decreased glucose sensitivity due to a breakdown of a mechanism by which neuro-mesenchymal interactions regulate energy homeostasis [174]. Finally, ILC3s also express RET in the intestine, which drives the production of IL-22 when recognizing environmental signals of

GDNF, produced by glial cells, localized in proximity to CPs and ILFs, in a MyD88-dependent manner [175].

These findings implicate the peripheral nervous system in the regulation of ILCs, creating an exciting new line of inquiry into the intricate interactions between the immune system, nervous system, and microbiota. Several studies have indicated an important connection between the microbiota and the functionality of the ENS as GF mice as well as antibiotic-treated mice show significant increases in intestinal transit times, indicating dysregulation of GI motility and alterations to enteric neurons when the microbiota is disrupted [176, 177]. Further, the microbiome-induced expression of Ahr in enteric neurons is critical for regulation of intestinal homeostasis. Cross talk between muscularis macrophages and enteric neurons has also been observed in which macrophages produce bone morphogenic protein 2 (Bmp2) and, in response, the neurons express colony-stimulating factor 1 (CSF1) to promote macrophage development [178]. Interestingly, this phenomenon seems to be directed by the microbiome and may depend on signaling through TLR4, as enteric neurons exhibit a substantial increase in CSF1 expression when cultured with LPS. Furthermore, muscularis macrophages expressing β_2AR have been shown to protect enteric neurons in mouse models of *Salmonella infection* [179, 180]. Based on these results, a potential route for further exploration may be to determine how muscularis macrophages or other immune cell types, such as ILCs, in the intestines may prevent enteric neuron death in the case of other bacterial, viral, or fungal infection as loss of enteric neurons has been observed in cases of lytic viral infections such as varicella zoster virus [181]. Although, as of yet, there is no direct evidence to show that these mechanisms will impact ILC functions, it is reasonable to hypothesize a cross-regulatory network in which ILCs help to maintain peripheral neuron functions while, at the same time, peripheral neurons act to regulate ILC functions. Moreover, although several mechanisms of the regulation of ILCs by neurons have been discovered, to date, little work has been done on how ILCs can regulate the function of neurons. This is an important contrast as it can also elucidate additional mechanisms by which ILCs utilize neurons as an intermediate partner to interact with and regulate other cell types as well as influence overall tissue homeostasis.

8.5 Concluding Remarks

Due to the exposure of mucosal barrier surfaces to the external environment, they are particularly vulnerable to pathogen colonization and, therefore, require complex immunological regulatory networks that promote immune tolerance to microbiota while still maintaining protective immunity against invading pathogens. ILCs have been shown to be key regulators of barrier tissue immunity through their interactions with both pathogenic microbes and the microbiota. Recent studies have discovered novel mechanisms by which these microbes modulate ILC function through direct and indirect mechanisms. In particular, a growing body of literature has begun to describe the interactions between ILCs and neurons in both the central nervous system and the peripheral nervous system. These interactions with neurons may endow them with unique properties and the ability to rapidly tune their responses to microbes or other environmental stimuli. This provides new opportunities to study the mechanisms by which the nervous system, immune system, and microbes, both commensal and pathogenic, form signaling networks to promote protective immunity and maintain tissue homeostasis. Finally, how these interactions may impact other systems, such as the metabolic or the endocrine systems, and conversely how other systems may impact these interactions are still enigmatic. Future work may focus on incorporating a more interdisciplinary approach to studying ILC functions and cross-regulatory networks that govern host–microbe interactions.

Acknowledgements We thank members of the Sonnenberg Laboratory for discussions and critical reading of the manuscript. Research in the Sonnenberg

Laboratory is supported by the National Institutes of Health (R01AI143842, R01AI123368, R01AI145989, R01AI162936, R21CA249274, and U01AI095608), the NIAID Mucosal Immunology Studies Team (MIST), the Searle Scholars Program, the American Asthma Foundation Scholar Award, an Investigators in the Pathogenesis of Infectious Disease Award from the Burroughs Wellcome Fund, a Wade F.B. Thompson/Cancer Research Institute (CRI) CLIP Investigator grant, the Meyer Cancer Center Collaborative Research Initiative, Linda and Glenn Greenberg, and the Roberts Institute for Research in IBD. G.F.S. is a CRI Lloyd J. Old STAR.

References

1. Sender R, Fuchs S, Milo R. Are we really vastly outnumbered? Revisiting the ratio of bacterial to host cells in humans. Cell. 2016;164(3):337–40.
2. Krajmalnik-Brown R, Ilhan ZE, Kang DW, DiBaise JK. Effects of gut microbes on nutrient absorption and energy regulation. Nutr Clin Pract. 2012;27(2): 201–14.
3. Rowland I, Gibson G, Heinken A, Scott K, Swann J, Thiele I, et al. Gut microbiota functions: metabolism of nutrients and other food components. Eur J Nutr. 2018;57(1):1–24.
4. Belkaid Y, Hand TW. Role of the microbiota in immunity and inflammation. Cell. 2014;157(1): 121–41.
5. Zheng D, Liwinski T, Elinav E. Interaction between microbiota and immunity in health and disease. Cell Res. 2020;30(6):492–506.
6. Fawkner-Corbett D, Antanaviciute A, Parikh K, Jagielowicz M, Geros AS, Gupta T, et al. Spatiotemporal analysis of human intestinal development at single-cell resolution. Cell. 2021;184(3):810–26.
7. Spits H, Artis D, Colonna M, Diefenbach A, Di Santo JP, Eberl G, et al. Innate lymphoid cells—a proposal for uniform nomenclature. Nat Rev Immunol. 2013;13(2):145–9.
8. Artis D, Spits H. The biology of innate lymphoid cells. Nature. 2015;517(7534):293–301.
9. Eberl G, Colonna M, Di Santo JP, AN MK. Innate lymphoid cells. Innate lymphoid cells: a new paradigm in immunology. Science. 2015;348(6237): aaa6566.
10. Vivier E, Artis D, Colonna M, Diefenbach A, Di Santo JP, Eberl G, et al. Innate lymphoid cells: 10 years on. Cell. 2018;174(5):1054–66.
11. Yokota Y, Mansouri A, Mori S, Sugawara S, Adachi S, Nishikawa S, et al. Development of peripheral lymphoid organs and natural killer cells depends on the helix-loop-helix inhibitor Id2. Nature. 1999;397(6721):702–6.
12. Klose CSN, Flach M, Mohle L, Rogell L, Hoyler T, Ebert K, et al. Differentiation of type 1 ILCs from a common progenitor to all helper-like innate lymphoid cell lineages. Cell. 2014;157(2):340–56.
13. Zhong C, Zheng M, Cui K, Martins AJ, Hu G, Li D, et al. Differential expression of the transcription factor GATA3 specifies lineage and functions of innate lymphoid cells. Immunity. 2020;52(1):83–95.
14. Constantinides MG, McDonald BD, Verhoef PA, Bendelac A. A committed precursor to innate lymphoid cells. Nature. 2014;508(7496):397–401.
15. Koues OI, Collins PL, Cella M, Robinette ML, Porter SI, Pyfrom SC, et al. Distinct gene regulatory pathways for human innate versus adaptive lymphoid cells. Cell. 2016;165(5):1134–46.
16. Shih HY, Sciume G, Mikami Y, Guo L, Sun HW, Brooks SR, et al. Developmental acquisition of regulomes underlies innate lymphoid cell functionality. Cell. 2016;165(5):1120–33.
17. Kiessling R, Klein E, Wigzell H. "Natural" killer cells in the mouse. I. Cytotoxic cells with specificity for mouse Moloney leukemia cells. Specificity and distribution according to genotype. Eur J Immunol. 1975;5(2):112–7.
18. Mebius RE, Rennert P, Weissman IL. Developing lymph nodes collect CD4+CD3− LTbeta+ cells that can differentiate to APC, NK cells, and follicular cells but not T or B cells. Immunity. 1997;7(4): 493–504.
19. Scoville SD, Mundy-Bosse BL, Zhang MH, Chen L, Zhang X, Keller KA, et al. A progenitor cell expressing transcription factor RORγt generates all human innate lymphoid cell subsets. Immunity. 2016;44(5):1140–50.
20. Lim AI, Li Y, Lopez-Lastra S, Stadhouders R, Paul F, Casrouge A, et al. Systemic human ILC precursors provide a substrate for tissue ILC differentiation. Cell. 2017;168(6):1086–100.
21. Bernink JH, Peters CP, Munneke M, te Velde AA, Meijer SL, Weijer K, et al. Human type 1 innate lymphoid cells accumulate in inflamed mucosal tissues. Nat Immunol. 2013;14(3):221–9.
22. Fuchs A, Vermi W, Lee JS, Lonardi S, Gilfillan S, Newberry RD, et al. Intraepithelial type 1 innate lymphoid cells are a unique subset of IL-12- and IL-15-responsive IFN-gamma-producing cells. Immunity. 2013;38(4):769–81.
23. Weizman OE, Adams NM, Schuster IS, Krishna C, Pritykin Y, Lau C, et al. ILC1 confer early host protection at initial sites of viral infection. Cell. 2017;171(4):795–808.
24. Klose CS, Kiss EA, Schwierzeck V, Ebert K, Hoyler T, d'Hargues Y, et al. A T-bet gradient controls the fate and function of CCR6-RORγt+ innate lymphoid cells. Nature. 2013;494(7436): 261–5.
25. Mjosberg J, Bernink J, Golebski K, Karrich JJ, Peters CP, Blom B, et al. The transcription factor GATA3 is essential for the function of human type 2 innate lymphoid cells. Immunity. 2012;37(4):649–59.

26. Halim TY, MacLaren A, Romanish MT, Gold MJ, McNagny KM, Takei F. Retinoic-acid-receptor-related orphan nuclear receptor alpha is required for natural helper cell development and allergic inflammation. Immunity. 2012;37(3):463–74.
27. Wong SH, Walker JA, Jolin HE, Drynan LF, Hams E, Camelo A, et al. Transcription factor RORalpha is critical for nuocyte development. Nat Immunol. 2012;13(3):229–36.
28. Fort MM, Cheung J, Yen D, Li J, Zurawski SM, Lo S, et al. IL-25 induces IL-4, IL-5, and IL-13 and Th2-associated pathologies in vivo. Immunity. 2001;15(6):985–95.
29. Moro K, Yamada T, Tanabe M, Takeuchi T, Ikawa T, Kawamoto H, et al. Innate production of T(H) 2 cytokines by adipose tissue-associated c-Kit(+) Sca-1(+) lymphoid cells. Nature. 2010;463(7280): 540–4.
30. Wilhelm C, Hirota K, Stieglitz B, Van Snick J, Tolaini M, Lahl K, et al. An IL-9 fate reporter demonstrates the induction of an innate IL-9 response in lung inflammation. Nat Immunol. 2011;12(11): 1071–7.
31. Monticelli LA, Sonnenberg GF, Abt MC, Alenghat T, Ziegler CG, Doering TA, et al. Innate lymphoid cells promote lung-tissue homeostasis after infection with influenza virus. Nat Immunol. 2011;12 (11):1045–54.
32. Halim TY, Krauss RH, Sun AC, Takei F. Lung natural helper cells are a critical source of Th2 cell-type cytokines in protease allergen-induced airway inflammation. Immunity. 2012;36(3):451–63.
33. Seehus CR, Kadavallore A, Torre B, Yeckes AR, Wang Y, Tang J, et al. Alternative activation generates IL-10 producing type 2 innate lymphoid cells. Nat Commun. 2017;8(1):1900.
34. Bando JK, Gilfillan S, Di Luccia B, Fachi JL, Secca C, Cella M, et al. ILC2s are the predominant source of intestinal ILC-derived IL-10. J Exp Med. 2020;217(2):e20191520.
35. Neill DR, Wong SH, Bellosi A, Flynn RJ, Daly M, Langford TK, et al. Nuocytes represent a new innate effector leukocyte that mediates type-2 immunity. Nature. 2010;464(7293):1367–70.
36. McKenzie AN. Type-2 innate lymphoid cells in asthma and allergy. Ann Am Thorac Soc. 2014;11 (Suppl 5):S263–70.
37. Gil-Cruz C, Perez-Shibayama C, Onder L, Chai Q, Cupovic J, Cheng HW, et al. Fibroblastic reticular cells regulate intestinal inflammation via IL-15-mediated control of group 1 ILCs. Nat Immunol. 2016;17(12):1388–96.
38. Kvestak D, Juranic Lisnic V, Lisnic B, Tomac J, Golemac M, Brizic I, et al. NK/ILC1 cells mediate neuroinflammation and brain pathology following congenital CMV infection. J Exp Med. 2021;218 (5):e20201503.
39. Park E, Patel S, Wang Q, Andhey P, Zaitsev K, Porter S, et al. Toxoplasma gondii infection drives conversion of NK cells into ILC1-like cells. Elife. 2019;8:e47605.
40. Hurst SD, Muchamuel T, Gorman DM, Gilbert JM, Clifford T, Kwan S, et al. New IL-17 family members promote Th1 or Th2 responses in the lung: in vivo function of the novel cytokine IL-25. J Immunol. 2002;169(1):443–53.
41. Owyang AM, Zaph C, Wilson EH, Guild KJ, McClanahan T, Miller HR, et al. Interleukin 25 regulates type 2 cytokine-dependent immunity and limits chronic inflammation in the gastrointestinal tract. J Exp Med. 2006;203(4):843–9.
42. Fallon PG, Ballantyne SJ, Mangan NE, Barlow JL, Dasvarma A, Hewett DR, et al. Identification of an interleukin (IL)-25-dependent cell population that provides IL-4, IL-5, and IL-13 at the onset of helminth expulsion. J Exp Med. 2006;203(4):1105–16.
43. Loser S, Smith KA, Maizels RM. Innate lymphoid cells in helminth infections-obligatory or accessory? Front Immunol. 2019;10:620.
44. von Moltke J, Ji M, Liang HE, Locksley RM. Tuft-cell-derived IL-25 regulates an intestinal ILC2-epithelial response circuit. Nature. 2016;529(7585): 221–5.
45. Gerbe F, Sidot E, Smyth DJ, Ohmoto M, Matsumoto I, Dardalhon V, et al. Intestinal epithelial tuft cells initiate type 2 mucosal immunity to helminth parasites. Nature. 2016;529(7585):226–30.
46. Howitt MR, Lavoie S, Michaud M, Blum AM, Tran SV, Weinstock JV, et al. Tuft cells, taste-chemosensory cells, orchestrate parasite type 2 immunity in the gut. Science. 2016;351(6279):1329–33.
47. Schneider C, O'Leary CE, von Moltke J, Liang HE, Ang QY, Turnbaugh PJ, et al. A metabolite-triggered tuft cell-ILC2 circuit drives small intestinal remodeling. Cell. 2018;174(2):271–84.
48. Nadjsombati MS, McGinty JW, Lyons-Cohen MR, Jaffe JB, DiPeso L, Schneider C, et al. Detection of succinate by intestinal tuft cells triggers a type 2 innate immune circuit. Immunity. 2018;49(1): 33–41.
49. Bornstein C, Nevo S, Giladi A, Kadouri N, Pouzolles M, Gerbe F, et al. Single-cell mapping of the thymic stroma identifies IL-25-producing tuft epithelial cells. Nature. 2018;559(7715):622–6.
50. Hung LY, Lewkowich IP, Dawson LA, Downey J, Yang Y, Smith DE, et al. IL-33 drives biphasic IL-13 production for noncanonical type 2 immunity against hookworms. Proc Natl Acad Sci USA. 2013;110(1): 282–7.
51. Yasuda K, Muto T, Kawagoe T, Matsumoto M, Sasaki Y, Matsushita K, et al. Contribution of IL-33-activated type II innate lymphoid cells to pulmonary eosinophilia in intestinal nematode-infected mice. Proc Natl Acad Sci USA. 2012;109(9):3451–6.
52. Frisbee AL, Saleh MM, Young MK, Leslie JL, Simpson ME, Abhyankar MM, et al. IL-33 drives group 2 innate lymphoid cell-mediated protection

during *Clostridium difficile* infection. Nat Commun. 2019;10(1):2712.
53. Zeng B, Shi S, Ashworth G, Dong C, Liu J, Xing F. ILC3 function as a double-edged sword in inflammatory bowel diseases. Cell Death Dis. 2019;10(4):315.
54. Abt MC, Lewis BB, Caballero S, Xiong H, Carter RA, Susac B, et al. Innate immune defenses mediated by two ILC subsets are critical for protection against acute *Clostridium difficile* infection. Cell Host Microbe. 2015;18(1):27–37.
55. Mirchandani AS, Besnard AG, Yip E, Scott C, Bain CC, Cerovic V, et al. Type 2 innate lymphoid cells drive CD4+ Th2 cell responses. J Immunol. 2014;192(5):2442–8.
56. Oliphant CJ, Hwang YY, Walker JA, Salimi M, Wong SH, Brewer JM, et al. MHCII-mediated dialog between group 2 innate lymphoid cells and CD4(+) T cells potentiates type 2 immunity and promotes parasitic helminth expulsion. Immunity. 2014;41(2):283–95.
57. Halim TY, Hwang YY, Scanlon ST, Zaghouani H, Garbi N, Fallon PG, et al. Group 2 innate lymphoid cells license dendritic cells to potentiate memory TH2 cell responses. Nat Immunol. 2016;17(1):57–64.
58. Bayes HK, Ritchie ND, Evans TJ. Interleukin-17 is required for control of chronic lung infection caused by *Pseudomonas aeruginosa*. Infect Immun. 2016;84(12):3507–16.
59. Chen K, Eddens T, Trevejo-Nunez G, Way EE, Elsegeiny W, Ricks DM, et al. IL-17 receptor signaling in the lung epithelium is required for mucosal chemokine gradients and pulmonary host defense against *K. pneumoniae*. Cell Host Microbe. 2016;20(5):596–605.
60. Ardain A, Domingo-Gonzalez R, Das S, Kazer SW, Howard NC, Singh A, et al. Group 3 innate lymphoid cells mediate early protective immunity against tuberculosis. Nature. 2019;570(7762):528–32.
61. Xiong H, Keith JW, Samilo DW, Carter RA, Leiner IM, Pamer EG. Innate lymphocyte/Ly6C(hi) monocyte crosstalk promotes *Klebsiella pneumoniae* clearance. Cell. 2016;165(3):679–89.
62. Zheng Y, Valdez PA, Danilenko DM, Hu Y, Sa SM, Gong Q, et al. Interleukin-22 mediates early host defense against attaching and effacing bacterial pathogens. Nat Med. 2008;14(3):282–9.
63. Cella M, Fuchs A, Vermi W, Facchetti F, Otero K, Lennerz JK, et al. A human natural killer cell subset provides an innate source of IL-22 for mucosal immunity. Nature. 2009;457(7230):722–5.
64. Satoh-Takayama N, Vosshenrich CA, Lesjean-Pottier S, Sawa S, Lochner M, Rattis F, et al. Microbial flora drives interleukin 22 production in intestinal NKp46+ cells that provide innate mucosal immune defense. Immunity. 2008;29(6):958–70.
65. Sonnenberg GF, Monticelli LA, Elloso MM, Fouser LA, Artis D. CD4(+) lymphoid tissue-inducer cells promote innate immunity in the gut. Immunity. 2011;34(1):122–34.
66. Xia P, Liu J, Wang S, Ye B, Du Y, Xiong Z, et al. WASH maintains NKp46(+) ILC3 cells by promoting AHR expression. Nat Commun. 2017;8:15685.
67. Guo X, Qiu J, Tu T, Yang X, Deng L, Anders RA, et al. Induction of innate lymphoid cell-derived interleukin-22 by the transcription factor STAT3 mediates protection against intestinal infection. Immunity. 2014;40(1):25–39.
68. Bernshtein B, Curato C, Ioannou M, Thaiss CA, Gross-Vered M, Kolesnikov M, et al. IL-23-producing IL-10Rα-deficient gut macrophages elicit an IL-22-driven proinflammatory epithelial cell response. Sci Immunol. 2019;4(36):6571.
69. Zigmond E, Bernshtein B, Friedlander G, Walker CR, Yona S, Kim KW, et al. Macrophage-restricted interleukin-10 receptor deficiency, but not IL-10 deficiency, causes severe spontaneous colitis. Immunity. 2014;40(5):720–33.
70. Aychek T, Mildner A, Yona S, Kim KW, Lampl N, Reich-Zeliger S, et al. IL-23-mediated mononuclear phagocyte crosstalk protects mice from *Citrobacter rodentium*-induced colon immunopathology. Nat Commun. 2015;6:6525.
71. Basu R, O'Quinn DB, Silberger DJ, Schoeb TR, Fouser L, Ouyang W, et al. Th22 cells are an important source of IL-22 for host protection against enteropathogenic bacteria. Immunity. 2012;37(6):1061–75.
72. Vely F, Barlogis V, Vallentin B, Neven B, Piperoglou C, Ebbo M, et al. Evidence of innate lymphoid cell redundancy in humans. Nat Immunol. 2016;17(11):1291–9.
73. Rankin LC, Girard-Madoux MJ, Seillet C, Mielke LA, Kerdiles Y, Fenis A, et al. Complementarity and redundancy of IL-22-producing innate lymphoid cells. Nat Immunol. 2016;17(2):179–86.
74. Song C, Lee JS, Gilfillan S, Robinette ML, Newberry RD, Stappenbeck TS, et al. Unique and redundant functions of NKp46+ ILC3s in models of intestinal inflammation. J Exp Med. 2015;212(11):1869–82.
75. van de Pavert SA, Ferreira M, Domingues RG, Ribeiro H, Molenaar R, Moreira-Santos L, et al. Maternal retinoids control type 3 innate lymphoid cells and set the offspring immunity. Nature. 2014;508(7494):123–7.
76. Scandella E, Bolinger B, Lattmann E, Miller S, Favre S, Littman DR, et al. Restoration of lymphoid organ integrity through the interaction of lymphoid tissue-inducer cells with stroma of the T cell zone. Nat Immunol. 2008;9(6):667–75.
77. Rossi SW, Kim MY, Leibbrandt A, Parnell SM, Jenkinson WE, Glanville SH, et al. RANK signals from CD4(+)3(−) inducer cells regulate development of Aire-expressing epithelial cells in the thymic medulla. J Exp Med. 2007;204(6):1267–72.
78. Zhou L, Chu C, Teng F, Bessman NJ, Goc J, Santosa EK, et al. Innate lymphoid cells support regulatory T cells in the intestine through interleukin-2. Nature. 2019;568(7752):405–9.

79. Sawa S, Cherrier M, Lochner M, Satoh-Takayama N, Fehling HJ, Langa F, et al. Lineage relationship analysis of RORγt+ innate lymphoid cells. Science. 2010;330(6004):665–9.
80. Ganal SC, Sanos SL, Kallfass C, Oberle K, Johner C, Kirschning C, et al. Priming of natural killer cells by nonmucosal mononuclear phagocytes requires instructive signals from commensal microbiota. Immunity. 2012;37(1):171–86.
81. Sawa S, Lochner M, Satoh-Takayama N, Dulauroy S, Berard M, Kleinschek M, et al. RORγt+ innate lymphoid cells regulate intestinal homeostasis by integrating negative signals from the symbiotic microbiota. Nat Immunol. 2011;12(4):320–6.
82. Vonarbourg C, Mortha A, Bui VL, Hernandez PP, Kiss EA, Hoyler T, et al. Regulated expression of nuclear receptor RORγt confers distinct functional fates to NK cell receptor-expressing RORγt(+) innate lymphocytes. Immunity. 2010;33(5):736–51.
83. Qiu J, Heller JJ, Guo X, Chen ZM, Fish K, Fu YX, et al. The aryl hydrocarbon receptor regulates gut immunity through modulation of innate lymphoid cells. Immunity. 2012;36(1):92–104.
84. Kiss EA, Vonarbourg C, Kopfmann S, Hobeika E, Finke D, Esser C, et al. Natural aryl hydrocarbon receptor ligands control organogenesis of intestinal lymphoid follicles. Science. 2011;334(6062): 1561–5.
85. Lee JS, Cella M, McDonald KG, Garlanda C, Kennedy GD, Nukaya M, et al. AHR drives the development of gut ILC22 cells and postnatal lymphoid tissues via pathways dependent on and independent of Notch. Nat Immunol. 2011;13(2):144–51.
86. Gomez de Aguero M, Ganal-Vonarburg SC, Fuhrer T, Rupp S, Uchimura Y, Li H, et al. The maternal microbiota drives early postnatal innate immune development. Science. 2016;351(6279): 1296–302.
87. Mishra A, Lai GC, Yao LJ, Aung TT, Shental N, Rotter-Maskowitz A, et al. Microbial exposure during early human development primes fetal immune cells. Cell. 2021;184(13):3394–409.
88. Bhattacharya N, Yuan R, Prestwood TR, Penny HL, DiMaio MA, Reticker-Flynn NE, et al. Normalizing microbiota-induced retinoic acid deficiency stimulates protective CD8(+) T cell-mediated immunity in colorectal cancer. Immunity. 2016;45(3): 641–55.
89. Bouskra D, Brezillon C, Berard M, Werts C, Varona R, Boneca IG, et al. Lymphoid tissue genesis induced by commensals through NOD1 regulates intestinal homeostasis. Nature. 2008;456(7221): 507–10.
90. Gury-BenAri M, Thaiss CA, Serafini N, Winter DR, Giladi A, Lara-Astiaso D, et al. The spectrum and regulatory landscape of intestinal innate lymphoid cells are shaped by the microbiome. Cell. 2016;166 (5):1231–46.
91. Lauzon NM, Mian F, MacKenzie R, Ashkar AA. The direct effects of Toll-like receptor ligands on human NK cell cytokine production and cytotoxicity. Cell Immunol. 2006;241(2):102–12.
92. Sivori S, Falco M, Della Chiesa M, Carlomagno S, Vitale M, Moretta L, et al. CpG and double-stranded RNA trigger human NK cells by Toll-like receptors: induction of cytokine release and cytotoxicity against tumors and dendritic cells. Proc Natl Acad Sci USA. 2004;101(27):10116–21.
93. Martinez J, Huang X, Yang Y. Direct TLR2 signaling is critical for NK cell activation and function in response to vaccinia viral infection. PLoS Pathog. 2010;6(3):e1000811.
94. Maggi L, Montaini G, Mazzoni A, Rossettini B, Capone M, Rossi MC, et al. Human circulating group 2 innate lymphoid cells can express CD154 and promote IgE production. J Allergy Clin Immunol. 2017;139(3):964–76.
95. Thio CL, Lai AC, Chi PY, Webster G, Chang YJ. Toll-like receptor 9-dependent interferon production prevents group 2 innate lymphoid cell-driven airway hyperreactivity. J Allergy Clin Immunol. 2019;144(3):682–97.
96. Crellin NK, Trifari S, Kaplan CD, Satoh-Takayama-N, Di Santo JP, Spits H. Regulation of cytokine secretion in human CD127(+) LTi-like innate lymphoid cells by Toll-like receptor 2. Immunity. 2010;33(5):752–64.
97. Xu H, Wang X, Lackner AA, Veazey RS. Type 3 innate lymphoid cell depletion is mediated by TLRs in lymphoid tissues of simian immunodeficiency virus-infected macaques. FASEB J. 2015;29 (12):5072–80.
98. Barrow AD, Martin CJ, Colonna M. The natural cytotoxicity receptors in health and disease. Front Immunol. 2019;10:909.
99. Glatzer T, Killig M, Meisig J, Ommert I, Luetke-Eversloh M, Babic M, et al. RORγt(+) innate lymphoid cells acquire a proinflammatory program upon engagement of the activating receptor NKp44. Immunity. 2013;38(6):1223–35.
100. Li S, Bostick JW, Ye J, Qiu J, Zhang B, Urban JF Jr, et al. Aryl hydrocarbon receptor signaling cell intrinsically inhibits intestinal group 2 innate lymphoid cell function. Immunity. 2018;49(5):915–28.
101. Miani M, Le Naour J, Waeckel-Enee E, Verma SC, Straube M, Emond P, et al. Gut microbiota-stimulated innate lymphoid cells support β-defensin 14 expression in pancreatic endocrine cells, preventing autoimmune diabetes. Cell Metab. 2018;28(4):557–72.
102. Lewis G, Wang B, Shafiei Jahani P, Hurrell BP, Banie H, Aleman Muench GR, et al. Dietary fiber-induced microbial short chain fatty acids suppress ILC2-dependent airway inflammation. Front Immunol. 2019;10:2051.

103. Sepahi A, Liu Q, Friesen L, Kim CH. Dietary fiber metabolites regulate innate lymphoid cell responses. Mucosal Immunol. 2021;14(2):317–30.
104. Chun E, Lavoie S, Fonseca-Pereira D, Bae S, Michaud M, Hoveyda HR, et al. Metabolite-sensing receptor Ffar2 regulates colonic group 3 innate lymphoid cells and gut immunity. Immunity. 2019;51(5):871–84.
105. Mortha A, Chudnovskiy A, Hashimoto D, Bogunovic M, Spencer SP, Belkaid Y, et al. Microbiota-dependent crosstalk between macrophages and ILC3 promotes intestinal homeostasis. Science. 2014;343(6178):1249288.
106. Zhou W, Sonnenberg GF. Activation and suppression of group 3 innate lymphoid cells in the gut. Trends Immunol. 2020;41(8):721–33.
107. Sonnenberg GF, Artis D. Innate lymphoid cell interactions with microbiota: implications for intestinal health and disease. Immunity. 2012;37(4):601–10.
108. Garrett WS, Lord GM, Punit S, Lugo-Villarino G, Mazmanian SK, Ito S, et al. Communicable ulcerative colitis induced by T-bet deficiency in the innate immune system. Cell. 2007;131(1):33–45.
109. Garrett WS, Gallini CA, Yatsunenko T, Michaud M, DuBois A, Delaney ML, et al. Enterobacteriaceae act in concert with the gut microbiota to induce spontaneous and maternally transmitted colitis. Cell Host Microbe. 2010;8(3):292–300.
110. Geremia A, Arancibia-Carcamo CV, Fleming MP, Rust N, Singh B, Mortensen NJ, et al. IL-23-responsive innate lymphoid cells are increased in inflammatory bowel disease. J Exp Med. 2011;208(6):1127–33.
111. Powell N, Walker AW, Stolarczyk E, Canavan JB, Gokmen MR, Marks E, et al. The transcription factor T-bet regulates intestinal inflammation mediated by interleukin-7 receptor+ innate lymphoid cells. Immunity. 2012;37(4):674–84.
112. Lee HC, Ziegler SF. Inducible expression of the proallergic cytokine thymic stromal lymphopoietin in airway epithelial cells is controlled by NFkappaB. Proc Natl Acad Sci USA. 2007;104(3):914–9.
113. Taylor BC, Zaph C, Troy AE, Du Y, Guild KJ, Comeau MR, et al. TSLP regulates intestinal immunity and inflammation in mouse models of helminth infection and colitis. J Exp Med. 2009;206(3):655–67.
114. Cultrone A, de Wouters T, Lakhdari O, Kelly D, Mulder I, Logan E, et al. The NF-kappaB binding site located in the proximal region of the TSLP promoter is critical for TSLP modulation in human intestinal epithelial cells. Eur J Immunol. 2013;43(4):1053–62.
115. Kato A, Favoreto S Jr, Avila PC, Schleimer RP. TLR3- and Th2 cytokine-dependent production of thymic stromal lymphopoietin in human airway epithelial cells. J Immunol. 2007;179(2):1080–7.
116. Price AE, Shamardani K, Lugo KA, Deguine J, Roberts AW, Lee BL, et al. A map of Toll-like receptor expression in the intestinal epithelium reveals distinct spatial, cell type-specific, and temporal patterns. Immunity. 2018;49(3):560–75.
117. Giacomin PR, Moy RH, Noti M, Osborne LC, Siracusa MC, Alenghat T, et al. Epithelial-intrinsic IKKα expression regulates group 3 innate lymphoid cell responses and antibacterial immunity. J Exp Med. 2015;212(10):1513–28.
118. Shaw MH, Kamada N, Kim YG, Nunez G. Microbiota-induced IL-1β, but not IL-6, is critical for the development of steady-state TH17 cells in the intestine. J Exp Med. 2012;209(2):251–8.
119. Castleman MJ, Dillon SM, Purba CM, Cogswell AC, Kibbie JJ, McCarter MD, et al. Commensal and pathogenic bacteria indirectly induce IL-22 but not IFNγ production from human colonic ILC3s via multiple mechanisms. Front Immunol. 2019;10:649.
120. Pickert G, Neufert C, Leppkes M, Zheng Y, Wittkopf N, Warntjen M, et al. STAT3 links IL-22 signaling in intestinal epithelial cells to mucosal wound healing. J Exp Med. 2009;206(7):1465–72.
121. McGee HM, Schmidt BA, Booth CJ, Yancopoulos GD, Valenzuela DM, Murphy AJ, et al. IL-22 promotes fibroblast-mediated wound repair in the skin. J Investig Dermatol. 2013;133(5):1321–9.
122. Goto Y, Obata T, Kunisawa J, Sato S, Ivanov II, Lamichhane A, et al. Innate lymphoid cells regulate intestinal epithelial cell glycosylation. Science. 2014;345(6202):1254009.
123. Pickard JM, Maurice CF, Kinnebrew MA, Abt MC, Schenten D, Golovkina TV, et al. Rapid fucosylation of intestinal epithelium sustains host-commensal symbiosis in sickness. Nature. 2014;514(7524):638–41.
124. Pham TA, Clare S, Goulding D, Arasteh JM, Stares MD, Browne HP, et al. Epithelial IL-22RA1-mediated fucosylation promotes intestinal colonization resistance to an opportunistic pathogen. Cell Host Microbe. 2014;16(4):504–16.
125. Vaishnava S, Yamamoto M, Severson KM, Ruhn KA, Yu X, Koren O, et al. The antibacterial lectin RegIIIgamma promotes the spatial segregation of microbiota and host in the intestine. Science. 2011;334(6053):255–8.
126. Lehmann FM, von Burg N, Ivanek R, Teufel C, Horvath E, Peter A, et al. Microbiota-induced tissue signals regulate ILC3-mediated antigen presentation. Nat Commun. 2020;11(1):1794.
127. Rao A, Strauss O, Kokkinou E, Bruchard M, Tripathi KP, Schlums H, et al. Cytokines regulate the antigen-presenting characteristics of human circulating and tissue-resident intestinal ILCs. Nat Commun. 2020;11(1):2049.
128. Hepworth MR, Fung TC, Masur SH, Kelsen JR, McConnell FM, Dubrot J, et al. Immune tolerance. Group 3 innate lymphoid cells mediate intestinal

selection of commensal bacteria-specific CD4(+) T cells. Science. 2015;348(6238):1031–5.
129. Melo-Gonzalez F, Kammoun H, Evren E, Dutton EE, Papadopoulou M, Bradford BM, et al. Antigen-presenting ILC3 regulate T cell-dependent IgA responses to colonic mucosal bacteria. J Exp Med. 2019;216(4):728–42.
130. Goc J, Lv M, Bessman NJ, Flamar AL, Sahota S, Suzuki H, et al. Dysregulation of ILC3s unleashes progression and immunotherapy resistance in colon cancer. Cell. 2021;184(19):5015–30.
131. Joseph AM, Monticelli LA, Sonnenberg GF. Metabolic regulation of innate and adaptive lymphocyte effector responses. Immunol Rev. 2018;286 (1):137–47.
132. Helou DG, Shafiei-Jahani P, Lo R, Howard E, Hurrell BP, Galle-Treger L, et al. PD-1 pathway regulates ILC2 metabolism and PD-1 agonist treatment ameliorates airway hyperreactivity. Nat Commun. 2020;11(1):3998.
133. Fachi JL, Pral LP, Dos Santos JAC, Codo AC, de Oliveira S, Felipe JS, et al. Hypoxia enhances ILC3 responses through HIF-1α-dependent mechanism. Mucosal Immunol. 2021;14(4):828–41.
134. Di Luccia B, Gilfillan S, Cella M, Colonna M, Huang SC. ILC3s integrate glycolysis and mitochondrial production of reactive oxygen species to fulfill activation demands. J Exp Med. 2019;216(10):2231–41.
135. Mundt S, Mrdjen D, Utz SG, Greter M, Schreiner B, Becher B. Conventional DCs sample and present myelin antigens in the healthy CNS and allow parenchymal T cell entry to initiate neuroinflammation. Sci Immunol. 2019;4(31):8380.
136. Hickey WF, Kimura H. Perivascular microglial cells of the CNS are bone marrow-derived and present antigen in vivo. Science. 1988;239(4837):290–2.
137. Ransohoff RM, Kivisakk P, Kidd G. Three or more routes for leukocyte migration into the central nervous system. Nat Rev Immunol. 2003;3(7):569–81.
138. Brioschi S, Wang WL, Peng V, Wang M, Shchukina I, Greenberg ZJ, et al. Heterogeneity of meningeal B cells reveals a lymphopoietic niche at the CNS borders. Science. 2021;373(6553):9277.
139. Cugurra A, Mamuladze T, Rustenhoven J, Dykstra T, Beroshvili G, Greenberg ZJ, et al. Skull and vertebral bone marrow are myeloid cell reservoirs for the meninges and CNS parenchyma. Science. 2021;373 (6553):7844.
140. Aspelund A, Antila S, Proulx ST, Karlsen TV, Karaman S, Detmar M, et al. A dural lymphatic vascular system that drains brain interstitial fluid and macromolecules. J Exp Med. 2015;212(7): 991–9.
141. Louveau A, Smirnov I, Keyes TJ, Eccles JD, Rouhani SJ, Peske JD, et al. Structural and functional features of central nervous system lymphatic vessels. Nature. 2015;523(7560):337–41.
142. Ma Q, Ineichen BV, Detmar M, Proulx ST. Outflow of cerebrospinal fluid is predominantly through lymphatic vessels and is reduced in aged mice. Nat Commun. 2017;8(1):1434.
143. Miljkovic D, Jevtic B, Stojanovic I, Dimitrijevic M. ILC3, a central innate immune component of the gut–brain axis in multiple sclerosis. Front Immunol. 2021;12:657622.
144. Ma Q, Xing C, Long W, Wang HY, Liu Q, Wang RF. Impact of microbiota on central nervous system and neurological diseases: the gut–brain axis. J Neuroinflamm. 2019;16(1):53.
145. Bettelli E, Sullivan B, Szabo SJ, Sobel RA, Glimcher LH, Kuchroo VK. Loss of T-bet, but not STAT1, prevents the development of experimental autoimmune encephalomyelitis. J Exp Med. 2004;200(1): 79–87.
146. Lovett-Racke AE, Rocchini AE, Choy J, Northrop SC, Hussain RZ, Ratts RB, et al. Silencing T-bet defines a critical role in the differentiation of autoreactive T lymphocytes. Immunity. 2004;21(5): 719–31.
147. Kwong B, Rua R, Gao Y, Flickinger J Jr, Wang Y, Kruhlak MJ, et al. T-bet-dependent NKp46(+) innate lymphoid cells regulate the onset of TH17-induced neuroinflammation. Nat Immunol. 2017;18(10): 1117–27.
148. Sanmarco LM, Wheeler MA, Gutierrez-Vazquez C, Polonio CM, Linnerbauer M, Pinho-Ribeiro FA, et al. Gut-licensed IFNgamma(+) NK cells drive LAMP1 (+)TRAIL(+) anti-inflammatory astrocytes. Nature. 2021;590(7846):473–9.
149. Hirose S, Jahani PS, Wang S, Jaggi U, Tormanen K, Yu J, et al. Type 2 innate lymphoid cells induce CNS demyelination in an HSV-IL-2 mouse model of multiple sclerosis. iScience. 2020;23(10):101549.
150. Russi AE, Walker-Caulfield ME, Ebel ME, Brown MA. Cutting edge: c-Kit signaling differentially regulates type 2 innate lymphoid cell accumulation and susceptibility to central nervous system demyelination in male and female SJL mice. J Immunol. 2015;194(12):5609–13.
151. Gadani SP, Smirnov I, Smith AT, Overall CC, Kipnis J. Characterization of meningeal type 2 innate lymphocytes and their response to CNS injury. J Exp Med. 2017;214(2):285–96.
152. Fung ITH, Sankar P, Zhang Y, Robison LS, Zhao X, D'Souza SS, et al. Activation of group 2 innate lymphoid cells alleviates aging-associated cognitive decline. J Exp Med. 2020;217(4):e20190915.
153. Godinho-Silva C, Domingues RG, Rendas M, Raposo B, Ribeiro H, da Silva JA, et al. Light-entrained and brain-tuned circadian circuits regulate ILC3s and gut homeostasis. Nature. 2019;574(7777): 254–8.
154. Wang Q, Robinette ML, Billon C, Collins PL, Bando JK, Fachi JL, et al. Circadian rhythm-dependent and circadian rhythm-independent impacts of the molecular clock on type 3 innate lymphoid cells. Sci Immunol. 2019;4(40):7501.

155. Teng F, Goc J, Zhou L, Chu C, Shah MA, Eberl G, et al. A circadian clock is essential for homeostasis of group 3 innate lymphoid cells in the gut. Sci Immunol. 2019;4(40):1215.
156. Ordovas-Montanes J, Rakoff-Nahoum S, Huang S, Riol-Blanco L, Barreiro O, von Andrian UH. The regulation of immunological processes by peripheral neurons in homeostasis and disease. Trends Immunol. 2015;36(10):578–604.
157. Chapman LF, Goodell H. The participation of the nervous system in the inflammatory reaction. Ann N Y Acad Sci. 1964;116:990–1017.
158. Moriyama S, Brestoff JR, Flamar AL, Moeller JB, Klose CSN, Rankin LC, et al. β2-adrenergic receptor-mediated negative regulation of group 2 innate lymphoid cell responses. Science. 2018;359 (6379):1056–61.
159. Nagashima H, Mahlakoiv T, Shih HY, Davis FP, Meylan F, Huang Y, et al. Neuropeptide CGRP limits group 2 innate lymphoid cell responses and constrains type 2 inflammation. Immunity. 2019;51 (4):682–95.
160. Xu H, Ding J, Porter CBM, Wallrapp A, Tabaka M, Ma S, et al. Transcriptional atlas of intestinal immune cells reveals that neuropeptide α-CGRP modulates group 2 innate lymphoid cell responses. Immunity. 2019;51(4):696–708.
161. Wallrapp A, Burkett PR, Riesenfeld SJ, Kim SJ, Christian E, Abdulnour RE, et al. Calcitonin gene-related peptide negatively regulates alarmin-driven type 2 innate lymphoid cell responses. Immunity. 2019;51(4):709–23.
162. Sui P, Wiesner DL, Xu J, Zhang Y, Lee J, Van Dyken S, et al. Pulmonary neuroendocrine cells amplify allergic asthma responses. Science. 2018;360(6393):8546.
163. Talbot J, Hahn P, Kroehling L, Nguyen H, Li D, Littman DR. Feeding-dependent VIP neuron-ILC3 circuit regulates the intestinal barrier. Nature. 2020;579(7800):575–80.
164. Seillet C, Luong K, Tellier J, Jacquelot N, Shen RD, Hickey P, et al. The neuropeptide VIP confers anticipatory mucosal immunity by regulating ILC3 activity. Nat Immunol. 2020;21(2):168–77.
165. Talbot S, Abdulnour RE, Burkett PR, Lee S, Cronin SJ, Pascal MA, et al. Silencing nociceptor neurons reduces allergic airway inflammation. Neuron. 2015;87(2):341–54.
166. Nussbaum JC, Van Dyken SJ, von Moltke J, Cheng LE, Mohapatra A, Molofsky AB, et al. Type 2 innate lymphoid cells control eosinophil homeostasis. Nature. 2013;502(7470):245–8.
167. Cardoso V, Chesne J, Ribeiro H, Garcia-Cassani B, Carvalho T, Bouchery T, et al. Neuronal regulation of type 2 innate lymphoid cells via neuromedin U. Nature. 2017;549(7671):277–81.
168. Klose CSN, Mahlakoiv T, Moeller JB, Rankin LC, Flamar AL, Kabata H, et al. The neuropeptide neuromedin U stimulates innate lymphoid cells and type 2 inflammation. Nature. 2017;549(7671):282–6.
169. Wallrapp A, Riesenfeld SJ, Burkett PR, Abdulnour RE, Nyman J, Dionne D, et al. The neuropeptide NMU amplifies ILC2-driven allergic lung inflammation. Nature. 2017;549(7672):351–6.
170. Wang S, Zhang Y, Wang Y, Ye P, Li J, Li H, et al. Amphiregulin confers regulatory T cell suppressive function and tumor invasion via the EGFR/GSK-3β/Foxp3 axis. J Biol Chem. 2016;291(40):21085–95.
171. Yan Y, Ramanan D, Rozenberg M, McGovern K, Rastelli D, Vijaykumar B, et al. Interleukin-6 produced by enteric neurons regulates the number and phenotype of microbe-responsive regulatory T cells in the gut. Immunity. 2021;54(3):499–513.
172. Darby M, Roberts LB, Mackowiak C, Chetty A, Tinelli S, Schnoeller C, et al. ILC3-derived acetylcholine promotes protease-driven allergic lung pathology. J Allergy Clin Immunol. 2021;147(4): 1513–6.
173. Chu C, Parkhurst CN, Zhang W, Zhou L, Yano H, Arifuzzaman M, et al. The ChAT-acetylcholine pathway promotes group 2 innate lymphoid cell responses and anti-helminth immunity. Sci Immunol. 2021;6(57):3218.
174. Cardoso F, Klein Wolterink RGJ, Godinho-Silva C, Domingues RG, Ribeiro H, da Silva JA, et al. Neuro-mesenchymal units control ILC2 and obesity via a brain-adipose circuit. Nature. 2021;597(7876):410–4.
175. Ibiza S, Garcia-Cassani B, Ribeiro H, Carvalho T, Almeida L, Marques R, et al. Glial-cell-derived neuroregulators control type 3 innate lymphoid cells and gut defence. Nature. 2016;535(7612):440–3.
176. Anitha M, Vijay-Kumar M, Sitaraman SV, Gewirtz AT, Srinivasan S. Gut microbial products regulate murine gastrointestinal motility via Toll-like receptor 4 signaling. Gastroenterology. 2012;143(4):1006–16.
177. Obata Y, Castano A, Boeing S, Bon-Frauches AC, Fung C, Fallesen T, et al. Neuronal programming by microbiota regulates intestinal physiology. Nature. 2020;578(7794):284–9.
178. Muller PA, Koscso B, Rajani GM, Stevanovic K, Berres ML, Hashimoto D, et al. Crosstalk between muscularis macrophages and enteric neurons regulates gastrointestinal motility. Cell. 2014;158 (2):300–13.
179. Matheis F, Muller PA, Graves CL, Gabanyi I, Kerner ZJ, Costa-Borges D, et al. Adrenergic signaling in muscularis macrophages limits infection-induced neuronal loss. Cell. 2020;180(1):64–78.
180. Gabanyi I, Muller PA, Feighery L, Oliveira TY, Costa-Pinto FA, Mucida D. Neuro-immune interactions drive tissue programming in intestinal macrophages. Cell. 2016;164(3):378–91.
181. Holland-Cunz S, Goppl M, Rauch U, Bar C, Klotz M, Schafer KH. Acquired intestinal aganglionosis after a lytic infection with varicella-zoster virus. J Pediatr Surg. 2006;41(3):e29–31.

Interaction Between Innate Lymphoid Cells and the Nervous System

9

Yuanyue Zhang, Rachel Grazda, and Qi Yang

Abstract

The interaction between the immune system and the nervous system remains an intriguing enigma. Recent studies indicate that innate lymphoid cells (ILCs), a unique family of innate effector cells, participate in intense cross talk with the nervous system. In the mucosal barrier sites, ILCs have been found to co-localize with neurons, nerves, glial cell projectors, and neuroendocrine cells. The cross talk between ILCs and peripheral nervous system orchestrates mucosal homeostasis and immunity. In addition, the barrier tissues of the central nervous system (CNS) also provide conductive microenvironment for ILC development and maintenance. Activities of CNS-associated ILCs impact the outcome of various CNS disorders. In this chapter, we review and discuss the intricate and bidirectional interaction between ILCs and nervous system.

Y. Zhang · Q. Yang (✉)
Department of Immunology & Microbial Disease, Albany Medical College, Albany, NY, USA
e-mail: qy117@rwjms.rutgers.edu

R. Grazda
Department of Immunology & Microbial Disease, Albany Medical College, Albany, NY, USA

Rutgers Child Health Institute of New Jersey, New Brunswick, NJ, USA

Keywords

Innate lymphoid cells · Neuroimmunology · The enteric nervous system · Meningeal immunity · Choroid plexus

9.1 Introduction

The bidirectional cross talk between the nervous system and immune system has attracted increasing attention in the recent years. In the gastrointestinal tract, the activity of immune cells plays a crucial role in the gut–brain axis, bridging the communication between the central nervous system, the enteric nervous system, and the microbiome [1–3]. In the lung and in the skin, orchestration between immune cells and neuronal network allows coordinated and rapid responses to pathogens and other noxious stimuli [4, 5]. In the brain and associated structures, the activities of immune cells impact blood-brain barrier integrity, neuroinflammation, cognition, and behavior [6, 7].

Advances in the last few decades have greatly expanded our knowledge of immune cell diversity. The discovery and characterization of the innate lymphoid cell (ILC) family have provided revolutionary insights into our understanding of lymphocyte differentiation and function [8]. ILCs lack clonally distributed antigen receptors, but

© The Author(s), under exclusive license to Springer Nature Singapore Pte Ltd. 2022

X.-H. Sun (ed.), *Innate Lymphoid Cells*, Advances in Experimental Medicine and Biology 1365,
https://doi.org/10.1007/978-981-16-8387-9_9

transcriptionally and functionally mirror T cells [8]. Yet unlike traditional T cells, ILCs already acquire effector function during early development and may constitutively produce effector molecules even at homeostasis. Notably, with the exception of circulating NK cells, the majority of cytokine-producing helper ILCs are tissue-resident lymphocytes that are enriched in mucosal barrier sites and other non-lymphoid vital organs [9]. Interestingly, recent work indicates co-location of tissue-resident ILCs with neurons, nerves, glia cells, and neuroendocrine cells [1]. Studies on the intriguing neuro-ILC circuits at mucosal barriers have provided an exciting glimpse into the close cross talk between the immune and the nervous system.

Unlike the peripheral nervous system, the central nervous system (CNS) is relatively immune privileged and is largely devoid of immune cells save for microglia in the parenchyma. Yet, mounting evidence indicates important roles for adaptive and innate lymphocytes in regulating CNS homeostasis and function. The brain barrier regions, including blood-brain barrier, blood-cerebrospinal fluid (CSF) barrier, and blood-meninges barrier, are the hub of neuroimmune interaction in the CNS [6, 7]. Numerous trafficking and tissue-resident innate and adaptive immune cell subsets have been detected in and around vessels and capillaries that lie within Virchow–Robin space, choroid plexus, and meninges. Recent work has highlighted pivotal roles for both trafficking blood-borne lymphocytes and noncirculating brain barrier-resident lymphocytes in regulating brain and spinal cord homeostasis and function [6, 7].

In this chapter, we discuss the bidirectional communication between innate lymphoid cells and nervous system. We review current knowledge about the interaction between ILCs and peripheral nervous system and how they orchestrate mucosal immunity. We also outline the role of circulating and tissue-resident ILCs in regulating CNS homeostasis and inflammation.

9.2 Regulation of Innate Lymphoid Cells by the Nervous System

9.2.1 The Peripheral Nervous System at Mucosal Barriers

Mucosal barriers, such as the gut and the lung, are highly innervated organs. The mammalian gastrointestinal tract harbors numerous neurons and glia cells in the ganglia at the myenteric plexus and the submucosal plexus [10]. The gut is densely innervated by both the intrinsic enteric nervous system and the extrinsic sympathetic and parasympathetic nervous system [10]. The lung nervous plexus consists of nerve fibers at the submucosal regions, and parasympathetic ganglionic neurons have also been found along trachea bifurcations and main bronchus [11]. In addition, pulmonary neuroendocrine cells (PNEC), rare innervated airway epithelial cells, cluster into neuroepithelial bodies at bronchial branch points [12, 13]. Recent work indicates that many tissue-resident innate lymphoid cells are located in proximity to neurons, nerves, glial cell projectors, and neuroendocrine cells in the mucosal barriers, indicating close communication between ILCs and peripheral nervous system [1].

9.2.2 Regulation of Group 2 Innate Lymphoid Cell (ILC2) Activity by the Peripheral Nervous System

ILC2s, a major subset of ILCs, are critically involved in host defense, tissue repair, and asthma pathogenesis. ILC2s are enriched in mucosal barrier sites, such as lung, gut, and skin. They reside in highly innervated connective tissue around major vessels in the mucosal barriers [14]. Mounting evidence suggests an intriguing role for the neuronal network in regulating ILC2 activity by providing both activating and inhibitory signals. In the gastrointestinal tract, ILC2s co-localize with both cholinergic enteric neurons and adrenergic sympathetic neurons [15, 16]. Co-localization of ILC2s with cholinergic parasympathetic neurons and SNAP-25^{+} nerve fibers has also been found in the lung [16, 17]. In addition, a subset of ILC2s localize in proximity to pulmonary neuroendocrine cells (PNEC), a highly specialized subset of epithelial cells that secrete neurotransmitters [18]. Tissue-resident ILC2s express high amounts of surface receptors to, and directly respond to, a variety of neuropeptides and neurotransmitters.

ILC2s are the main cell subset that expresses NMUR1 in the lungs and intestines of NMUR1 reporter mice [15, 16, 19]. NMU produced by cholinergic neurons potently and rapidly activates ILC2 [15, 16, 19]. NMU alone is sufficient to activate ILC2, via ERK1/2 activation and a calcineurin/NFAT cascade [15, 16, 19]. NMU may also amplify ILC2 responses to IL-25 and IL-33 [19]. Genetic deletion of NMUR1 led to impaired innate type 2 responses for helminth clearance in mouse models of *N. brasiliensis* infection [15, 16]. The absence of NMUR1 signaling also reduced inflammatory ILC2 responses during allergic airway inflammation [19].

In contrast, NMB, another neuromedin family of neuropeptide, inhibits lung and intestinal ILC2 activation via NMB/NMBR interaction [20]. The presence of basophils enhances NMBR expression by ILC2s, thus repressing ILC2 responses during helminth infection [20]. Although the precise mechanisms by which NMBR/NMB signaling represses ILC2s are yet to be further elucidated, this neuropeptide receptor-mediated cross talk between ILC2 and basophils might represent an important negative feedback pathway that mammalians utilize to restrict excessive type 2 inflammation [20].

ILC2s also express β2 adrenoreceptor, and their activities are negatively regulated by norepinephrine produced by sympathetic neurons and nerves during helminth infection and allergic airway inflammation [17]. Deletion of β2 adrenoreceptor increased intestinal ILC2 responses in the intestine and in the lung, leading to accelerated helminth clearance and exacerbated type 2 airway inflammation [17]. Treatment of β2 adrenoreceptor agonist inhibited ILC2 responses during *Alternaria* allergen-induced airway inflammation [17].

ILC2s also respond to many other neuropeptides such as the vasoactive intestinal peptide (VIP), CALCA-encoding calcitonin gene-related peptide (CGRP), and acetylcholine. ILC2s express mRNA of VIP receptor type 1 (VPAC1) and VIP receptor type 2 (VPAC2) [21]. Exposure to VIP may enhance the homeostatic production of IL-5 by intestinal and lung ILC2s [21]. Because VIP controls circadian behaviors, the regulation of ILC2 activation by VIP might contribute to the circadian variation of ILC2 activity [21]. ILC2s also express CGRP receptors [18, 22]. CGRP appears to have complicated effects on ILC2 responses [18, 22]. CGRP enhances IL-5 production but restricts IL-13 production, in concert of IL-33 and UMU [18, 22]. Of note, CGFP can also be produced by ILC2s under certain conditions, indicating potential autocrine regulation [23].

Acetylcholine regulation of ILC2s has also been documented. mRNA expression for both muscarinic and nicotinic Ach receptors was detected in intestinal ILC2s [24, 25]. Activated ILC2s upregulated choline acetyltransferase (ChAT) expression, and released acetylcholine in vivo [24, 25]. Deletion of ILC2-intrinsic ChAT led to reduced ILC2 responses and defective helminth expulsion, indicating an intriguing feedback loop underlined by autocrine-regulatory mechanisms [24, 25]. Interestingly, other studies indicate that treatment with exogenous α7nAChR agonist instead attenuated ILC2 responses in IL-33 and *Alternaria* allergen-induced airway inflammation, suggesting that exogenous α7nAChR may positively regulate ILC2 activity [26, 27]. It is possible that distinct Ach receptors differentially control ILC2 activities under diverse physiological conditions.

In the lung, many ILC2s are also found to co-localize with pulmonary neuroendocrine cells in the bronchial branch points [18]. The abundance of PNECs is increased in human asthmatics. Mice lacking PNEC cells exhibited enhanced ILC2 proliferation and cytokine production in the OVA-induced allergic airway inflammation mouse model [18]. Baseline levels of ILC2 abundance and activity remained comparable between PNEC-deficient mice and wild-type mice. Induced deletion of PNECs in adult mice also reduced ILC2 responses in house dust mite-induced airway inflammation [18]. Thus, neuroendocrine cells might provide an important niche to support ILC2 activity during allergic airway inflammation.

9.2.3 Regulation of ILC3 Response by the Neuronal Network

Group 3 innate lymphoid cells (ILC3s) are another major ILC subset in the mucosal barriers. ILC3s are enriched in the intestinal lamina propria, cryptopatches (CP), and isolated lymphoid follicles (ILFs). $CCR6^+$ ILC3s express high levels of VIPR2, and localize in close proximity to the projections of enteric neurons that express high amounts of VIP [28, 29]. Feeding induces VIP release from enteric neurons, activating VIPR2 signaling in ILC3s [28, 29]. Recent work reported both positive and negative regulation of ILC3 activity by VIP/VIPR2 signaling [28, 29]. ILC3 from *Rorgt*^{cre} *Vipr2*f/f mice or mice with VIP inhibition exhibited reduced IL-22 production after feeding, indicating that VIP might repress ILC3 activity in response to feeding [29]. Nevertheless, another study suggested that exposure to exogenous VIP enhanced IL-22 production by ILC3s in vitro and in vivo, indicating that VIP can also enhance ILC3 activity [28]. Deletion or inhibition of VIP signaling might lead to impaired or enhanced ILC3 responses in different disease states, leading to altered tissue inflammation and disease susceptibility [28, 29]. The distinctive role of VIP on ILC3s in different studies might be related with differences in VIP signaling strength and microenvironmental context such as microbiota variation [28, 29]. In addition to neurons and nerves, enteric glial cells are also important components of the neuroimmune circuits at the mucosal barriers. A large percentage of $CCR6^+$ ILC3s also express high amounts of *Ret, a receptor for glial* cell line-derived neurotrophic factor (GDNF)-family ligands (GFL) [30]. Ret^+ ILC3s aggregate around the projections of enteric glial cells in the cryptic patch and isolated lymphoid follicles [30]. Enteric glial cells sense environmental cues to produce GFL through Myd88-dependent mechanisms [30]. Enteric glial cell-derived GFL positively regulates ILC3 function, by stimulating ILC3s to produce IL-22 that promotes epithelial homeostasis and repair [30]. P38 MAPK/ERK-AKT cascade and STAT3 mediate GFL/Ret signaling-induced ILC3 activation [30].

ILC3 activity may also be regulated by the vagus nerve and its main neuro-mediator acetylcholine [31]. Acetylcholine upregulates the biosynthesis of immuno-resolvents such as PCTR1 in ILC3s [31]. Vagotomy leads to altered peritoneal macrophage responses and decreased resistance to *E. coli* infection, which can be restored by treatment with PCTR1 or replenishment of ILC3s [31].

9.2.4 Regulation of NK Activity by the Neurotransmitters

Regulation of NK cell, the cytotoxic innate lymphocytes, by sympathetic neurotransmitters has been documented by multiple studies. NK cells express high amounts of β2-adrenoreceptor receptors. β2-Adrenoreceptor receptor signals repress IFNγ production by liver NK cells, thereby enhancing susceptibility to MCMV infection in mouse models [32]. β2-Adrenergic stimulation also leads to reduced NK cell numbers in a mouse model of acute stress [33]. Exposure to adrenaline or isoprenaline reduces cytotoxic activity of human NK cells in vitro [34–37]. Nevertheless, a few other reports also suggest evidence for positive regulation of NK cell activity by β2-adrenoreceptor signaling [38–40]. The discrepancy might be related with variations in the microenvironmental factors and in the assays used to assess NK cell activity. β2-Adrenoreceptor activation also alters human NK cell adhesion to endothelial cells, thus affecting NK cell migration and circulation [34, 41, 42].

NK cells express dopamine receptors. Exposure to dopamine upregulated expression of D5R on human NK cells and inhibited human NK cell production and IFNγ production [43]. In mouse models, stimulation of dopamine receptors by D2-like agonists repressed NK cell cytotoxicity [44, 45]. NK cell activities were reduced in hypodopaminergic rats. However, the same studies reported that D1-like DR stimulation may instead enhance NK cell activity and that different D2 agonists may have antagonizing effects [44, 45]. In addition, chemical sympathectomy by injection of 6OH-DA in adult rats led to reduced number of NK cells in the blood and spleen after 7 days of treatment [46]. 6OH-DA treatment at birth, however, resulted in increased cytotoxic activity of NK cells [47, 48]. Together, while increasing evidence suggests that sympathetic neurotransmitters are critically involved in the regulation of NK cell activity, the precise pathways are yet to be better elucidated.

9.3 Regulation of the Central Nervous System by Innate Lymphoid Cells

9.3.1 Immunity in the CNS Barriers

The CNS is a delicate system that is highly sensitive to changes in the cytokine and immune cell milieu. Recent work has shed light on critical roles of various immune cells in regulating CNS homeostasis and function. Of note, the brain and spinal cord parenchyma are relatively immune-privileged structures that are devoid of immune cells other than microglia at homeostasis. Nevertheless, a variety of non-microglial immune cells are present with relative abundance in the blood-CNS interface, including blood-CNS barriers, blood-meningeal barriers, and blood-CSF barriers [6, 7]. In particular, the dura venous sinus of meninges harbors numerous circulating and tissue-resident myeloid and lymphoid immune cell subsets [49]. Many meningeal immune cells possess a unique ontogeny, partly because the calvaria provides a bone marrow niche for immune cell development [50, 51]. Immune cells have also been identified in choroid plexus; their similarities and differences with meningeal immune cells are also beginning to be revealed [52–55].

9.3.2 Regulation of CNS Function by NK Cells

NK cells and ILC2 are the two major subsets of innate lymphoid cells in the brain barriers [53]. Brain-associated NK cells express high amounts of cytotoxic molecules including granzymes, perforins, and cathepsins [56]. Chemokines such as *Ccl3* and *Ccl4* are also expressed with relative abundance in brain-associated NK cells [56]. Expression of cytokine genes such as *Ifng* and *Tnf* is minimal in brain-associated NK cells at homeostasis, but may be upregulated during brain infection and inflammation [56]. NK cells play complicated multiple-faceted roles in various brain disorders.

9.3.2.1 NK Cells and Brain Tumors

Known for their ability to kill malignant cells and to mediate antibody-dependent cellular cytotoxicity (ADCC), NK cells are believed to play an important role in the surveillance of brain tumors [57]. Tumor-infiltrating NK cells have been detected in glioma and meningiomas [58–60]. Several studies indicate that an activated NK cell profile is associated with reduced disease severity and a better prognosis in glioma [61–64]. NK cells can recognize and lyse glioma cells, particularly immature glioblastoma cells with stem cell-like properties [65, 66]. NK cell-mediated killing of neuroblastoma cells involves natural cytotoxicity receptors NKp46 and NKp30 [67, 68]. The engagement of NKp44 expressed by NK cells with PDGF-DD expressed by glioblastoma also stimulates NK cells to produce IFNγ and TNFα that inhibit tumor growth [69]. Susceptibility to NK cell-mediated killing is influenced by the expression of poliovirus receptor (PVR) on neuroblasts [68]. Recognition of PVR by the DNAX accessory molecule-1 (DNAM-1) expressed by NK cells facilitates neuroblastoma killing [68]. Neuroblasts also express B7-H3 that inhibits NK cell-mediated cytotoxicity [70]. In addition, malignant gliomas express high amounts of galectin-1 that suppress NK immune surveillance [71]. These knowledges gained from basic research provide important information to improve NK cell-based immunotherapy to combat brain cancer.

9.3.2.2 NK Cells and Brain Infections

NK cells are extensively implicated in brain infections. NK cells are important players in antiviral responses. NK cells may directly eradiate infected cells via cytotoxic activity, and may also be induced to produce the antiviral cytokine IFNγ. Increased susceptibility to viral encephalitis has been observed in various individuals with genetic mutations that affect NK cell numbers or function [72, 73]. Deficient NK cell responses underlined impaired virus clearance and increased susceptibly to Theiler's virus-induced demyelinating disease in mouse models [74]. Depletion of NK cells significantly led to increased mortality in a mouse model of acute virulent Semliki Forest virus-induced encephalitis [75]. Increased cytotoxic activity of NK cells was associated with decreased CNS lesions in macaques infected with neurovirulent SIV [75]. NK cells also played an important role in virus clearance during HSV-1-induced encephalitis [76]. NK cells may also limit neuro-invasion of bacteria such as *Listeria* bacteria [77].

NK cells might also affect the outcome of brain infections through mechanisms independently of pathogen clearance. One study reported that depletion of NK cells exacerbated morbidity in mice infected with the neurotropic JHM strain of mouse hepatitis virus, without affecting virus titers [78]. In contrast, another report indicated that depletion of NK cells improved the CNS outcome of Venezuelan equine encephalitis, indicating that NK cells might play a pathogenic role in this model [79]. In addition, NK cells were the major source of IFNγ that enhanced susceptibility to *Streptococcus pneumoniae* meningitis in mouse models [80]. The proinflammatory activity of NK cells might also contribute to severe cerebral malaria [81, 82]. Thus, NK cells may play a double-edged role in CNS infection. Both insufficient and dysregulated NK cell activities might disrupt the delicate balance between immune defense and disease tolerance in CNS infection.

9.3.2.3 NK Cells and Brain Ischemic Stroke

Recent studies have also extensively explored NK cell activity in patients and mouse models of ischemic stroke. NK cells infiltrated the lesions of postmortem samples of ischemic stroke patients [83, 84]. NK cells contributed to local neuroinflammation and exacerbated brain infarction in mouse models of MCAO [83, 84]. NK cells also promoted blood-brain barrier disruption and brain edema in a mouse model of ischemic hemorrhage, by killing cerebral endothelial cells and producing neutrophil-recruiting chemokines [85]. Nevertheless, protective role of NK cells in ischemic stroke and post-stroke conditions has also been suggested by several other studies. One recent study indicates that CXCL12-induced NK cell infiltration restricts stroke lesions in photothrombotic-induced stroke [86]. In addition, the presence of NK cells protected lethal *Listeria* infection in post-stroke mouse models [87]. These studies together highlight double-edged roles of NK cells in controlling neuroinflammation and immune defense in the CNS.

Interestingly, ischemia stroke is also associated with spleen atrophy and decreased numbers of NK cells in the circulation [87, 88]. Catecholaminergic and hypothalamic-pituitary-adrenal axis activation contributed to spleen atrophy and contraction of peripheral NK cells [87]. Reversal of this repression might enhance resistance to post-ischemic pneumonia [87].

9.3.2.4 NK Cells in Multiple Sclerosis (MS) and Experimental Autoimmune Encephalomyelitis

Studies with multiple sclerosis models indicate striking immunoregulatory function for NK cells. While $CD56^{dim}$ NK cells are the predominant subset of NK cells in the peripheral blood of healthy individuals, clinical reports suggest that $CD56^{bright}$ NK cells with enhanced immunoregulatory function are enriched in the CSF of patients with various neuroimmunological diseases including MS [89, 90]. A large-scale clinical study with relapsing-remitting MS patients from Australia indicates that increased proportion of CD56bright NK cells is associated with stable magnetic resonate imaging (MRI) [91]. In MS patients, $CD56^{bright}$ NK cells localize in proximity to T cells, and express high amounts of granzyme K that induces caspase-independent apoptosis of activated T cells [92, 93]. Modulation of IL-2Rα signaling by daclizumab enhanced the activity of $CD56^{hi}$ regulatory NK cells and alleviated CNS inflammation in MS patients [92, 93]. Multiple studies with mouse model of EAE also indicate that depletion of NK cells alleviates neuroinflammation and disease progression of EAE [94–96]. NK cells restrict IL-23 production by microglia and inhibit myelin-reactive Th17 responses in mouse models of EAE [96]. NK cells also produce acetylcholine that represses CNS infiltration of proinflammatory monocytes/macrophages [97]. Interaction between the MHCI molecule Qa-I expressed by autoreactive T cells and the inhibitory receptor NKG2A expressed by NK cells protects some autoreactive T cells from lysis by NK cells in mouse EAE models [98, 99]. Inhibition of NKG2A reduces T cell infiltration and microglial activation in EAE [98]. Nevertheless, recent work indicates that NK cells may impair CNS recovery in later stages of EAE by inducing the apoptosis of neural stem cells in the subventricular zone [100]. Thus, a deeper dissection into NK cell heterogeneity and the roles of specific NK cell subsets in distinct disease stages might help better understand the multiple-faceted function of NK cells in CNS disorders.

9.3.2.5 NK Cells and Neurodegeneration

Deviation of NK cell activity has been reported in patients with neurodegenerative diseases, such as Parkinson's disease (PD) and Alzheimer's disease (AD) [101–111]. However, the results of these clinical studies widely vary, possibly due to patient heterogeneity and distinct in vitro methods used to assess NK cell function [101–111]. NK cells have been found to play a protective role in a mouse model of PD [112]. Depletion of NK cells increased α-syn pathogenies and exacerbated motor symptoms in M83 Tg mice with intrastriatal injection of preformed fibrils (PPF) α-syn [112]. Nevertheless, depletion of NK cells did not affect Aβ pathologies or microglia uptake of Aβ in a triple-transgenic mouse model of AD (3xTg AD) [56]. In contrast, NK cell depletion alleviated microglia inflammation, stimulated neurogenesis, and improved cognitive function in 3xTg-AD mice, indicating a proinflammatory role of NK cells in these mice [56]. Together, these studies highlight the complicated context-dependent function of NK cells in regulating neuroinflammation and neurodegeneration.

9.3.3 Regulation of CNS Function by Group 2 Innate Lymphoid Cells

In addition to NK cells, ILC2s are another major subset of innate lymphoid cells in CNS-associated tissue. Function and regulation of ILC2 in CNS homeostasis and inflammation are beginning to be revealed.

9.3.3.1 ILC2 and CNS Injury

ILC2s are enriched along the meningeal dural venous sinus [14, 23]. Fibroblast-like stromal cells near the large and intermediate vessels might provide an important niche to support ILC2 survival and proliferation [14]. Meningeal ILC2s exhibit a distinct transcriptional profile compared to ILC2s in other anatomic locations such as the lung [23]. A neuroprotective role has been reported for meningeal ILC2s in a mouse model of spinal cord contusion [23]. Meningeal ILC2s responded to spinal cord injury by upregulating the expression of CGRP and other molecules involved in neuroprotection [23]. Transfer of ILC2 into ILC2-deficient mice improved CNS recovery [23].

ILC2 responses have also been examined in traumatic brain injury (TBI) models [113]. ILC2s were present in the dura and increased in the CSF of TBI patients [113]. AMP-activated protein kinase (AMPK) activation induced IL-10-

producing regulatory ILC2 that might help improve neurological outcomes in a controlled cortical impact mouse model of experimental TBI [113].

ILC2s also accumulated in brain-associated tissue in neonate mouse models of hypoxia ischemia, but their presence did not alter the outcome of brain injury in this model [114].

9.3.3.2 ILC2s and Brain Infection

ILC2s may play both anti-inflammatory and proinflammatory roles in brain infections. An immune-regulatory role for ILC2s has also been observed in a mouse model of cerebral malaria [115]. Transfer of ILC2s promoted the responses of regulatory T cells and reparative microglia and enhanced the protective effects of IL-33 in mice with cerebral malaria [115]. In contrast, ILC2s exacerbated demyelination in a mouse model of HSV-IL-2-induced CNS demyelination [116]. Potential roles for ILC2 in other CNS infectious disorders remain to be deciphered.

9.3.3.3 ILC2s and Multiple Sclerosis

Function and regulation of ILC2s have also been examined in mouse models of EAE. Differential ILC2 responses between sexes might contribute to sex-dimorphic effects on EAE susceptibility [117, 118]. Male-specific IL-33 expanded ILC2 that promoted protective Th2 response in EAE mice with myelin peptide immunization [117, 118]. Female mice lacked robust ILC2 responses, which might underline elevated encephalitogenic TH17 responses and increased disease severity [117, 118]. These studies together highlight immune-regulatory function for ILC2 in EAE. Nevertheless, ILC2s exacerbated CNS damage in an HSV-IL-2-induced EAE model [116]. Mechanisms underlying the potentially multiple-faceted roles for ILC2s in regulatory CNS inflammation remain to be better elucidated.

ILC3s, another major ILC subset, are barely detectable in the CNS or associated tissue at homeostasis. A subset of Rorγt+ ILC that expressed IL-17 and IFNγ infiltrated the inflamed CNS tissue in mouse model of EAE [119]. The presence of this ILC subset did not influence disease outcome [119].

9.3.3.4 ILC2s and Aging

Aging is a complicated process that is paradoxically associated with both a deterioration of the adaptive immune function (immunosenescence) and increased levels of inflammation (inflammaging). Interestingly, aging led to accumulation of ILC2s in CP of mice and humans [53]. CP ILC2s exhibited enhanced proliferative and functional capability compared to meningeal ILC2 in aged mice [53]. Activation of ILC2s was associated with improved cognitive function and reduced neuroinflammation in aged mice [53]. Potential roles of ILC2s in aging-associated CNS disorders such as neurodegenerative diseases remain to be deciphered.

Together, results from these studies suggest complicated multiple-faceted roles for innate lymphoid cells in regulating CNS homeostasis and inflammation. The interaction between ILCs and other immune and nonimmune cells in CNS-associated tissue, the heterogeneity of CNS-associated ILCs, and the precise effector function of ILCs in CNS disorders warrant further investigation.

9.4 CNS Lymphatics

The recent discovery of the dura meningeal lymphatics has uncovered the secrets of brain drain and revolutionized our understanding of immune surveillance in the CNS [120]. CNS is generally considered an immune-privileged system that lacks classical lymphatic system. However, recent work uncovers that functional lymphatic vessels line the dural sinuses of the meninges [120]. Meningeal lymphatics promote drainage of cerebrospinal fluid contents, and allow migration of CNS-associated lymphocytes to the draining lymph nodes [120, 121]. Meningeal lymphatics play an essential role in maintaining brain homeostasis, and its dysfunction exacerbates aging and Alzheimer's disease pathology-induced cognitive impairment [122]. Meningeal lymphatic ablation also led to increased inflammatory microglial responses and behavior deficits during anti-Aβ immunotherapy in mouse models of AD [123]. Meningeal lymphatic dysfunction exacerbated neuroinflammation in TBI mouse models, but reduced development of encephalitogenic T cells in EAE mouse models [121, 124]. Meningeal lymphatic vessels might also be important players in brain tumor drainage and immunity [125]. These studies together indicate critical roles for meningeal lymphatics in regulating CNS immunity. The interaction between tissue-resident and trafficking innate lymphocytes and meningeal lymphatics is an intriguing topic for future investigation.

9.5 Summary

Recent advance has provided an exciting glimpse into the intriguing interaction between ILCs and nervous system. Nevertheless, our understanding of ILC-neuro interaction is still at an infancy stage. How ILCs interact with the neuronal network in various vital organs and the importance of such interactions in human health and diseases are yet incompletely understood. The precise roles of ILCs in transmitting the messages to and from the nervous system remain largely unexplored. A better understanding of ILCs in regulating CNS physiology and function requires more extensive dissection into the heterogeneity and the multiple-faceted function of this unique family of immune cells. We expect that new and exciting insights will be uncovered in the next few years, as answers to these questions unfold.

References

1. Chu C, Artis D, Chiu IM. Neuro-immune interactions in the tissues. Immunity. 2020;52(3):464–74.
2. Abdel-Haq R, Schlachetzki JCM, Glass CK, Mazmanian SK. Microbiome-microglia connections via the gut-brain axis. J Exp Med. 2019;216(1): 41–59.
3. Huh JR, Veiga-Fernandes H. Neuroimmune circuits in inter-organ communication. Nat Rev Immunol. 2020;20(4):217–28.
4. Kabata H, Artis D. Neuro-immune crosstalk and allergic inflammation. J Clin Invest. 2019;129(4): 1475–82.
5. Blake KJ, Jiang XR, Chiu IM. Neuronal regulation of immunity in the skin and lungs. Trends Neurosci. 2019;42(8):537–51.
6. Alves de Lima K, Rustenhoven J, Kipnis J. Meningeal immunity and its function in maintenance of the central nervous system in health and disease. Annu Rev Immunol. 2020;38:597–620.
7. Rua R, McGavern DB. Advances in meningeal immunity. Trends Mol Med. 2018;24(6):542–59.
8. Yang Q, Bhandoola A. The development of adult innate lymphoid cells. Curr Opin Immunol. 2016;39:114–20.
9. Gasteiger G, Fan X, Dikiy S, Lee SY, Rudensky AY. Tissue residency of innate lymphoid cells in lymphoid and nonlymphoid organs. Science. 2015;350(6263):981–5.
10. Furness JB. The enteric nervous system and neurogastroenterology. Nat Rev Gastroenterol Hepatol. 2012;9(5):286–94.
11. Weigand LA, Myers AC. Synaptic and membrane properties of parasympathetic ganglionic neurons innervating mouse trachea and bronchi. Am J Physiol Lung Cell Mol Physiol. 2010;298(4):L593–9.
12. Kuo CS, Krasnow MA. Formation of a neurosensory organ by epithelial cell slithering. Cell. 2015;163(2): 394–405.
13. Noguchi M, Sumiyama K, Morimoto M. Directed migration of pulmonary neuroendocrine cells toward airway branches organizes the stereotypic location of neuroepithelial bodies. Cell Rep. 2015;13(12): 2679–86.
14. Dahlgren MW, Jones SW, Cautivo KM, Dubinin A, Ortiz-Carpena JF, Farhat S, et al. Adventitial stromal cells define group 2 innate lymphoid cell tissue niches. Immunity. 2019;50(3):707–22 e6.
15. Cardoso V, Chesne J, Ribeiro H, Garcia-Cassani B, Carvalho T, Bouchery T, et al. Neuronal regulation of type 2 innate lymphoid cells via neuromedin U. Nature. 2017;549(7671):277–81.
16. Klose CSN, Mahlakoiv T, Moeller JB, Rankin LC, Flamar AL, Kabata H, et al. The neuropeptide neuromedin U stimulates innate lymphoid cells and type 2 inflammation. Nature. 2017;549(7671):282–6.
17. Moriyama S, Brestoff JR, Flamar AL, Moeller JB, Klose CSN, Rankin LC, et al. beta2-adrenergic receptor-mediated negative regulation of group 2 innate lymphoid cell responses. Science. 2018;359(6379): 1056–61.
18. Sui P, Wiesner DL, Xu J, Zhang Y, Lee J, Van Dyken S, et al. Pulmonary neuroendocrine cells amplify allergic asthma responses. Science. 2018;360(6393):8546.
19. Wallrapp A, Riesenfeld SJ, Burkett PR, Abdulnour RE, Nyman J, Dionne D, et al. The neuropeptide NMU amplifies ILC2-driven allergic lung inflammation. Nature. 2017;549(7672):351–6.

20. Inclan-Rico JM, Ponessa JJ, Valero-Pacheco N, Hernandez CM, Sy CB, Lemenze AD, et al. Basophils prime group 2 innate lymphoid cells for neuropeptide-mediated inhibition. Nat Immunol. 2020;21(10):1181–93.
21. Nussbaum JC, Van Dyken SJ, von Moltke J, Cheng LE, Mohapatra A, Molofsky AB, et al. Type 2 innate lymphoid cells control eosinophil homeostasis. Nature. 2013;502(7470):245–8.
22. Nagashima H, Mahlakoiv T, Shih HY, Davis FP, Meylan F, Huang Y, et al. Neuropeptide CGRP limits group 2 innate lymphoid cell responses and constrains type 2 inflammation. Immunity. 2019;51 (4):682–95 e6.
23. Gadani SP, Smirnov I, Smith AT, Overall CC, Kipnis J. Characterization of meningeal type 2 innate lymphocytes and their response to CNS injury. J Exp Med. 2017;214(2):285–96.
24. Chu C, Parkhurst CN, Zhang W, Zhou L, Yano H, Arifuzzaman M, et al. The ChAT-acetylcholine pathway promotes group 2 innate lymphoid cell responses and anti-helminth immunity. Sci Immunol. 2021;6(57):3218.
25. Roberts LB, Schnoeller C, Berkachy R, Darby M, Pillaye J, Oudhoff MJ, et al. Acetylcholine production by group 2 innate lymphoid cells promotes mucosal immunity to helminths. Sci Immunol. 2021;6(57):0359.
26. Yuan F, Jiang L, Li Q, Sokulsky L, Wanyan Y, Wang L, et al. A selective alpha7 nicotinic acetylcholine receptor agonist, PNU-282987, attenuates ILC2s activation and Alternaria-induced airway inflammation. Front Immunol. 2020;11:598165.
27. Galle-Treger L, Suzuki Y, Patel N, Sankaranarayanan I, Aron JL, Maazi H, et al. Nicotinic acetylcholine receptor agonist attenuates ILC2-dependent airway hyperreactivity. Nat Commun. 2016;7:13202.
28. Seillet C, Luong K, Tellier J, Jacquelot N, Shen RD, Hickey P, et al. The neuropeptide VIP confers anticipatory mucosal immunity by regulating ILC3 activity. Nat Immunol. 2020;21(2):168–77.
29. Talbot J, Hahn P, Kroehling L, Nguyen H, Li D, Littman DR. Feeding-dependent VIP neuron-ILC3 circuit regulates the intestinal barrier. Nature. 2020;579(7800):575–80.
30. Ibiza S, Garcia-Cassani B, Ribeiro H, Carvalho T, Almeida L, Marques R, et al. Glial-cell-derived neuroregulators control type 3 innate lymphoid cells and gut defence. Nature. 2016;535(7612):440–3.
31. Dalli J, Colas RA, Arnardottir H, Serhan CN. Vagal regulation of group 3 innate lymphoid cells and the immunoresolvent PCTR1 controls infection resolution. Immunity. 2017;46(1):92–105.
32. Wieduwild E, Girard-Madoux MJ, Quatrini L, Laprie C, Chasson L, Rossignol R, et al. β2-adrenergic signals downregulate the innate immune response and reduce host resistance to viral infection. J Exp Med. 2020;217(4):e20190554.
33. Kanemi O, Zhang X, Sakamoto Y, Ebina M, Nagatomi R. Acute stress reduces intraparenchymal lung natural killer cells via beta-adrenergic stimulation. Clin Exp Immunol. 2005;139(1):25–34.
34. Whalen MM, Bankhurst AD. Effects of beta-adrenergic receptor activation, cholera toxin and forskolin on human natural killer cell function. Biochem J. 1990;272(2):327–31.
35. Rosenne E, Sorski L, Shaashua L, Neeman E, Matzner P, Levi B, et al. In vivo suppression of NK cell cytotoxicity by stress and surgery: glucocorticoids have a minor role compared to catecholamines and prostaglandins. Brain Behav Immun. 2014;37:207–19.
36. Theorell J, Gustavsson AL, Tesi B, Sigmundsson K, Ljunggren HG, Lundback T, et al. Immunomodulatory activity of commonly used drugs on Fc-receptor-mediated human natural killer cell activation. Cancer Immunol Immunother. 2014;63(6):627–41.
37. Takamoto T, Hori Y, Koga Y, Toshima H, Hara A, Yokoyama MM. Norepinephrine inhibits human natural killer cell activity in vitro. Int J Neurosci. 1991;58(1–2):127–31.
38. Tarr AJ, Powell ND, Reader BF, Bhave NS, Roloson AL, Carson WE 3rd, et al. beta-Adrenergic receptor mediated increases in activation and function of natural killer cells following repeated social disruption. Brain Behav Immun. 2012;26(8):1226–38.
39. Hellstrand K, Hermodsson S, Strannegard O. Evidence for a beta-adrenoceptor-mediated regulation of human natural killer cells. J Immunol. 1985;134(6):4095–9.
40. Glac W, Borman A, Badtke P, Stojek W, Orlikowska A, Tokarski J. Amphetamine enhances natural killer cytotoxic activity via beta-adrenergic mechanism. J Physiol Pharmacol. 2006;57(Suppl 11):125–32.
41. Benschop RJ, Schedlowski M, Wienecke H, Jacobs R, Schmidt RE. Adrenergic control of natural killer cell circulation and adhesion. Brain Behav Immun. 1997;11(4):321–32.
42. Schedlowski M, Hosch W, Oberbeck R, Benschop RJ, Jacobs R, Raab HR, et al. Catecholamines modulate human NK cell circulation and function via spleen-independent beta 2-adrenergic mechanisms. J Immunol. 1996;156(1):93–9.
43. Mikulak J, Bozzo L, Roberto A, Pontarini E, Tentorio P, Hudspeth K, et al. Dopamine inhibits the effector functions of activated NK cells via the upregulation of the D5 receptor. J Immunol. 2014;193(6):2792–800.
44. Nozaki H, Hozumi K, Nishimura T, Habu S. Regulation of NK activity by the administration of bromocriptine in haloperidol-treated mice. Brain Behav Immun. 1996;10(1):17–26.
45. Zhao W, Huang Y, Liu Z, Cao BB, Peng YP, Qiu YH. Dopamine receptors modulate cytotoxicity of natural killer cells via cAMP-PKA-CREB signaling pathway. PLoS One. 2013;8(6):e65860.

46. Pacheco-Lopez G, Niemi MB, Kou W, Bildhauser A, Gross CM, Goebel MU, et al. Central catecholamine depletion inhibits peripheral lymphocyte responsiveness in spleen and blood. J Neurochem. 2003;86(4):1024–31.
47. Reder A, Checinski M, Chelmicka-Schorr E. The effect of chemical sympathectomy on natural killer cells in mice. Brain Behav Immun. 1989;3(2):110–8.
48. Dishman RK, Hong S, Soares J, Edwards GL, Bunnell BN, Jaso-Friedmann L, et al. Activity-wheel running blunts suppression of splenic natural killer cell cytotoxicity after sympathectomy and footshock. Physiol Behav. 2000;71(3–4):297–304.
49. Rustenhoven J, Drieu A, Mamuladze T, de Lima KA, Dykstra T, Wall M, et al. Functional characterization of the dural sinuses as a neuroimmune interface. Cell. 2021;184(4):1000–16 e27.
50. Brioschi S, Wang WL, Peng V, Wang M, Shchukina I, Greenberg ZJ, et al. Heterogeneity of meningeal B cells reveals a lymphopoietic niche at the CNS borders. Science. 2021;373(6553): eabf9277.
51. Cugurra A, Mamuladze T, Rustenhoven J, Dykstra T, Beroshvili G, Greenberg ZJ, et al. Skull and vertebral bone marrow are myeloid cell reservoirs for the meninges and CNS parenchyma. Science. 2021;373 (6553):eabf7844.
52. Baruch K, Ron-Harel N, Gal H, Deczkowska A, Shifrut E, Ndifon W, et al. CNS-specific immunity at the choroid plexus shifts toward destructive Th2 inflammation in brain aging. Proc Natl Acad Sci. 2013;110(6):2264–9.
53. Fung ITH, Sankar P, Zhang Y, Robison LS, Zhao X, D'Souza SS, et al. Activation of group 2 innate lymphoid cells alleviates aging-associated cognitive decline. J Exp Med. 2020;217(4):e20190915.
54. Strominger I, Elyahu Y, Berner O, Reckhow J, Mittal K, Nemirovsky A, et al. The choroid plexus functions as a niche for T-cell stimulation within the central nervous system. Front Immunol. 2018;9: 1066.
55. Baruch K, Schwartz M. CNS-specific T cells shape brain function via the choroid plexus. Brain Behav Immun. 2013;34:11–6.
56. Zhang Y, Fung ITH, Sankar P, Chen X, Robison LS, Ye L, et al. Depletion of NK cells improves cognitive function in the Alzheimer disease mouse model. J Immunol. 2020;205(2):502–10.
57. Sedgwick AJ, Ghazanfari N, Constantinescu P, Mantamadiotis T, Barrow AD. The role of NK cells and innate lymphoid cells in brain cancer. Front Immunol. 2020;11:1549.
58. Yang I, Han SJ, Sughrue ME, Tihan T, Parsa AT. Immune cell infiltrate differences in pilocytic astrocytoma and glioblastoma: evidence of distinct immunological microenvironments that reflect tumor biology. J Neurosurg. 2011;115(3):505–11.
59. Domingues PH, Teodosio C, Ortiz J, Sousa P, Otero A, Maillo A, et al. Immunophenotypic identification and characterization of tumor cells and infiltrating cell populations in meningiomas. Am J Pathol. 2012;181(5):1749–61.
60. Domingues P, Gonzalez-Tablas M, Otero A, Pascual D, Miranda D, Ruiz L, et al. Tumor infiltrating immune cells in gliomas and meningiomas. Brain Behav Immun. 2016;53:1–15.
61. Zhu C, Zou C, Guan G, Guo Q, Yan Z, Liu T, et al. Development and validation of an interferon signature predicting prognosis and treatment response for glioblastoma. Onco Targets Ther. 2019;8(9): e1621677.
62. Lu J, Li H, Chen Z, Fan L, Feng S, Cai X, et al. Identification of 3 subpopulations of tumor-infiltrating immune cells for malignant transformation of low-grade glioma. Cancer Cell Int. 2019;19: 265.
63. Vauleon E, Tony A, Hamlat A, Etcheverry A, Chiforeanu DC, Menei P, et al. Immune genes are associated with human glioblastoma pathology and patient survival. BMC Med Genet. 2012;5:41.
64. Zhong QY, Fan EX, Feng GY, Chen QY, Gou XX, Yue GJ, et al. A gene expression-based study on immune cell subtypes and glioma prognosis. BMC Cancer. 2019;19(1):1116.
65. Castriconi R, Daga A, Dondero A, Zona G, Poliani PL, Melotti A, et al. NK cells recognize and kill human glioblastoma cells with stem cell-like properties. J Immunol. 2009;182(6):3530–9.
66. Haspels HN, Rahman MA, Joseph JV, Gras Navarro A, Chekenya M. Glioblastoma stem-like cells are more susceptible than differentiated cells to natural killer cell lysis mediated through killer immunoglobulin-like receptors-human leukocyte antigen ligand mismatch and activation receptor-ligand interactions. Front Immunol. 2018;9:1345.
67. Sivori S, Parolini S, Marcenaro E, Castriconi R, Pende D, Millo R, et al. Involvement of natural cytotoxicity receptors in human natural killer cell-mediated lysis of neuroblastoma and glioblastoma cell lines. J Neuroimmunol. 2000;107(2):220–5.
68. Castriconi R, Dondero A, Corrias MV, Lanino E, Pende D, Moretta L, et al. Natural killer cell-mediated killing of freshly isolated neuroblastoma cells: critical role of DNAX accessory molecule-1-poliovirus receptor interaction. Cancer Res. 2004;64(24): 9180–4.
69. Barrow AD, Edeling MA, Trifonov V, Luo J, Goyal P, Bohl B, et al. Natural killer cells control tumor growth by sensing a growth factor. Cell. 2018;172(3):534–48 e19.
70. Castriconi R, Dondero A, Augugliaro R, Cantoni C, Carnemolla B, Sementa AR, et al. Identification of 4Ig-B7-H3 as a neuroblastoma-associated molecule that exerts a protective role from an NK cell-mediated lysis. Proc Natl Acad Sci. 2004;101(34):12640–5.
71. Baker GJ, Chockley P, Yadav VN, Doherty R, Ritt M, Sivaramakrishnan S, et al. Natural killer cells eradicate galectin-1-deficient glioma in the

absence of adaptive immunity. Cancer Res. 2014;74 (18):5079–90.
72. Biron CA, Byron KS, Sullivan JL. Severe herpesvirus infections in an adolescent without natural killer cells. N Engl J Med. 1989;320(26):1731–5.
73. Almerigogna F, Fassio F, Giudizi MG, Biagiotti R, Manuelli C, Chiappini E, et al. Natural killer cell deficiencies in a consecutive series of children with herpetic encephalitis. Int J Immunopathol Pharmacol. 2011;24(1):231–8.
74. Chastain EM, Getts DR, Miller SD. Deficient natural killer dendritic cell responses underlay the induction of Theiler's virus-induced autoimmunity. MBio. 2015;6(4):e01175.
75. Alsharifi M, Lobigs M, Simon MM, Kersten A, Muller K, Koskinen A, et al. NK cell-mediated immunopathology during an acute viral infection of the CNS. Eur J Immunol. 2006;36(4):887–96.
76. Adler H, Beland JL, Del-Pan NC, Kobzik L, Sobel RA, Rimm IJ. In the absence of T cells, natural killer cells protect from mortality due to HSV-1 encephalitis. J Neuroimmunol. 1999;93(1–2):208–13.
77. Jin Y, Dons L, Kristensson K, Rottenberg ME. Neural route of cerebral Listeria monocytogenes murine infection: role of immune response mechanisms in controlling bacterial neuroinvasion. Infect Immun. 2001;69(2):1093–100.
78. Khanolkar A, Hartwig SM, Haag BA, Meyerholz DK, Epping LL, Haring JS, et al. Protective and pathologic roles of the immune response to mouse hepatitis virus type 1: implications for severe acute respiratory syndrome. J Virol. 2009;83(18):9258–72.
79. Taylor K, Kolokoltsova O, Patterson M, Poussard A, Smith J, Estes DM, et al. Natural killer cell mediated pathogenesis determines outcome of central nervous system infection with Venezuelan equine encephalitis virus in C3H/HeN mice. Vaccine. 2012;30(27): 4095–105.
80. Mitchell AJ, Yau B, McQuillan JA, Ball HJ, Too LK, Abtin A, et al. Inflammasome-dependent IFN-gamma drives pathogenesis in Streptococcus pneumoniae meningitis. J Immunol. 2012;189(10):4970–80.
81. Hansen DS, Bernard NJ, Nie CQ, Schofield L. NK cells stimulate recruitment of CXCR3+ T cells to the brain during *Plasmodium berghei*-mediated cerebral malaria. J Immunol. 2007;178(9):5779–88.
82. Hansen DS, Evans KJ, D'Ombrain MC, Bernard NJ, Sexton AC, Buckingham L, et al. The natural killer complex regulates severe malarial pathogenesis and influences acquired immune responses to *Plasmodium berghei* ANKA. Infect Immun. 2005;73(4): 2288–97.
83. Gan Y, Liu Q, Wu W, Yin JX, Bai XF, Shen R, et al. Ischemic neurons recruit natural killer cells that accelerate brain infarction. Proc Natl Acad Sci. 2014;111(7):2704–9.
84. Zhang Y, Gao Z, Wang D, Zhang T, Sun B, Mu L, et al. Accumulation of natural killer cells in ischemic brain tissues and the chemotactic effect of IP-10. J Neuroinflammation. 2014;11:79.
85. Li Z, Li M, Shi SX, Yao N, Cheng X, Guo A, et al. Brain transforms natural killer cells that exacerbate brain edema after intracerebral hemorrhage. J Exp Med. 2020;217(12):e20200213.
86. Wang S, de Fabritus L, Kumar PA, Werner Y, Siret C, Simic M, et al. Brain endothelial CXCL12 attracts protective natural killer cells during ischemic stroke. bioRxiv. 2021; https://doi.org/10.1101/2021.02.18.431426.
87. Liu Q, Jin WN, Liu Y, Shi K, Sun H, Zhang F, et al. Brain ischemia suppresses immunity in the periphery and brain via different neurogenic innervations. Immunity. 2017;46(3):474–87.
88. Jiang C, Kong W, Wang Y, Ziai W, Yang Q, Zuo F, et al. Changes in the cellular immune system and circulating inflammatory markers of stroke patients. Oncotarget. 2017;8(2):3553–67.
89. Han S, Lin YC, Wu T, Salgado AD, Mexhitaj I, Wuest SC, et al. Comprehensive immunophenotyping of cerebrospinal fluid cells in patients with neuroimmunological diseases. J Immunol. 2014;192 (6):2551–63.
90. Rodriguez-Martin E, Picon C, Costa-Frossard L, Alenda R, Sainz de la Maza S, Roldan E, et al. Natural killer cell subsets in cerebrospinal fluid of patients with multiple sclerosis. Clin Exp Immunol. 2015;180(2):243–9.
91. Caruana P, Lemmert K, Ribbons K, Lea R, Lechner-Scott J. Natural killer cell subpopulations are associated with MRI activity in a relapsing-remitting multiple sclerosis patient cohort from Australia. Mult Scler. 2017;23(11):1479–87.
92. Gross CC, Schulte-Mecklenbeck A, Runzi A, Kuhlmann T, Posevitz-Fejfar A, Schwab N, et al. Impaired NK-mediated regulation of T-cell activity in multiple sclerosis is reconstituted by IL-2 receptor modulation. Proc Natl Acad Sci. 2016;113(21): E2973–82.
93. Jiang W, Chai NR, Maric D, Bielekova B. Unexpected role for granzyme K in CD56bright NK cell-mediated immunoregulation of multiple sclerosis. J Immunol. 2011;187(2):781–90.
94. Xu W, Fazekas G, Hara H, Tabira T. Mechanism of natural killer (NK) cell regulatory role in experimental autoimmune encephalomyelitis. J Neuroimmunol. 2005;163(1–2):24–30.
95. Zhang B, Yamamura T, Kondo T, Fujiwara M, Tabira T. Regulation of experimental autoimmune encephalomyelitis by natural killer (NK) cells. J Exp Med. 1997;186(10):1677–87.
96. Hao J, Liu R, Piao W, Zhou Q, Vollmer TL, Campagnolo DI, et al. Central nervous system (CNS)-resident natural killer cells suppress Th17 responses and CNS autoimmune pathology. J Exp Med. 2010;207(9):1907–21.
97. Jiang W, Li D, Han R, Zhang C, Jin WN, Wood K, et al. Acetylcholine-producing NK cells attenuate

CNS inflammation via modulation of infiltrating monocytes/macrophages. Proc Natl Acad Sci. 2017;114(30):E6202–11.
98. Leavenworth JW, Schellack C, Kim HJ, Lu L, Spee P, Cantor H. Analysis of the cellular mechanism underlying inhibition of EAE after treatment with anti-NKG2A F(ab')2. Proc Natl Acad Sci. 2010;107(6):2562–7.
99. Lu L, Ikizawa K, Hu D, Werneck MB, Wucherpfennig KW, Cantor H. Regulation of activated CD4+ T cells by NK cells via the Qa-1-NKG2A inhibitory pathway. Immunity. 2007;26(5):593–604.
100. Liu Q, Sanai N, Jin WN, La Cava A, Van Kaer L, Shi FD. Neural stem cells sustain natural killer cells that dictate recovery from brain inflammation. Nat Neurosci. 2016;19(2):243–52.
101. Solerte SB, Fioravanti M, Pascale A, Ferrari E, Govoni S, Battaini F. Increased natural killer cell cytotoxicity in Alzheimer's disease may involve protein kinase C dysregulation. Neurobiol Aging. 1998;19(3):191–9.
102. Solana C, Tarazona R, Solana R. Immunosenescence of natural killer cells, inflammation, and Alzheimer's disease. Int J Alzheimers Dis. 2018;2018:3128758.
103. Schindowski K, Peters J, Gorriz C, Schramm U, Weinandi T, Leutner S, et al. Apoptosis of CD4+ T and natural killer cells in Alzheimer's disease. Pharmacopsychiatry. 2006;39(6):220–8.
104. Prolo P, Chiappelli F, Angeli A, Dovio A, Perotti P, Pautasso M, et al. Physiologic modulation of natural killer cell activity as an index of Alzheimer's disease progression. Bioinformation. 2007;1(9):363–6.
105. Martins LC, Rocha NP, Torres KC, Dos Santos RR, Franca GS, de Moraes EN, et al. Disease-specific expression of the serotonin-receptor 5-HT(2C) in natural killer cells in Alzheimer's dementia. J Neuroimmunol. 2012;251(1–2):73–9.
106. Masera RG, Prolo P, Sartori ML, Staurenghi A, Griot G, Ravizza L, et al. Mental deterioration correlates with response of natural killer (NK) cell activity to physiological modifiers in patients with short history of Alzheimer's disease. Psychoneuroendocrinology. 2002;27(4):447–61.
107. Solerte SB, Cravello L, Ferrari E, Fioravanti M. Overproduction of IFN-gamma and TNF-alpha from natural killer (NK) cells is associated with abnormal NK reactivity and cognitive derangement in Alzheimer's disease. Ann N Y Acad Sci. 2000;917:331–40.
108. Solerte SB, Fioravanti M, Severgnini S, Locatelli M, Renzullo M, Pezza N, et al. Enhanced cytotoxic response of natural killer cells to interleukin-2 in Alzheimer's disease. Dementia. 1996;7(6):343–8.
109. Mihara T, Nakashima M, Kuroiwa A, Akitake Y, Ono K, Hosokawa M, et al. Natural killer cells of Parkinson's disease patients are set up for activation: a possible role for innate immunity in the pathogenesis of this disease. Parkinsonism Relat Disord. 2008;14(1):46–51.
110. Cen L, Yang C, Huang S, Zhou M, Tang X, Li K, et al. Peripheral lymphocyte subsets as a marker of Parkinson's disease in a Chinese population. Neurosci Bull. 2017;33(5):493–500.
111. Niwa F, Kuriyama N, Nakagawa M, Imanishi J. Effects of peripheral lymphocyte subpopulations and the clinical correlation with Parkinson's disease. Geriatr Gerontol Int. 2012;12(1):102–7.
112. Earls RH, Menees KB, Chung J, Gutekunst CA, Lee HJ, Hazim MG, et al. NK cells clear alpha-synuclein and the depletion of NK cells exacerbates synuclein pathology in a mouse model of alpha-synucleinopathy. Proc Natl Acad Sci. 2020;117(3):1762–71.
113. Baban B, Braun M, Khodadadi H, Ward A, Alverson K, Malik A, et al. AMPK induces regulatory innate lymphoid cells after traumatic brain injury. JCI Insight. 2021;6(1):e126766.
114. Zelco A, Rocha-Ferreira E, Nazmi A, Ardalan M, Chumak T, Nilsson G, et al. Type 2 innate lymphoid cells accumulate in the brain after hypoxia-Ischemia but do not contribute to the development of preterm brain injury. Front Cell Neurosci. 2020;14:249.
115. Besnard AG, Guabiraba R, Niedbala W, Palomo J, Reverchon F, Shaw TN, et al. IL-33-mediated protection against experimental cerebral malaria is linked to induction of type 2 innate lymphoid cells, M2 macrophages and regulatory T cells. PLoS Pathog. 2015;11(2):e1004607.
116. Hirose S, Jahani PS, Wang S, Jaggi U, Tormanen K, Yu J, et al. Type 2 innate lymphoid cells induce CNS demyelination in an HSV-IL-2 mouse model of multiple sclerosis. iScience. 2020;23(10):101549.
117. Russi AE, Ebel ME, Yang Y, Brown MA. Male-specific IL-33 expression regulates sex-dimorphic EAE susceptibility. Proc Natl Acad Sci. 2018;115(7):E1520–9.
118. Russi AE, Walker-Caulfield ME, Ebel ME, Brown MA. Cutting edge: c-Kit signaling differentially regulates type 2 innate lymphoid cell accumulation and susceptibility to central nervous system demyelination in male and female SJL mice. J Immunol. 2015;194(12):5609–13.
119. Hatfield JK, Brown MA. Group 3 innate lymphoid cells accumulate and exhibit disease-induced activation in the meninges in EAE. Cell Immunol. 2015;297(2):69–79.
120. Louveau A, Smirnov I, Keyes TJ, Eccles JD, Rouhani SJ, Peske JD, et al. Structural and functional features of central nervous system lymphatic vessels. Nature. 2015;523(7560):337–41.
121. Louveau A, Herz J, Alme MN, Salvador AF, Dong MQ, Viar KE, et al. CNS lymphatic drainage and neuroinflammation are regulated by meningeal lymphatic vasculature. Nat Neurosci. 2018;21(10):1380–91.

122. Da Mesquita S, Louveau A, Vaccari A, Smirnov I, Cornelison RC, Kingsmore KM, et al. Functional aspects of meningeal lymphatics in ageing and Alzheimer's disease. Nature. 2018;560(7717):185–91.
123. Da Mesquita S, Papadopoulos Z, Dykstra T, Brase L, Farias FG, Wall M, et al. Meningeal lymphatics affect microglia responses and anti-Aβ immunotherapy. Nature. 2021;593(7858):255–60.
124. Bolte AC, Dutta AB, Hurt ME, Smirnov I, Kovacs MA, McKee CA, et al. Meningeal lymphatic dysfunction exacerbates traumatic brain injury pathogenesis. Nat Commun. 2020;11(1):4524.
125. Hu X, Deng Q, Ma L, Li Q, Chen Y, Liao Y, et al. Meningeal lymphatic vessels regulate brain tumor drainage and immunity. Cell Res. 2020;30(3):229–43.

Development of αβ T Cells with Innate Functions

10

José Alberola-Ila

Abstract

Although we mostly think of αβ T cells as components of the adaptive immune system, a number of them differentiate into alternative lineages. These lineages express TCRs with limited diversity, and functionally bridge the gap between innate and adaptive immunity. They tend to be tissue resident, and mount potent cytokine responses very rapidly after activation, and their development and functional maturation are strongly influenced by the microbiome. Here, we compare the development pathways and interactions with the microbiome of natural killer T (NKT) cells and mucosal-associated invariant T (MAIT cells), the two best studied "innate-like" αβ T cell populations.

Keywords

Innate lymphocytes · Microbiome · Development · MAIT · NKT

J. Alberola-Ila (✉)
Arthritis and Clinical Immunology, Oklahoma Medical Research Foundation (OMRF), Oklahoma City, OK, USA

Department of Microbiology and Immunology, University of Oklahoma Health Sciences Center (OUHSC), Oklahoma City, OK, USA

Department of Cell Biology, University of Oklahoma Health Sciences Center (OUHSC), Oklahoma City, OK, USA
e-mail: alberolaj@omrf.org

10.1 Introduction

Mucosal and barrier sites such as the skin, lungs, and digestive tract are major contact points with the external environment. As such, they are colonized by diverse commensal microbial species, whose cooperative coexistence with their host is critical to maintain homeostasis and to shape normal immune responses. Mucosal barriers are particularly enriched for "atypical lymphocytes," defined as lymphocytes expressing no antigen receptors (innate lymphoid cells) or antigen receptors with a limited repertoire. These include γδ T cells and two lineages of αβ T cells, natural killer T (NKT) cells and mucosal-associated invariant T (MAIT cells). These lineages bridge the gap between innate and adaptive immunity. They tend to be tissue resident, and mount potent cytokine responses very rapidly after activation, and their development and functional maturation are strongly influenced by the microbiome. Here, we compare the development pathways and interactions with the microbiome of natural killer T (NKT) cells and mucosal-associated invariant T (MAIT cells), the two best studied "innate-like" αβ T cell populations.

NKT cells recognize lipid-based antigens, presented by the β2M-associated MHC class I-like molecule CD1d, and are separated into two broad classes: type 1, also called invariant, and type 2 NKT cells. Invariant NKT cells recognize the prototypic NKT cell lipid antigen,

© The Author(s), under exclusive license to Springer Nature Singapore Pte Ltd. 2022

X.-H. Sun (ed.), *Innate Lymphoid Cells*, Advances in Experimental Medicine and Biology 1365,
https://doi.org/10.1007/978-981-16-8387-9_10

α-galactosylceramide (α-GalCer), and express a CD1d-restricted semi-invariant αβ TCR comprising an invariant TCR α chain (Vα14-Jα18 in mice or Vα24-Jα18 in humans) in combination with certain TCRβ chains (using Vβ8.2, Vβ7, or Vβ2 in mice, and Vβ11 in humans). NKT cells in mice can be $CD4^+$ or DN (CD4-CD8-), and generally have a "memory" or "activated" phenotype ($CD69^+CD62L^-CD44^{hi}$) [1, 2]. NKT cells are relatively abundant in mice, representing between 1% and 3% of T cells in most tissues, and up to 50% of T cells in the liver [3]. NKT cells are in general less frequent in humans, although they are highly variable [1]. Humans appear to have greater numbers of type 2 NKT cells, which express diverse TCRs that confer broader lipid antigen specificities [1, 4] but because these cells are difficult to identify, little is known about how they develop.

MAIT cells recognize riboflavin (vitamin B2) metabolic derivatives, which include 5-(2-oxopropylideneamino)-6-D-ribitylaminouracil (5-OP-RU), presented in the context of MHC class I-like molecule MR1 [5], and express predominantly a semi-invariant αβ TCR comprising an invariant α chain (Vα19–Jα33 in mice and Vα7.2–Jα33 in humans) in combination with certain TCRβ chains (using Vβ8 or Vβ6 in mice, or Vβ2 or Vβ13 in humans). MAIT cells are typically rare (<1% of T cells) in mouse tissues [6], although they are enriched in barrier tissues such as skin and lung [7]. They are more abundant in humans, usually representing >45% of liver lymphocytes and 2–5% of the T cells in human blood [7, 8]. As with CD1d-restricted NKT cell, recent evidence suggests that additional types of MR1-restricted cells using different TCRs can be identified (reviewed in [9]).

The roles NKT and MAIT cells play in health and disease have been the focus of intense work and the subject of excellent recent reviews [1, 10–13]. Briefly, these cells play an important role in maintaining homeostasis in mucosal tissues in close interaction with the microbiota [7, 14–16]. Furthermore, these cells participate in immune responses against bacteria, viruses, and other pathogens [17, 18] and play important roles in autoimmune, allergic, and inflammatory diseases such as type 1 diabetes asthma and Crohn's disease [19–22].

10.2 Development in the Thymus

10.2.1 Generation of Precursor NKT and MAIT Cells

Both NKT and MAITs develop from the same precursor as conventional αβ T cells, $CD4^+CD8^+$ DP thymocytes, but instead of being selected by thymic epithelial cells present in the cortex, they are selected by other DP thymocytes that express their ligands, CD1d and MR1, respectively [23–28].

Because of the distal location of the Vα14 chain used in the NKT TCR, and the orderly sequence of rearrangements in the TCRα locus [29], any decrease in the half-life of DP thymocytes results in a significant block in NKT development. This effect has been described in mice deficient in RORγ_T, c-Myb, and HEB, all of which regulate the expression of Bcl-x_L [30–34]. Since Vα19 is located at the extreme 5′ end of the TCRα locus, a similar constrain likely applies to MAIT development. This has not yet been experimentally demonstrated, in part because tetramers specific for murine MAITs [35] have been only recently made widely available through the NIH tetramer facility. However, in support of this possibility, it has been shown that RORγt- and c-Myb-deficient mice lack Vα19 transcripts [31, 36], and MAIT cells and NKT cells are virtually absent in humans lacking functional RORγ and RORγt [37].

10.2.2 Positive Selection of Innate αβ T Cells

The nature of the ligand(s) that induce positive selection of NKTs and MAITs is still unclear. In the case of NKTs several candidate glycolipids have been identified, including isoglobotrihexosylceramide (iGb3) [38], although mice deficient for iGb3 synthase have normal levels of NKT cells [39]. The peroxisome-derived lysophospholipid antigens plasmalogen lysophosphatidylethanolamine (pLPE) and lysophosphatidic acid (eLPA) appear to be important for NKT cell development, although small residual populations of NKT cells were still present in mice that were unable to synthesize these antigens, suggesting that they are not an absolute requirement [40]. Interestingly, mouse thymocytes express trace amounts of α-GalCer and α-GluCer [41, 42], although it remains to be determined whether mammalian-derived α-linked glycolipids are involved in NKT cell development. Development of NKT seems to be mostly independent of microbiota-derived antigens, because their numbers are not drastically altered in the thymus from germ-free mice [43]. In contrast, MAIT cell numbers in the thymus are dramatically decreased in germ-free mice, although a few can be detected [44, 45]. This suggests that self-antigen(s) presented by MR1 contribute directly to the positive selection of some MAIT cells. However, colonization with *E. coli* that can produce MAIT antigens (ribD competent) increases the numbers of HSA^{hi} MAIT precursors in the thymus [45] suggesting that *de novo* positive selection can also be mediated by ligands derived from microbiota. Interestingly, it has been shown that the cognate 5-OP-RU MAIT cell antigen could be detected within the thymus 1 h after application to the intact skin of mice and was presented by DP thymocytes, dendritic cells, and thymic epithelial cells [45], suggesting that there could be easy availability of microbiota-derived MR1 ligands in the thymus.

As opposed to conventional αβ T cells that are positively selected by interactions with MHC molecules expressed on thymic epithelial cells (TECs), NKTs and MAITs are selected by interactions with other DP thymocytes. A recent report used mixed bone marrow chimeras, where MR1 expression is restricted to hematopoietic or epithelial cells, to demonstrate that T cell precursors expressing the semi-invariant MAIT TCR can also be selected by thymic epithelial cells (TECs), but in this case they do not acquire the "innate" characteristics of MAITs, including PLZF expression [46], reinforcing the idea that selection by other DP induces signals important for the development of αβ T cells with innate characteristics. However, in these experiments not all the MAIT precursors selected by hematopoietic cells become *bona fide* $PLZF^+$ MAITs either, suggesting that there is some limiting niche in the thymus.

10.2.3 Signals that Control Positive Selection of Innate αβ T Cells

It is thought that selection by DP cells imparts the unique developmental program of NKT cells by the cooperative signaling through the TCR and the SLAM-SAP signaling pathways [47] and, although initial reports suggested that SAP was dispensable for MAIT development [28], this is also the case for MAITs [45, 47, 48]. In fact, forced expression of MHC I in DP thymocytes results in the generation of a population of $PLZF^+$ innate T cells in the thymus that is also SAP dependent, and seems to compete with NKTs for a not well-defined niche in the thymus [49]. This developmental program is characterized by the induction of PLZF which is a master regulator of NKT cell development [50, 51].

Numerous pathways downstream the TCR contribute to both conventional αβ T cells and NKT development, including activation of the Ras/MAPK cascade, increased intracellular calcium levels, and subsequent activation of calcineurin and activation of itk family kinases [52–58]. These pathways are probably also required for MAIT development, although most of them have not yet been directly tested in these cells. However, there are some interesting differences in the contribution of these signaling pathways to conventional vs. innate αβ T cells. For example, Egr1 and Egr2 play a quantitatively similar role during conventional αβ T cell development [59, 60], but Egr2 plays a much more central role during NKT cell development, including induction of PLZF [56], and only experiments using competitive mixed bone marrow chimeras and double Egr1-Egr2 knockouts revealed a minor role for Egr1 [57]. Whether this difference is due to different targets of these closely related transcription factors, or due to their different kinetics of expression, has not been clearly demonstrated. Current evidence, based on expression levels of Egr transcription factors [61] and Nur77-GFP reporter mice [62], indicates that the TCR-derived signals that drive positive selection of NKT cells are stronger than those that select conventional αβ T cells.

The SLAM family consists of several related proteins, Ly108 (Slamf1), CD48 (Slamf2), 2B4 (Slamf4), Ly9 (CD229, Slamf3), CD84 (Slamf5), NTB-A (Slamf6), Cracc (Slamf7), BLAME (Slamf8), and SF2001 (Slamf9). These molecules are expressed on many hematopoietic cells, and their function can be activating on inhibitory signals, depending on the cell where they are expressed and on their interactions with SAP adaptors and inhibitory molecules such as SHP-1 and SHIP-1 [63, 64]. DP thymocytes express high levels of SLAMF1, 2, 3, 5, and 6, and lower levels of SLAMF4, 7, 8, and 9 [31, 64, 65], but SLAMF1 and SLAMF6 are the only ones that seem to play a role in NKT cell development. Interestingly their function is redundant. Single knockouts have a partial phenotype [66, 67], but the combined lack of SLAMF1 and SLAMF6 causes a dramatic decrease in NKT cell numbers [47]. Similarly, NOD mice, a mouse strain used as a model for autoimmune diabetes, have decreased levels of NKT cells [68]. This phenotype was mapped to the SLAM locus [69], and a comparison of the expression pattern of SLAM family members in DP thymocytes showed a decrease in the expression of SLAMF1 and SLAMF6, but no other SLAM family members [65].

More recent experiments have confirmed these initial reports using combined knockouts of different SLAMs, and shown that the defects present in mice lacking SLAM family receptors seem to spare stage 0 NKT cells [70–73]. In contrast, studies in mice lacking SAP indicated that this defect is due to a block in positive selection at the most immature stage of NKT cell development, stage 0 [74–77]. MAIT cell development was not characterized in the SLAM-deficient models, but it is also affected in SAP knockout mice [48], although the defect seems to be later than in NKT cells.

Interestingly, a recent study using a complete knockout of the SLAM locus [78] showed that loss of SLAM family expression leads to higher TCR signaling in developing NKT cells, as measured by higher Nur77 and Egr2 levels, and to decreased numbers of mature NKT cells due to increased cell death. This suggests that inhibitory signals provided by SLAMs attenuate TCR signal strength after positive selection to promote NKT cell development, as opposed to previous studies proposing that SLAMs complemented TCR signaling to support NKT cell development [79].

10.2.4 Stages of Development

Early experiments proposed a four-stage developmental program in the thymus (S0–S3). NKT cells progress from the most immature stage S0 ($CD24^+CD44^-NK1.1^-$) to losing CD24 expression (S1) and then upregulating CD44 (S2) and NK1.1 (S3), so the mature NKTs in the thymus were defined as $CD24^-CD44^+NK1.1^+$ [80, 81]. Although this model reflects the trajectory of the most common NKT cell present in the thymus of C57BL/6 mice, it fails to account for a number of recent discoveries. The major problem with this model was the identification of different subsets of NKT cells that produced preferentially IL-17 or IL-4, were present in the periphery, and did not express NK1.1. These cells were also identified in the thymus, making it clear that not all the S2 cells are transitional immature NKTs [82–84]. Therefore, this original model has been replaced by a new functional classification of NKT cells into three terminally differentiated subsets, based on the expression pattern of characteristic cytokines and transcription factors [85–87]. In this model, NKT cells arise from a common progenitor designated as NKT0 cells ($Egr2^{hi}CD24^+$) and differentiate into NKT1, NKT2, or NKT17 cell subsets. NKT1 cells ($PLZF^{lo}Tbet^+$) produce predominantly interferon gamma (IFNγ) upon stimulation and express NK1.1. NKT2 cells ($PLZF^{hi}TGATA\text{-}3^{hi}$) produce IL-4, and NKT17 ($PLZF^{int}ROR\gamma t^+$) produce

IL-17. This original classification has been reinforced by transcriptome analysis of these thymic populations [88–90], although these experiments have also revealed additional levels of heterogeneity within these subsets. Recent experiments identified a small CCR7$^+$ PLZF^{hi} CD44^{lo} CD24^{lo} subset that can give rise to all effector subsets and seems to preferentially exit the thymus and seed the periphery [91]. It remains to be conclusively determined whether the majority of peripheral NKT cells exit the thymus with a predetermined effector program, or if their final differentiation takes place in the tissue where they home. Subset representation is very tissue specific, but this is compatible with both possibilities.

Although characterization of the stages of MAIT development has been slower, due to their rarity and the lack of reagents capable of specifically identifying them in mice, the development of MR1-5-OP-RU tetramers that specifically bind MAIT cells in both humans and mice [35] and the use of the NKT cell paradigm as a template led to the definition of a similar three-stage pathway for MAIT cell development [44], where cells progress from MR1–5-OP-RU tet$^+$ CD24$^+$CD44$^-$ cells (S1) to MR1–5-OP-RU tet$^+$ CD24$^-$CD44$^-$ (S2), and then MR1–5-OP-RU tet$^+$ CD24$^-$CD44$^+$ (S3). Within this mature, S3 stage, two distinct populations of MAIT cells have been identified: MAIT17 (PLZF^{int} RORγt$^+$) cells that secrete IL-17 upon stimulation and are the major population, and MAIT1 (T-bet$^+$ PLZF^{lo}) cells that predominantly produce IFNγ following activation [6, 44, 46, 48]. Interestingly, there does not appear to be a population of "MAIT2" cells equivalent to NKT2 cells. There is also a CCR7$^+$ subset of MAITs that are immature and may represent a similar developmental stage as the CCR7$^+$ NKTs [91], although this has not been demonstrated.

The mechanisms that regulate the generation of the different effector subsets are not completely understood, although there is evidence of contributions from both the TCR signals and environmental cues. Shortly after the description of the different NKT effector subsets [82–84] our group reported that small alterations in E protein activity during NKT positive selection resulted in changes in effector subset distribution, with a decrease in NKT1 and increase in NKT2 and NKT17 [92]. Since E protein activity during positive selection is controlled by the upregulation of Id2 and Id3 downstream the Ras/MAPK/Egr cascade [93], we interpreted these results as evidence that strong TCR signals favored NKT1 differentiation. However, a couple of recent papers using the SKG mouse model, where TCR signaling is weakened because of a hypomorphic ZAP70 allele, showed an abrogation in NKT2 and, to a lesser extent, NKT17 cell development while not reducing NKT1 cell development, suggesting that high signal strength is necessary for NKT2 and NKT17 development [94, 95]. Additionally, mice deficient in the Src homology 2 domain-containing phosphatase 1 (Shp1) showed an increase of NKT2 and NKT17 cells [96]. Although the authors found no evidence that this effect was due to alterations of TCR or SLAM signaling, Shp1 was previously identified as a negative regulator of TCR signaling by targeting ZAP-70 [97]. ZAP70-dependent TCR signaling is also important for MAIT cell development because SKG mice have a drastic reduction in thymic MAIT cells, although there was not a reported differential effect on MAIT1 versus MAIT17 [48].

Interestingly, a recent study showed that genetic manipulation in the levels of PLZF could control NKT effector differentiation [98], with higher levels required for the generation of NKT2. This could be interpreted as corroborating evidence for stronger signals being required for NKT1 differentiation, but it must be remembered that the earliest CCR7$^+$ NKT precursors express uniformly high PLZF levels.

Since, as discussed above, positive selection occurs in the cortex, it would seem likely that the impact of the TCR signal intensity would be evident in the early NKT populations, like the CCR7$^+$ PLZF^{hi} CD44^{lo} CD24^{lo}, but there is to date no evidence supporting any bias to distinct subsets at this stage. It is possible that the signals that impact effector differentiation are delivered in the medulla and are mediated by other cell types rather than by DP thymocytes. In this case, it is also important to consider the impact of other environmental signals that are known to promote and sustain development and homeostasis of these effector populations, like IL-15 for NKT1 [99] and IL25 for NKT2 [83, 100]. This leaves open the question of what regulates the differentiation of the effector subsets from the immature CCR7$^+$ precursors that preferentially exit the thymus [91]. Here, the role of different niches in different organs may be more relevant than the initial TCR strength. Furthermore, any model that tries to explain what determines the differentiation of NKT precursors into different effectors needs to consider the very different distribution of NKT subsets among different mouse strains [86, 101, 102]. It seems likely that other environmental factors are also major drivers of effector choice during NKT development, or differentiation in the periphery.

If TCR signal strength is a driving force directing subset differentiation, a question is what generates this gradient. One possible explanation is different specificity of the semi-invariant NKT TCR, driven by the limited diversity of the Vβ chains it can use [103–105]. It is therefore interesting that the different NKT subsets have differences in TCR beta usage [86, 94]. For example NKT2 expresses preferentially Vβ7 [86, 94], a Vβ chain that confers higher avidity binding to CD1d [105]. In fact, increasing ligand density by driving CD1d expression in the thymic medulla results in partial negative selection of NKTs, and changes the Vβ distribution of the remaining population, decreasing Vβ7 and Vβ8 representation [106]. However, in our experiments, where there were major changes in effector differentiation, we could not observe major changes in the Vβ repertoire of the thymic NKTs [72].

Although MAIT cells have been less studied, there does not seem to be a difference in Vβ usage between MAIT1 and MAIT 17 [107], and both subsets seem affected in ZAP70 hypomorph SKG mice [48].

10.3 Colonization of Peripheral Tissues, and Cross Talk with the Microbiome

It has been known for a number of years that NK1.1$^-$ stage 2 immature NKT cells can exit the thymus starting at day 5 post-birth [80, 81], and that many of the mature effectors present in the thymus are long-term thymic resident cells [108]. More recent work identified the migratory NKT as CCR7$^+$ CD44$^+$ cells and showed that they could differentiate into all three major effector subsets in the periphery [91]. Furthermore, parabiosis experiments showed that NKT cells are mostly tissue-resident cells in homeostasis [109, 110], and this includes thymic NKT cells [91]. This is also the case for MAITs [107].

These experiments suggest that most peripheral NKT cells develop from a small population of CCR7$^+$ NKT precursors that exit the thymus. It is unclear how much the maintenance of the peripheral pool depends on constant replenishment from thymic cells, or whether this changes during infection or chronic inflammation. Another unresolved issue is whether the cells that exit the thymus and colonize the periphery are already committed to an effector program, or if the type of effector they become is influenced by their final location. In the case of NKTs, there are specialized subsets that can be found in specific peripheral location, but not in the thymus, such as adipose tissue NKT10 [110–112] and NKT-FH [113]. It is likely that these subsets are derived from immature NKTs that home in these locations. Alternatively, they must derive from other differentiated effector types. Neither possibility has been experimentally tested.

For the more common NKT1, NKT2, and NKT17 subsets, it is not clear whether they exit the thymus as mature effector subsets and then home preferentially into different organs [108], or whether they differentiate in the tissue. Some evidence shows that the TCR repertoire, and possibly the effector program, of NKT cells is shaped by their tissue of residence [114]. However, single-cell analysis of the transcriptome and chromatin landscape of NKT subsets from the thymus and different peripheral tissues show that the subsets are similar, regardless of the location [115], with the possible exception of lung-resident NKT cells that seem to have more distinct features. Interestingly, these programs are highly overlapping between NKT and MAITs [107], but it is unclear whether the programs get pre-established in the thymus.

As we mentioned before, development of NKT seems to be mostly independent of microbiota-derived antigens, because their numbers are not drastically altered in the thymus from germ-free mice [43]. However, in germ-free animals NKT cells are decreased in peripheral tissues, such as the spleen and liver, and are hyporesponsive to lipid antigen stimulation [116]. These defects are normalized by monocolonization with bacteria expressing NKT antigens in adult mice [116]. In contrast, mucosal tissues (lung, colon) of germ-free mice contain increased numbers of NKT cells. This is associated with increased responses in experimental models of colitis and airway hyperresponsiveness [117, 118]. Interestingly, these alterations can be corrected by normalization of the microbiota, but only during the first 2 weeks of life [117, 118]. The same effect can be achieved by monocolonization with *B. fragilis* or treatment with *B. fragilis*-derived sphingolipid antigens, but only in the gut [118]. This tissue specificity suggests that specific microbes could regulate NKT recruitment and proliferation in distinct organs. Comparison of NKT populations from specific pathogen-free mice obtained from different vendors, which have different microbiota compositions, showed differences in the frequency, Vβ7 usage, and tumor necrosis factor production of NKT cells [116]. These differences were abolished by co-housing the offspring, emphasizing again the impact that different microbial products have on NKT populations early in life.

A recent paper has added an additional layer of complexity to the relationships between NKT cells and microbiota early in life [119]. This work confirmed that NKT cells migrate to and proliferate in the colon, but not the spleen, in the early days after birth (day 4 to day 11 post-birth), and then focused on the role of macrophages present in the gut at this stage. It has been shown that early in life the colon is populated by distinct macrophages of embryonic origin that progressively give way to bone marrow-derived macrophages [120, 121]. Depletion of macrophages in the second week of life, using a transgenic mouse model (MM^{DTR}) in which the combined expression of Lyz2 and Csf1r allows the precise expression of diphtheria toxin receptor (DTR) at the surface of macrophages in vivo, showed a specific effect on NKT cell accumulation in the colon, small intestine, lungs, and skin (other cell types, including MAITs, were not affected). This effect could be reproduced in $Plavp^{-/-}$ mice which have a defect in embryonic-derived macrophages due to an inadequacy of macrophage progenitor egress from the fetal liver [122], but not in $Ccr2^{-/-}$ mice, in which bone marrow macrophages are unable to migrate to the colon [123], suggesting that embryonic-derived, but not bone marrow-derived, macrophages induce NKT cell accumulation in the colon [119]. Interestingly, they observed a substantial increase in embryonic- but not bone marrow-derived macrophages in germ-free mice compared to SPF mice at day 15 of life [119]. This suggests that normal microbiota represses colonic macrophage levels during early life, and, in turn, these embryonic macrophages regulate NKT expansion.

In contrast to NKT cells, MAIT cell numbers are dramatically decreased in the thymus of germ-free mice, although a few can be detected [26, 44, 45], suggesting that microbiota-derived antigens contribute significantly to thymic selection and/or expansion of MAIT cells. In fact, the cognate 5-OP-RU MAIT cell antigen could be detected within the thymus 1 h after application to the intact skin of mice and was presented by DP thymocytes, dendritic cells, and thymic epithelial cells [45]. Accumulation of MAITs in peripheral tissues, including lung, skin, and small intestine, happens after the second week of life, later than NKT cells [124], but, as described with NKT cells, there seems to be a "window of opportunity" for this commensal effects. Colonization of germ-free mice with riboflavin-synthesizing bacteria in the first few weeks rescued the development of MAIT cells that home to tissues such as the skin, but this effect was only partial on adult mice [45, 124]. As mentioned above, this effect of the microbiota on MAIT cell development and colonization of the periphery is independent of embryonic macrophages [119].

10.4 Conclusion

Although we have now a better understanding of the mechanisms that regulate the development and function of NKT and MAITs, there are a number of outstanding issues that will require additional work in the future.

During development, it is still not clear what the natural ligands that induce positive selection are, and the role of microbiota-derived ligands needs to be further explored, especially in the case of MAITs, where they seem to play a critical role. We do not quite understand yet what mechanisms control the development of the different NKT and MAIT subsets, although there is evidence of contributions from both the TCR signals and environmental cues, or to what extent these decisions are made in the thymus versus in peripheral tissues, where the immature $CCR7^+$ precursors that preferentially exit the thymus home.

Similarly, it is unclear how much the maintenance of the peripheral pool depends on constant replenishment from thymic cells versus self-maintaining peripheral populations, or how these homeostatic mechanisms change during infection or chronic inflammation. The contribution of microbiota to the recruitment, maintenance, and effector differentiation in different peripheral tissues is also an emerging area of research that will increase our understanding of these populations, and their role in tissue homeostasis and immune responses.

Acknowledgements J.A.I. is funded by NIH R01 AI129458.

References

1. Godfrey DI. The burgeoning family of unconventional T cells. Nat Immunol. 2015;16:1114–23.
2. Kawano T, Cui J, Koezuka Y, et al. CD1d-restricted and TCR-mediated activation of valpha14 NKT cells by glycosylceramides. Science. 1997;278:1626–9.
3. Slauenwhite D, Johnston B. Regulation of NKT cell localization in homeostasis and infection. Front Immunol. 2015;6:255.
4. Pellicci DG, Uldrich AP. Unappreciated diversity within the pool of CD1d-restricted T cells. Semin Cell Dev Biol. 2018;84:835–44.
5. Kjer-Nielsen L, Patel O, Corbett AJ, et al. MR1 presents microbial vitamin B metabolites to MAIT cells. Nature. 2012;491:717–23.
6. Rahimpour A, Koay HF, Enders A, et al. Identification of phenotypically and functionally heterogeneous mouse mucosal-associated invariant T cells using MR1 tetramers. J Exp Med. 2015;212:1095–108.
7. Nel I, Bertrand L, Toubal A, Lehuen A. MAIT cells, guardians of skin and mucosa. Mucosal Immunol. 2021;14:803–14.
8. Dusseaux M. Human MAIT cells are xenobiotic-resistant, tissue-targeted, CD161hi IL-17-secreting T cells. Blood. 2011;117:1250–9.
9. Gherardin NA, McCluskey J, Rossjohn J, Godfrey DI. The diverse family of MR1-restricted T cells. J Immunol. 2018;201:2862–71.
10. Iwamura C, Nakayama T. Role of CD1d- and MR1-restricted T cells in asthma. Front Immunol. 2018;9:1942.
11. Toubal A, Lehuen A. Role of MAIT cells in metabolic diseases. Mol Immunol. 2021;130:142–7.
12. LaMarche NM, Kohlgruber AC, Brenner MB. Innate T cells govern adipose tissue biology. J Immunol. 2018;201:1827–34.
13. Godfrey DI, Koay HF, McCluskey J, Gherardin NA. The biology and functional importance of MAIT cells. Nat Immunol. 2019;20:1110–28.
14. Lin Q, Kuypers M, Philpott DJ, Mallevaey T. The dialogue between unconventional T cells and the microbiota. Mucosal Immunol. 2020;13:867–76.
15. Cox JR, Cruickshank SM, Saunders AE. Maintenance of barrier tissue integrity by unconventional lymphocytes. Front Immunol. 2021;12:670471.
16. Seo GY, Giles DA, Kronenberg M. The role of innate lymphoid cells in response to microbes at mucosal surfaces. Mucosal Immunol. 2020;13:399–412.
17. Juno JA, Keynan Y, Fowke KR. Invariant NKT cells: regulation and function during viral infection. PLoS Pathog. 2012;8:e1002838.
18. Gebhardt T, Palendira U, Tscharke DC, Bedoui S. Tissue-resident memory T cells in tissue homeostasis, persistent infection, and cancer surveillance. Immunol Rev. 2018;283:54–76.

19. Nel I, Beaudoin L, Lehuen A. MAIT cells in type 1 diabetes mouse models. Methods Mol Biol. 2020;2098:283–97.
20. Rouxel O, Da Silva J, Beaudoin L, et al. Cytotoxic and regulatory roles of mucosal-associated invariant T cells in type 1 diabetes. Nat Immunol. 2017;18: 1321–31.
21. Magalhaes I, Kiaf B, Lehuen A. iNKT and MAIT cell alterations in diabetes. Front Immunol. 2015;6:341.
22. Burrello C, Pellegrino G, Giuffrè MR, et al. Mucosa-associated microbiota drives pathogenic functions in IBD-derived intestinal iNKT cells. Life Sci Alliance. 2019;2:e201800229.
23. Wei DG, Lee H, Park SH, et al. Expansion and long-range differentiation of the NKT cell lineage in mice expressing CD1d exclusively on cortical thymocytes. J Exp Med. 2005;202:239–48.
24. Bendelac A. Positive selection of mouse NK1+ T cells by CD1-expressing cortical thymocytes. J Exp Med. 1995;182:2091–6.
25. Gapin L, Matsuda JL, Surh CD, Kronenberg M. NKT cells derive from double-positive thymocytes that are positively selected by CD1d. Nat Immunol. 2001;2: 971–8.
26. Treiner E, Duban L, Bahram S, et al. Selection of evolutionarily conserved mucosal-associated invariant T cells by MR1. Nature. 2003;422:164–9.
27. Seach N, Guerri L, Le Bourhis L, et al. Double-positive thymocytes select mucosal-associated invariant T cells. J Immunol. 2013;191:6002–9.
28. Martin E, Treiner E, Duban L, et al. Stepwise development of MAIT cells in mouse and human. PLoS Biol. 2009;7:e54.
29. Krangel MS. T cell development: better living through chromatin. Nat Immunol. 2007;8:687–94.
30. Egawa T, Eberl G, Taniuchi I, et al. Genetic evidence supporting selection of the Valpha14i NKT cell lineage from double-positive thymocyte precursors. Immunity. 2005;22:705–16.
31. Hu T, Simmons A, Yuan J, Bender TP, Alberola-Ila J. The transcription factor c-Myb primes CD4+CD8+ immature thymocytes for selection into the iNKT lineage. Nat Immunol. 2010;11:435–41.
32. Bezbradica JS, Hill T, Stanic AK, Van Kaer L, Joyce S. Commitment toward the natural T (iNKT) cell lineage occurs at the CD4+8+ stage of thymic ontogeny. Proc Natl Acad Sci U S A. 2005;102:5114–9.
33. D'Cruz LM, Knell J, Fujimoto JK, Goldrath AW. An essential role for the transcription factor HEB in thymocyte survival, Tcra rearrangement and the development of natural killer T cells. Nat Immunol. 2010;11:240–9.
34. Yuan J, Crittenden RB, Bender TP. C-Myb promotes the survival of CD4+CD8+ double-positive thymocytes through upregulation of Bcl-xL. J Immunol. 2010;184:2793–804.
35. Reantragoon R, Corbett AJ, Sakala IG, et al. Antigen-loaded MR1 tetramers define T cell receptor heterogeneity in mucosalassociated invariant T cells. J Exp Med. 2013;210:2305–20.
36. Guo J, Hawwari A, Li H, et al. Regulation of the TCRalpha repertoire by the survival window of CD4 (+)CD8(+) thymocytes. Nat Immunol. 2002;3:469–76.
37. Okada S, Markle JG, Deenick EK, et al. Impairment of immunity to Candida and Mycobacterium in humans with bi-allelic RORC mutations. Science. 2015;349:606–13.
38. Zhou D, Mattner J, Cantu C, et al. Lysosomal glycosphingolipid recognition by NKT cells. Science. 2004;306:1786–9.
39. Porubsky S, Speak AO, Luckow B, Cerundolo V, et al. Normal development and function of invariant natural killer T cells in mice with isoglobotrihexosylceramide (iGb3) deficiency. Proc Natl Acad Sci U S A. 2007;104:5977–82.
40. Facciotti F, Ramanjaneyulu GS, Lepore M, et al. Peroxisome-derived lipids are self antigens that stimulate invariant natural killer T cells in the thymus. Nat Immunol. 2012;13:474–80.
41. Kain L, Webb B, Anderson BL, et al. The identification of the endogenous ligands of natural killer T cells reveals the presence of mammalian alpha-linked glycosylceramides. Immunity. 2014;41:543–54.
42. Kain L, Costanzo A, Webb B, et al. Endogenous ligands of natural killer T cells are alpha-linked glycosylceramides. Mol Immunol. 2015;68:94–7.
43. Burns CM, Sakaguchi K, Appella E, Ashwell JD. CD45 regulation of tyrosine phosphorylation and enzyme activity of src family kinases. J Biol Chem. 1994;269:13594–600.
44. Koay HF, Gherardin NA, Enders A, et al. A three-stage intrathymic development pathway for the mucosal-associated invariant T cell lineage. Nat Immunol. 2016;17:1300–11.
45. Legoux F, Bellet D, Daviaud C, et al. Microbial metabolites control the thymic development of mucosal-associated invariant T cells. Science. 2019;366:494–9.
46. Legoux F, Gilet J, Procopio E, Echasserieau K, Bernardeau K, Lantz O. Molecular mechanisms of lineage decisions in metabolite-specific T cells. Nat Immunol. 2019;20:1244–55.
47. Griewank K, Borowski C, Rietdijk S, et al. Homotypic interactions mediated by Slamf1 and Slamf6 receptors control NKT cell lineage development. Immunity. 2007;27:751–62.
48. Koay HF, Su S, Amann-Zalcenstein D, et al. A divergent transcriptional landscape underpins the development and functional branching of MAIT cells. Sci Immunol. 2019;4:eaay6039.
49. Georgiev H, Peng C, Huggins MA, Jameson SC, Hogquist KA. Classical MHC expression by DP thymocytes impairs the selection of non-classical MHC restricted innate-like T cells. Nat Commun. 2021;12:2308.

50. Savage AK, Constantinides MG, Han J, et al. The transcription factor PLZF directs the effector program of the NKT cell lineage. Immunity. 2008;29:1–13.
51. Kovalovsky D, Uche OU, Eladad S, et al. The BTB-zinc finger transcriptional regulator PLZF controls the development of invariant natural killer T cell effector functions. Nat Immunol. 2008;9:1055–64.
52. Alberola-Ila J, Hernandez-Hoyos G. The Ras/MAPK cascade and the control of positive selection. Immunol Rev. 2003;191:79–96.
53. Neilson JR, Winslow MM, Hur EM, Crabtree GR. Calcineurin B1 is essential for positive but not negative selection during thymocyte development. Immunity. 2004;20:255–66.
54. Prince AL, Yin CC, Enos ME, Felices M, Berg LJ. The Tec kinases Itk and Rlk regulate conventional versus innate T-cell development. Immunol Rev. 2009;228:115–31.
55. Readinger JA, Mueller KL, Venegas AM, Horai R, Schwartzberg PL. Tec kinases regulate T-lymphocyte development and function: new insights into the roles of Itk and Rlk/Txk. Immunol Rev. 2009;228:93–114.
56. Lazarevic V, Zullo AJ, Schweitzer MN, et al. The gene encoding early growth response 2, a target of the transcription factor NFAT, is required for the development and maturation of natural killer T cells. Nat Immunol. 2009;10:306–13.
57. Hu T, Gimferrer I, Simmons A, Wiest D, Alberola-Ila J. The Ras/MAPK pathway is required for generation of iNKT cells. PLoS One. 2011;6:e19890.
58. Felices M, Berg LJ. The Tec kinases Itk and Rlk regulate NKT cell maturation, cytokine production, and survival. J Immunol. 2008;180:3007–18.
59. Bettini M, Xi H, Milbrandt J, Kersh GJ. Thymocyte development in early growth response gene 1-deficient mice. J Immunol. 2002;169:1713–20.
60. Lauritsen JP, Kurella S, Lee SY, et al. Egr2 is required for Bcl-2 induction during positive selection. J Immunol. 2008;181:7778–85.
61. Seiler MP, Mathew R, Liszewski MK, et al. Elevated and sustained expression of the transcription factors Egr1 and Egr2 controls NKT lineage differentiation in response to TCR signaling. Nat Immunol. 2012;13:264–71.
62. Moran AE, Holzapfel KL, Xing Y, et al. T cell receptor signal strength in Treg and iNKT cell development demonstrated by a novel fluorescent reporter mouse. J Exp Med. 2011;208:1279–89.
63. Veillette A. SLAM-family receptors: immune regulators with or without SAP-family adaptors. Cold Spring Harb Perspect Biol. 2010;2:a002469.
64. Cannons JL, Tangye SG, Schwartzberg PL. SLAM family receptors and SAP adaptors in immunity. Annu Rev Immunol. 2010;29:665–705.
65. Jordan MA, Fletcher JM, Pellicci D, Baxter AG. Slamf1, the NKT cell control gene Nkt1. J Immunol. 2007;178:1618–27.
66. Wang N, Satoskar A, Faubion W, et al. The cell surface receptor SLAM controls T cell and macrophage functions. J Exp Med. 2004;199:1255–64.
67. Howie D, Laroux FS, Morra M, et al. Cutting edge: the SLAM family receptor Ly108 controls T cell and neutrophil functions. J Immunol. 2005;174:5931–5.
68. Baxter AG, Kinder SJ, Hammond KJ, Scollay R, Godfrey DI. Association between alphabetaTCR +CD4-CD8- T-cell deficiency and IDDM in NOD/Lt mice. Diabetes. 1997;46:572–82.
69. Esteban LM, Tsoutsman T, Jordan MA, et al. Genetic control of NKT cell numbers maps to major diabetes and lupus loci. J Immunol. 2003;171:2873–8.
70. Chen S, Cai C, Li Z, et al. Dissection of SAP-dependent and SAP-independent SLAM family signaling in NKT cell development and humoral immunity. J Exp Med. 2017;214:475–89.
71. Huang B, Gomez-Rodriguez J, Preite S, Garrett LJ, Harper UL, Schwartzberg PL. CRISPR-mediated triple knockout of SLAMF1, SLAMF5 and SLAMF6 supports positive signaling roles in NKT cell development. PLoS One. 2016;11:e0156072.
72. Hu JK, Crampton JC, Locci M, Crotty S. CRISPR-mediated Slamf1Δ/Δ Slamf5Δ/Δ Slamf6Δ/Δ triple gene disruption reveals NKT cell defects but not T follicular helper cell defects. PLoS One. 2016;11: e0156074.
73. De Calisto J, Wang N, Wang G, Yigit B, Engel P, Terhorst C. SAP-dependent and -independent regulation of innate T cell development involving SLAMF receptors. Front Immunol. 2014;5:186.
74. Kageyama R, Cannons JL, Zhao F, et al. The receptor Ly108 functions as a SAP adaptor-dependent on-off switch for T cell help to B cells and NKT cell development. Immunity. 2012;36:986–1002.
75. Pasquier B, Yin L, Fondaneche MC, et al. Defective NKT cell development in mice and humans lacking the adapter SAP, the X-linked lymphoproliferative syndrome gene product. J Exp Med. 2005;201:695–701.
76. Nichols KE, Hom J, Gong SY, et al. Regulation of NKT cell development by SAP, the protein defective in XLP. Nat Med. 2005;11:340–5.
77. Chung B, Aoukaty A, Dutz J, Terhorst C, Tan R. Signaling lymphocytic activation molecule-associated protein controls NKT cell functions. J Immunol. 2005;174:3153–7.
78. Lu Y, Zhong MC, Qian J, et al. SLAM receptors foster iNKT cell development by reducing TCR signal strength after positive selection. Nat Immunol. 2019;20:447–57.
79. Dutta M, Kraus ZJ, Gomez-Rodriguez J, et al. A role for Ly108 in the induction of promyelocytic zinc finger transcription factor in developing thymocytes. J Immunol. 2013;190:2121–8.
80. Benlagha K, Kyin T, Beavis A, Teyton L, Bendelac A. A thymic precursor to the NK T cell lineage. Science. 2002;296:553–5.

81. Pellicci DG, Hammond KJ, Uldrich AP, Baxter AG, Smyth MJ, Godfrey DI. A natural killer T (NKT) cell developmental pathway iInvolving a thymus-dependent NK1.1(−)CD4(+) CD1d-dependent precursor stage. J Exp Med. 2002;195:835–44.
82. Terashima A, Watarai H, Inoue S, et al. A novel subset of mouse NKT cells bearing the IL-17 receptor B responds to IL-25 and contributes to airway hyperreactivity. J Exp Med. 2008;205:2727–33.
83. Watarai H, Nakagawa R, Omori-Miyake M, Dashtsoodol N, Taniguchi M. Methods for detection, isolation and culture of mouse and human invariant NKT cells. Nat Protoc. 2008;3:70–8.
84. Michel ML, Keller AC, Paget C, et al. Identification of an IL-17-producing NK1.1(neg) iNKT cell population involved in airway neutrophilia. J Exp Med. 2007;204:995–1001.
85. Constantinides MG, Bendelac A. Transcriptional regulation of the NKT cell lineage. Curr Opin Immunol. 2013;25:161–7.
86. Lee YJ, Holzapfel KL, Zhu J, Jameson SC, Hogquist KA. Steady-state production of IL-4 modulates immunity in mouse strains and is determined by lineage diversity of iNKT cells. Nat Immunol. 2013;14:1146–54.
87. Engel I, Kronenberg M. Transcriptional control of the development and function of valpha14i NKT cells. Curr Top Microbiol Immunol. 2014;381:51–81.
88. Engel I, Seumois G, Chavez L, et al. Innate-like functions of natural killer T cell subsets result from highly divergent gene programs. Nat Immunol. 2016;17:728–39.
89. Georgiev H, Ravens I, Benarafa C, Förster R, Bernhardt G. Distinct gene expression patterns correlate with developmental and functional traits of iNKT subsets. Nat Commun. 2016;7:13116.
90. Lee YJ, Starrett GJ, Lee ST, et al. Lineage-specific effector signatures of invariant NKT cells are shared amongst γδ T, innate lymphoid, and Th cells. J Immunol. 2016;197:1460–70.
91. Wang H, Hogquist KA. CCR7 defines a precursor for murine iNKT cells in thymus and periphery. elife. 2018;7:e34793.
92. Hu T, Wang H, Simmons A, et al. Increased level of E protein activity during invariant NKT development promotes differentiation of invariant NKT2 and invariant NKT17 subsets. J Immunol. 2013;191: 5065–73.
93. Bain G, Cravatt CB, Loomans C, Alberola-Ila J, Hedrick SM, Murre C. Regulation of the helix-loop-helix proteins, E2A and Id3, by the Ras-ERK MAPK cascade. Nat Immunol. 2001;2:165–71.
94. Tuttle KD, Krovi SH, Zhang J, et al. TCR signal strength controls thymic differentiation of iNKT cell subsets. Nat Commun. 2018;9:2650.
95. Zhao M, Svensson MND, Venken K, et al. Altered thymic differentiation and modulation of arthritis by invariant NKT cells expressing mutant ZAP70. Nat Commun. 2018;9:2627.
96. Cruz Tleugabulova M, Zhao M, Lau I, et al. The protein phosphatase Shp1 regulates invariant NKT cell effector differentiation independently of TCR and Slam signaling. J Immunol. 2019;202:2276–86.
97. Plas DR, Johnson R, Pingel JT, et al. Direct regulation of ZAP-70 by SHP-1 in T cell antigen receptor signaling. Science. 1996;272:1173–6.
98. Park JY, DiPalma DT, Kwon J, Fink J, Park JH. Quantitative difference in PLZF protein expression determines iNKT lineage fate and controls innate CD8 T cell generation. Cell Rep. 2019;27:2548–2557.e4.
99. Gordy LE, Bezbradica JS, Flyak AI, et al. IL-15 regulates homeostasis and terminal maturation of NKT cells. J Immunol. 2011;187:6335–45.
100. Miller CN, Proekt I, von Moltke J, et al. Thymic tuft cells promote an IL-4-enriched medulla and shape thymocyte development. Nature. 2018;559:627–31.
101. Rymarchyk SL, Lowenstein H, Mayette J, et al. Widespread natural variation in murine natural killer T-cell number and function. Immunology. 2008;125: 331–43.
102. Hammond KJ, Pellicci DG, Poulton LD, et al. CD1d-restricted NKT cells: an interstrain comparison. J Immunol. 2001;167:1164–73.
103. Cameron G, Pellicci DG, Uldrich AP, et al. Antigen specificity of type I NKT cells is governed by TCR β-chain diversity. J Immunol. 2015;195:4604–14.
104. Clancy-Thompson E, Chen GZ, Tyler PM, et al. Monoclonal invariant NKT (iNKT) cell mice reveal a role for both tissue of origin and the TCR in development of iNKT functional subsets. J Immunol. 2017;199:159–71.
105. Schümann J, Voyle RB, Wei BY, MacDonald HR. Cutting edge: influence of the TCR V beta domain on the avidity of CD1d:alpha-galactosylceramide binding by invariant V alpha 14 NKT cells. J Immunol. 2003;170:5815–9.
106. Chun T, Page MJ, Gapin L, et al. CD1d-expressing dendritic cells but not thymic epithelial cells can mediate negative selection of NKT cells. J Exp Med. 2003;197:907–18.
107. Salou M, Legoux F, Gilet J, et al. A common transcriptomic program acquired in the thymus defines tissue residency of MAIT and NKT subsets. J Exp Med. 2019;216:133–51.
108. Lee YJ, Wang H, Starrett GJ, Phuong V, Jameson SC, Hogquist KA. Tissue-specific distribution of iNKT cells impacts their cytokine response. Immunity. 2015;43:566–78.
109. Thomas SY, Scanlon ST, Griewank KG, et al. PLZF induces an intravascular surveillance program mediated by long-lived LFA-1-ICAM-1 interactions. J Exp Med. 2011;208:1179–88.
110. Lynch L, Michelet X, Zhang S, et al. Regulatory iNKT cells lack expression of the transcription factor PLZF and control the homeostasis of Treg cells and macrophages in adipose tissue. Nat Immunol. 2015;16:85–95.

111. Lynch L, Nowak M, Varghese B, et al. Adipose tissue invariant NKT cells protect against diet-induced obesity and metabolic disorder through regulatory cytokine production. Immunity. 2012;37:574–87.
112. Sag D, Krause P, Hedrick CC, Kronenberg M, Wingender G. IL-10-producing NKT10 cells are a distinct regulatory invariant NKT cell subset. J Clin Invest. 2014;124:3725–40.
113. Chang PP, Barral P, Fitch J, et al. Identification of Bcl-6-dependent follicular helper NKT cells that provide cognate help for B cell responses. Nat Immunol. 2011;13:35–43.
114. Jimeno R, Lebrusant-Fernandez M, Margreitter C, et al. Tissue-specific shaping of the TCR repertoire and antigen specificity of iNKT cells. elife. 2019;8:e51663.
115. Murray MP, Engel I, Seumois G, et al. Transcriptome and chromatin landscape of iNKT cells are shaped by subset differentiation and antigen exposure. Nat Commun. 2021;12:1446.
116. Wingender G, Stepniak D, Krebs P, et al. Intestinal microbes affect phenotypes and functions of invariant natural killer T cells in mice. Gastroenterology. 2012;143:418–28.
117. Olszak T, An D, Zeissig S, et al. Microbial exposure during early life has persistent effects on natural killer T cell function. Science. 2012;336:489–93.
118. An D, Oh SF, Olszak T, et al. Sphingolipids from a symbiotic microbe regulate homeostasis of host intestinal natural killer T cells. Cell. 2014;156:123–33.
119. Gensollen T, Lin X, Zhang T, et al. Embryonic macrophages function during early life to determine invariant natural killer T cell levels at barrier surfaces. Nat Immunol. 2021;22:699–710.
120. Ginhoux F, Guilliams M. Tissue-resident macrophage ontogeny and homeostasis. Immunity. 2016;44:439–49.
121. Bain CC, Bravo-Blas A, Scott CL, et al. Constant replenishment from circulating monocytes maintains the macrophage pool in the intestine of adult mice. Nat Immunol. 2014;15:929–37.
122. Rantakari P, Jäppinen N, Lokka E, et al. Fetal liver endothelium regulates the seeding of tissue-resident macrophages. Nature. 2016;538:392–6.
123. Serbina NV, Pamer EG. Monocyte emigration from bone marrow during bacterial infection requires signals mediated by chemokine receptor CCR2. Nat Immunol. 2006;7:311–7.
124. Constantinides MG, Link VM, Tamoutounour S, et al. MAIT cells are imprinted by the microbiota in early life and promote tissue repair. Science. 2019;366:eaax6624.

The Road from Mouse to Human ILCs: A Perspective of Understanding the Roles of ILCs in Disease

11

Hergen Spits

Abstract

The chapters of this book give a comprehensive overview of the current state of knowledge of ILCs. Most of this knowledge stems from studies in mouse models. Translation to the human situation is not always straightforward because of differences between human and mouse ILCs and the microenvironments in which these ILCs are operating. Nonetheless, these mouse studies formed the basis for investigations in human diseases using state-of-the art technologies which are beginning to provide an understanding of the role of ILCs in inflammatory diseases in humans. This perspective discusses gaps in our knowledge about human ILCs and what type of studies may be done to resolve these.

Keywords

Innate immunity · Innate lymphoid cells · Inflammatory diseases · Cancer immunology

Although NK cells and lymphoid tissue inducer (LTi) cells were discovered in 1974 [1] and 1997 [2], respectively, and hints of the existence of other ILC family members were published in 2001 [3] and 2006 [4], it was not before the appearance of a number of papers describing what is now known as ILC3 in 2009 [5] and type 2 ILCs in 2010 [6, 7] that it was realized that ILCs form a family of cell types that showed similarity with the T helper cell subsets [8] (see also Chap. 1 of this volume). In that paper the existence of ILC1 was predicted which was indeed confirmed with their identification of ILC1s in humans [9, 10] and in mice [11]. The chapters in this book provide a comprehensive snapshot of the still rapidly expanding state of knowledge of these cells. Following the early publications describing the discovery of these cells profound complexities emerged such as differences in so-called central and peripheral, tissue-specific, developmental paths of ILC subsets as pointed out in several chapters. In addition, ILC subsets have been found to be highly plastic in that mature ILC subsets can change their phenotype and cytokine production profiles in response to alterations in the tissue microenvironment ([12], reviewed in [13]). The classification of ILCs in three canonical groups 1, 2, and 3 based on (the level of) expression of cytokines and lineage-determining transcription factors was a useful guideline in the field [14]. More recently it has been proposed to define five canonical populations, NK cells, ILC1, ILC2, ILC3, and LTi cells [15]. However, it turns out not to be that simple to place ILCs in clear categories. For instance, the distinction between NK cells and ILC1 is blurred because of the many shared cell surface antigens [16]. The presence of

H. Spits (✉)
Department of Experimental Immunology of the Amsterdam University Medical Centers, University of Amsterdam, Amsterdam, The Netherlands
e-mail: Hergen.spits@amsterdamumc.nl

© The Author(s), under exclusive license to Springer Nature Singapore Pte Ltd. 2022

X.-H. Sun (ed.), *Innate Lymphoid Cells*, Advances in Experimental Medicine and Biology 1365,
https://doi.org/10.1007/978-981-16-8387-9_11

several intermediate cell types between ILC1 and ILC3 further complicates their classification [17]. In addition, ILC2 turns out to appear in different flavors with distinct functions; sedentary natural and migratory inflammatory ILC2s have been described [18, 19] and ILC2s have been found that produce IL-10 and much less IL-5 and have suppressor activity [20–22]. Moreover, whereas in the mouse LTi cells have been well defined and found to follow a different path of development than ILC3, in humans defining a clear distinction between LTi cells and ILC3 has been problematic although neuropilin (NRP)1 which is expressed on mouse LTi cells [23, 24] and on a proportion of human ILC3 [23, 24] might be a useful marker for human LTi cells. Indeed NRP1+ is also abundantly expressed on human fetal ILC3 at a gestational age where lymph nodes are formed [25]. However, it has so far been impossible to prove that NRP1 is restricted to LTi cells in humans. Also, there is presently no evidence that in humans LTi cells develop as a lineage distinct from ILC3.

Chapter 2 gives a comprehensive overview of the development of ILC. Understandably our knowledge of ILC development is based mainly on studies in mouse models. Much is now known about the developmental pathways from precursors in fetal liver and bone marrow. Based on these data hierarchical models could be built in which precursor cells gradually become committed to particular ILC types. Understanding ILC development is complicated by observations that deletion of certain transcription factors which are needed for the development of bone marrow precursors to a particular ILC type does not affect those ILCs in tissues as outlined in Chap. 2. This has led to the hypothesis that the process of ILC-poiesis in tissues can take place independently from the so-called central ILC development from bone marrow precursors [26, 27]. In this model ILC precursors home into tissues which then form a substrate for the development of these precursors in mature ILCs. The distinct tissue microenvironments may then be responsible for the differences in requirements for transcription factors for ILC development. Future research should further validate this concept and should give insight into the importance of central versus tissue-specific ILC development for establishing the overall ILC cellularity.

Although the ILC system is clearly evolutionarily conserved, data obtained in mouse systems cannot be translated one on one to humans. For instance, fetal development is very different in mice and humans and therefore findings with respect to fetal versus postnatal development of ILCs cannot be applied to humans. Careful analysis of human fetal tissues at distinct gestational ages may give clues as to whether and how different waves of ILC development occur in humans. In contrast to mouse ILC development our knowledge of differentiation of these cells in humans is still fragmentary. Several studies have shown that ILC precursors are included within CD34 + CD45RA+ precursor cells which can reside not only in bone marrow but also in lymph nodes [28] and in the thymus ([29] Chap. 3). Moreover, human peripheral blood contains ILC precursors which are CD34 negative [27, 30]. Such precursors were also found in different organs which would support the concept of organ-specific ILC-poiesis [26]. But CD34+ ILC precursors (ILCp) are also present in tonsils [28] and possibly in other organs as well which raises the possibility that the committed ILC precursors in those organs are derived from CD34+ cells rather than from peripheral ILC precursors that home in the tonsil. Indeed, the precursors described by Lim et al. were shown to develop from CD34+ progenitors in adoptive transfer experiments in human immune system mice [26]. The anatomical location where ILCp develop from CD34 + CD45+ precursors and the underlying mechanisms of this differentiation require further studies.

Obviously, the requirement for transcription factors and cytokines is much harder to investigate in humans than in mice but investigations on rare human subjects with gene defects have confirmed the essential roles of RORγt in the development of ILC3 [31] and of gamma common receptor (also known as IL2R γ chain) in the development of all ILCs [32]. The importance of GATA3 for the development of human ILC2 was elucidated using RNAi-mediated reduction of

GATA3 [33]. Enforced expression of Id2 into human CD34+ precursors induced the development of ILCs [29]. Recently a platform was published for in vitro development of all human ILC subsets from CD34 + CD45RA+ cells [34] which opens the possibility to investigate the effects of deletion or mutations of certain transcription factors in precursors using CRISPR/Cas9 on the development of ILC precursors. The availability of these tools will enable in-depth investigations into the mechanisms of human ILC development. In addition, pseudotime algorithms applied to single-cell RNA-seq data will provide more detailed information about the developmental trajectories of ILCs in the peripheral blood and tissues.

A landmark paper in 2015 indicated that under resting homeostatic conditions ILCs are tissue resident [35]. Close interactions between ILCs and stromal cells are essential drivers of ILC function. Chapter 5 summarizes our knowledge about this issue. Also, here most of our knowledge comes from mouse studies as stromal cell populations are better defined in mice and it will be challenging to verify the principles of interactions of ILC subsets with stroma in humans. This will require more comprehensive definition of stromal cell types and their functional activities in humans. For instance, the important finding that Tuft cells in the gut are producing IL-25 [36] has yet to be confirmed in humans. Structural differences in tissue of mice and humans may result in different compositions of ILC subpopulations. The human skin is structurally quite distinct from mouse. The consequences of differences in skin thickness and numbers of hair follicles and the presence of a cutaneous muscle layer in mouse skin for skin ILCs have yet to be mapped. In mice the great majority of ILCs in the lung are ILC2; this subset forms a minority of ILCs both in fetal and adult human lung. The reason for these differences has yet to be figured out.

Whereas ILCs under homeostatic conditions are tissue resident, studies both in mice and more recently in humans have shown that under certain conditions ILC2s can migrate out of tissues [18, 19]. The results of experiments in mice were interpreted to mean that there are two subsets of ILC2s, one sedentary population of natural ILC2 which responds to IL-33 and another subset which was called inflammatory (i)ILC2 which responds to IL-25 and is migratory. Data in humans suggest, however, that migratory ILC2s with a same inflammatory imprint as mouse iILC2 and expressing CD45RO are derived from nonactivated CD45RA+ ILC2 which may be equivalent to mouse natural ILC2 [37]. Thus, whereas it is now clear that ILCs are not just tissue resident [38], the relationship between sedentary and migratory ILCs has yet to be firmly established.

It has now been accepted in the field that ILCs are not only beneficial but also involved in inflammatory diseases as pointed out in Chaps. 6 and 7. Like other cell types ILCs are part of circuits that are normally implicated in productive immune reactions but when dysregulated can cause inflammatory diseases. Epithelial cells present in barrier tissues secrete factors needed for homeostasis of multiple cell types including ILC subsets ([39, 40] Chap. 8). Disturbances in the interactions of epithelial cells and ILCs and other cell types may result in inflammatory circuits. Chapter 6 describes examples of involvement of ILC2s in type 2 inflammatory circuits in type 2 inflammatory diseases like allergic asthma and chronic rhinosinusitis. Respiratory viruses and allergenic microorganisms as house dust mite induce IL-33 which together with other factors as IL-25 and TSLP can activate ILC2s to generate an inflammatory milieu which if uncontrolled can lead to chronic inflammation and pathology. Other inflammatory circuits are described in Chap. 7 involving ILC1 and ILC3s in IBD. These chapters also discuss a variety of models that have aided mechanistic studies on the roles of ILCs in inflammatory diseases. Using Rag-deficient [no T and B cells] and Rag−/− gamma common−/− mice (no T, B, NK cells and ILCs) and in adoptive transfer experiments it has been firmly established that ILCs can be pathogenic. However, it is still challenging to determine the relative roles of adaptive and innate lymphocytes in inflammatory diseases in unmanipulated mice with a normal immune system. Because mouse disease models are rarely true mimics of human diseases and the distribution of ILC2 in mice and humans is strikingly different, clinical translation of the results of mechanistic studies in experimental models to human disease remains problematic. On the other hand, as described in Chaps. 6 and 7 careful analysis of ex vivo-isolated ILC subpopulations combined with new molecular and imaging technologies is now enabling translational researchers to monitor changes in the composition and functional activities of immune cells including ILCs in disease. Expansion of such studies should provide more precise information about the critical disease-causing cellular interactions and the role ILCs play in pathology.

How ILCs function in antitumor immune reactions or in promoting the growth of tumors is a very active field of research. As discussed in Chap. 4 , it is known for a long time that NK cells are involved in antitumor immune reactions. However, information on the function of ILCs in tumor immunity is yet limited. Whereas Chap. 4 discusses tumor-promoting and -suppressing effects of ILC1, recent studies have analyzed the function of other ILC subsets in tumors. Interestingly ILC2s contribute to the suppression of melanoma by recruiting eosinophils. ILC2s were found to express PD1 and a combination of PD1-blocking antibodies with IL-33-stimulated proliferation of ILC2 and a concomitant expansion of eosinophils [41]. However, in other tumor types such as bladder carcinoma and gastric cancer, ILC2 may have tumor-promoting effects. Understanding the complex interactions of ILCs with other immune cells and with tumor cells both in mouse models and in human cancer will be highly challenging but such research should eventually lead to novel therapeutic strategies.

Chapter 10 makes clear that there are many more cell types in the innate lymphocyte repertoire. NK-T cells and MAIT cells share many characteristics with ILCs and like ILCs they participate in the immune response against microorganisms and are also involved in allergic and inflammatory diseases. Surprisingly there is no information on whether and how the innate TCRαβ+ T cells and ILCs interact with each other and whether they are redundant. An interested study of Vely et al. reported that gamma common receptor-deficient patients who underwent bone marrow transplantation in the absence of myeloablation recover their T cells but not their ILCs [32]. The ILC deficiencies were not associated with susceptibility to disease. Perhaps the innate T cells may partly compensate for the ILC deficiencies in these patients. Understanding the relative importance of innate T cells and ILCs will be essential for the development of therapies targeting the innate lymphocyte system in life-threatening infections and inflammatory diseases.

References

1. Kiessling R, Klein E, Wigzell H. "Natural" killer cells in the mouse. I. Cytotoxic cells with specificity for mouse Moloney leukemia cells. Specificity and distribution according to genotype. Eur J Immunol. 1975;5 (2):112–7.
2. Mebius RE, Rennert P, Weissman IL. Developing lymph nodes collect CD4+CD3- LTbeta+ cells that can differentiate to APC, NK cells, and follicular cells but not T or B cells. Immunity. 1997;7(4): 493–504.
3. Fort MM, Cheung J, Yen D, Li J, Zurawski SM, Lo S, et al. IL-25 induces IL-4, IL-5, and IL-13 and Th2-associated pathologies in vivo. Immunity. 2001;15(6):985–95.
4. Fallon PG, Ballantyne SJ, Mangan NE, Barlow JL, Dasvarma A, Hewett DR, et al. Identification of an interleukin (IL)-25-dependent cell population that provides IL-4, IL-5, and IL-13 at the onset of helminth expulsion. J Exp Med. 2006;203(4):1105–16.
5. Cella M, Fuchs A, Vermi W, Facchetti F, Otero K, Lennerz JK, et al. A human natural killer cell subset provides an innate source of IL-22 for mucosal immunity. Nature. 2009;457(7230):722–5.
6. Moro K, Yamada T, Tanabe M, Takeuchi T, Ikawa T, Kawamoto H, et al. Innate production of T(H) 2 cytokines by adipose tissue-associated c-Kit(+)Sca-1(+) lymphoid cells. Nature. 2010;463(7280):540–4.
7. Neill DR, Wong SH, Bellosi A, Flynn RJ, Daly M, Langford TK, et al. Nuocytes represent a new innate effector leukocyte that mediates type-2 immunity. Nature. 2010;464(7293):1367–70.
8. Spits H, Di Santo JP. The expanding family of innate lymphoid cells: regulators and effectors of immunity and tissue remodeling. Nat Immunol. 2011;12(1): 21–7.
9. Bernink JH, Peters CP, Munneke M, te Velde AA, Meijer SL, Weijer K, et al. Human type 1 innate

lymphoid cells accumulate in inflamed mucosal tissues. Nat Immunol. 2013;14(3):221–9.
10. Fuchs A, Vermi W, Lee JS, Lonardi S, Gilfillan S, Newberry RD, et al. Intraepithelial type 1 innate lymphoid cells are a unique subset of IL-12- and IL-15-responsive IFN-gamma-producing cells. Immunity. 2013;38(4):769–81.
11. Klose CS, Flach M, Mohle L, Rogell L, Hoyler T, Ebert K, et al. Differentiation of type 1 ILCs from a common progenitor to all helper-like innate lymphoid cell lineages. Cell. 2014;157(2):340–56.
12. Cella M, Otero K, Colonna M. Expansion of human NK-22 cells with IL-7, IL-2, and IL-1beta reveals intrinsic functional plasticity. Proc Natl Acad Sci. 2010;107(24):10961–6.
13. Bal SM, Golebski K, Spits H. Plasticity of innate lymphoid cell subsets. Nat Rev Immunol. 2020;20 (9):552–65.
14. Spits H, Artis D, Colonna M, Diefenbach A, Di Santo JP, Eberl G, et al. Innate lymphoid cells--a proposal for uniform nomenclature. Nat Rev Immunol. 2013;13(2): 145–9.
15. Vivier E, Artis D, Colonna M, Diefenbach A, Di Santo JP, Eberl G, et al. Innate lymphoid cells: 10 years on. Cell. 2018;174(5):1054–66.
16. Spits H, Bernink JH, Lanier L. NK cells and type 1 innate lymphoid cells: partners in host defense. Nat Immunol. 2016;17(7):758–64.
17. Cella M, Gamini R, Secca C, Collins PL, Zhao S, Peng V, et al. Subsets of ILC3-ILC1-like cells generate a diversity spectrum of innate lymphoid cells in human mucosal tissues. Nat Immunol. 2019;20(8):980–91.
18. Huang Y, Guo L, Qiu J, Chen X, Hu-Li J, Siebenlist U, et al. IL-25-responsive, lineage-negative KLRG1 (hi) cells are multipotential 'inflammatory' type 2 innate lymphoid cells. Nat Immunol. 2015;16(2): 161–9.
19. Huang Y, Mao K, Chen X, Sun MA, Kawabe T, Li W, et al. S1P-dependent interorgan trafficking of group 2 innate lymphoid cells supports host defense. Science. 2018;359(6371):114–9.
20. Golebski K, Layhadi JA, Sahiner U, Steveling-Klein EH, Lenormand MM, Li RCY, et al. Induction of IL-10-producing type 2 innate lymphoid cells by allergen immunotherapy is associated with clinical response. Immunity. 2021;54(2):291–307 e7.
21. Morita H, Kubo T, Ruckert B, Ravindran A, Soyka MB, Rinaldi AO, et al. Induction of human regulatory innate lymphoid cells from group 2 innate lymphoid cells by retinoic acid. J Allergy Clin Immunol. 2019;143(6):2190–201 e9.
22. Seehus CR, Kadavallore A, Torre B, Yeckes AR, Wang Y, Tang J, et al. Alternative activation generates IL-10 producing type 2 innate lymphoid cells. Nat Commun. 2017;8(1):1900.
23. Robinette ML, Fuchs A, Cortez VS, Lee JS, Wang Y, Durum SK, et al. Transcriptional programs define molecular characteristics of innate lymphoid cell classes and subsets. Nat Immunol. 2015;16(3):306–17.
24. Shikhagaie MM, Bjorklund AK, Mjosberg J, Erjefalt JS, Cornelissen AS, Ros XR, et al. Neuropilin-1 Is expressed on lymphoid tissue residing LTi-like group 3 innate lymphoid cells and associated with ectopic lymphoid aggregates. Cell Rep. 2017;18(7):1761–73.
25. Liu C, Gong Y, Zhang H, Yang H, Zeng Y, Bian Z, et al. Delineating spatiotemporal and hierarchical development of human fetal innate lymphoid cells. Cell Res. 2021;31(10):1106–22.
26. Lim AI, Di Santo JP. ILC-poiesis: ensuring tissue ILC differentiation at the right place and time. Eur J Immunol. 2019;49(1):11–8.
27. Lim AI, Li Y, Lopez-Lastra S, Stadhouders R, Paul F, Casrouge A, et al. Systemic human ILC precursors provide a substrate for tissue ILC differentiation. Cell. 2017;168(6):1086–100 e10.
28. Scoville SD, Mundy-Bosse BL, Zhang MH, Chen L, Zhang X, Keller KA, et al. A progenitor cell expressing transcription factor RORgammat generates all human innate lymphoid cell subsets. Immunity. 2016;44(5):1140–50.
29. Nagasawa M, Germar K, Blom B, Spits H. Human CD5(+) innate lymphoid cells are functionally immature and their development from CD34(+) progenitor cells Is regulated by Id2. Front Immunol. 2017;8:1047.
30. Nagasawa M, Heesters BA, Kradolfer CMA, Krabbendam L, Martinez-Gonzalez I, de Bruijn MJW, et al. KLRG1 and NKp46 discriminate subpopulations of human CD117(+)CRTH2(−) ILCs biased toward ILC2 or ILC3. J Exp Med. 2019;216(8): 1762–76.
31. Okada S, Markle JG, Deenick EK, Mele F, Averbuch D, Lagos M, et al. IMMUNODEFICIENCIES. impairment of immunity to Candida and Mycobacterium in humans with bi-allelic RORC mutations. Science. 2015;349(6248):606–13.
32. Vely F, Barlogis V, Vallentin B, Neven B, Piperoglou C, Ebbo M, et al. Evidence of innate lymphoid cell redundancy in humans. Nat Immunol. 2016;17(11):1291–9.
33. Mjosberg J, Bernink J, Golebski K, Karrich JJ, Peters CP, Blom B, et al. The transcription factor GATA3 is essential for the function of human type 2 innate lymphoid cells. Immunity. 2012;37(4):649–59.
34. Hernandez DC, Juelke K, Muller NC, Durek P, Ugursu B, Mashreghi MF, et al. An in vitro platform supports generation of human innate lymphoid cells from CD34(+) hematopoietic progenitors that recapitulate ex vivo identity. Immunity. 2021;54(10): 2417–32.
35. Gasteiger G, Fan X, Dikiy S, Lee SY, Rudensky AY. Tissue residency of innate lymphoid cells in lymphoid and non-lymphoid organs. Science. 2015;350 (6263):981–5.
36. von Moltke J, Ji M, Liang HE, Locksley RM. Tuft-cell-derived IL-25 regulates an intestinal ILC2-epithelial response circuit. Nature. 2016;529(7585): 221–5.

37. van der Ploeg EK, Golebski K, van Nimwegen M, Fergusson JR, Heesters BA, Martinez-Gonzalez I, et al. Steroid-resistant human inflammatory ILC2s are marked by CD45RO and elevated in type 2 respiratory diseases. Sci Immunol. 2021;6(55):3489.
38. Huang Y, Mao K, Germain RN. Thinking differently about ILCs-Not just tissue resident and not just the same as CD4(+) T-cell effectors. Immunol Rev. 2018;286(1):160–71.
39. Akdis CA. Does the epithelial barrier hypothesis explain the increase in allergy, autoimmunity and other chronic conditions? Nat Rev Immunol. 2021;21(11):739–51.
40. Schulz-Kuhnt A, Neurath MF, Wirtz S, Atreya I. Innate lymphoid cells as regulators of epithelial integrity: therapeutic implications for inflammatory bowel diseases. Front Med (Lausanne). 2021;8:656745.
41. Jacquelot N, Seillet C, Wang M, Pizzolla A, Liao Y, Hediyeh-Zadeh S, et al. Blockade of the co-inhibitory molecule PD-1 unleashes ILC2-dependent antitumor immunity in melanoma. Nat Immunol. 2021;22(7):851–64.

GPSR Compliance

The European Union's (EU) General Product Safety Regulation (GPSR) is a set of rules that requires consumer products to be safe and our obligations to ensure this.

If you have any concerns about our products, you can contact us on ProductSafety@springernature.com

In case Publisher is established outside the EU, the EU authorized representative is:

Springer Nature Customer Service Center GmbH
Europaplatz 3
69115 Heidelberg, Germany

Batch number: 10732801

Printed by Printforce, the Netherlands